Tropical Mexico
The Ecotravellers' Wildlife Guide

Other Ecotravellers' Wildlife Guides

Costa Rica *by Les Beletsky*
Belize and Northern Guatemala *by Les Beletsky*

Forthcoming guides
Hawaii *by Les Beletsky*
Ecuador and the Galápagos *by David Pearson and Les Beletsky*

series editor Les Beletsky

Tropical Mexico
The Ecotravellers' Wildlife Guide

The Cancún region, Yucatán Peninsula, Oaxaca, Chiapas, and Tabasco

Les Beletsky

Illustrated by:
Priscilla Barrett (Plates 70–81),
David Beadle (Plates 20–69),
David Dennis (Plates 1–19),
John Myers (Figures 1–12) and
Colin Newman (Plates 82–104).

ACADEMIC PRESS
SAN DIEGO   LONDON   BOSTON
NEW YORK   SYDNEY   TOKYO   TORONTO

This book is printed on acid-free paper.

Academic Press
24–28 Oval Road, London NW1 7DX, UK

AP Natural World is published by
Academic Press
525 B Street, Suite 1900, San Diego
California 92101–4495, USA

ISBN: 0-12-084812-0

A catalogue record for this book is available from the British Library

Typeset by J&L Composition Ltd, Filey, North Yorkshire
Colour Separation by Tenon & Polert Colour Scanning Ltd
Printed in Hong Kong by Midas Printing Ltd
99 00 01 02 03 04 MD 9 8 7 6 5 4 3 2 1

Contents

Foreword

Throughout the world, wild places and wildlife are dwindling. Their conservation will require ever more intense protection, care, and management. We always value things more when we stand to lose them, and it is perhaps no coincidence that people are increasingly eager to experience unspoiled nature, and to see the great wildlife spectacles. Tourists are increasingly forsaking the package tour and the crowded beach, to wade through jungle streams, to dive on coral reefs, and to track elusive wildlife. But despite its increasing popularity, nature tourism is nothing new, and the attraction to the tourist is self evident – so why should a conservation organization like the Wildlife Conservation Society encourage it?

The answer is that nature tourism, if properly conducted, can contribute directly to the conservation of wild places and wildlife. If it does that, such tourism earns the sobriquet *ecotourism*. A defining quality of ecotourism is that people are actively encouraged to appreciate nature. If people experience wild areas, they can grow to appreciate their beauty, stability and integrity. And only if they do so, will people care about conserving these places. Before you can save nature, people need to know that it exists.

Another characteristic of ecotourism is that people tread lightly on the natural fabric of wild places. By their very definition, these are places with minimal human impact, so people must not destroy or degrade what they come to experience. Tourists need to take only photographs, leave only footprints – and ideally not even that. Wastes and pollution need to be minimized. Potential disturbance to animals and damage to vegetation must always be considered.

The third characteristic, and that which most clearly separates ecotourism from other forms of tourism, is that tourists actively participate in the conservation of the area. That participation can be direct. For instance, people or tour companies might pay fees or make contributions that support local conservation efforts, or tourists might volunteer to work on a project. More likely, the participation is indirect, with the revenues generated by the tourism entering the local economy. In this way, tourism provides an economic incentive to local communities to continue to conserve the area.

Ecotourists thus are likely to be relatively well informed about nature, and able to appreciate the exceptional nature of wild places. They are more likely to travel by canoe than cruise ship. They will be found staying at locally owned lodges rather than huge multi-national hotels. They will tend to travel to national parks and protected areas rather than to resorts. And they are more likely to contribute to conservation than detract from it.

The Wildlife Conservation Society was involved in promoting ecotourism since before the term was generally accepted. In the early 1960s, the Society (then known as the New York Zoological Society) studied how to use tourism to provide revenues for national park protection in Tanzania (then known as Tanganyika).

By the 1970s, the Society was actively involved in using tourism as a conservation strategy, focusing especially on Amboseli National Park in Kenya. The Mountain Gorilla Project in the Virunga mountains of Rwanda, a project started in the late 1970s, still remains one of the classic efforts to promote conservation through tourism. Today the Society continues to encourage tourism as a strategy from the lowland Amazonian forests to the savannas of East Africa.

We at the Wildlife Conservation Society believe that you will find these Ecotravellers' Wildlife Guides to be useful, educational introductions to the wildlife of many of the world's most spectacular ecotourism destinations.

John G. Robinson
Vice President and
Director of International Conservation
Wildlife Conservation Society

WCS
WILDLIFE CONSERVATION SOCIETY

● to sustain wildlife   ● to teach ecology   ● to inspire care for nature

The mission of the Wildlife Conservation Society, since its founding in 1895 as the New York Zoological Society, has been to save wildlife and inspire people to care about our nature heritage. Today, that mission is achieved through the world's leading international conservation program working in 53 nations to save endangered species and ecosystems, as well as through pioneering environmental education programs that reach more than two million schoolchildren in the New York metropolitan area and are used in 49 states and several nations, and through the nation's largest system of urban zoological facilities including the world famous Bronx Zoo.l WCS is working to make future generations inheritors, not just survivors after Bronx Zoo.

With 60 staff scientists and more than 100 research fellows, WCS has the largest professional field staff of any US-based international conservation organization. WCS's field programs benefit from the technical support of specialists based at WCS's Bronx Zoo headquarters in New York. The Field Veterinarian Program sends experts around the globe to assess wildlife health, develop monitoring techniques, and train local veterinarians. WCS's curatorial staff provides expertise in breeding endangered species in captivity. The Science Resource Center helps researchers assess data through computer mapping, statistical treatments, and cutting-edge genetic analysis. The Education Department writes primary and secondary school curricula that address conservation issues and hosts teacher-training workshops around the world.

WCS's strategy is to conduct comprehensive field studies to gather information on wildlife needs, train local conservation professionals to protect and manage wildlife and wild areas for the future, and advise on protected area creation, expansion, and management. Because WCS scientists are familiar with local conditions, they can effectively translate field data into conservation action and policies, and develop locally sustainable solutions to conflicts between humans and wildlife. An acknowledged leader in the field, the Wildlife Conservation Society forges productive relationships with governments, international agencies and local organizations.

To learn more about WCS's regional programs and membership opportunities, please see our pages in the back of this book. And please visit our website at **www.wcs.org**.

Preface

This book and others in the series are aimed at environmentally conscious travellers for whom some of the best parts of any trip are glimpses of animals in natural settings; at people who, when speaking of a journey, often remember days and locations by encountered wildlife: "That was where we watched the monkeys," and "That was the day we saw the snake eat the frog." The purpose of the book is to enhance enjoyment of a trip and enrich wildlife sightings by providing identifying information on several hundred of the most frequently encountered animals of the tropical regions of southern Mexico (chiefly the Yucatán Peninsula, Oaxaca, and Chiapas), along with up-to-date information on their ecology, behavior, and conservation. With color illustrations of 95 species of amphibians and reptiles, 55 mammals, and about 250 birds, this book truly includes most of the vertebrate land animals that visitors are likely to encounter. Also included is information on marine life, together with illustrations of 140 of the most commonly sighted sea creatures.

The idea to write these books grew out of my own travel experiences and frustrations. First and foremost, I found that I could not locate a single book to take along on a trip that would help me identify all the types of animals that really interested me – birds and mammals, amphibians and reptiles. There are bird field guides, which I've used, but they are often large books, featuring information on every bird species in a given country or region. For Mexico, for instance, field guides, some of them heavy as bricks, detail more than a thousand birds, most of which are rarely seen. If I wanted to be able to identify mammals, I needed to carry another book. For "herps" – amphibians and reptiles – I was a bit astonished to learn, no good, small book existed that might permit me to identify these animals during my travels. Thus, the idea: create a single guide book that travellers could carry to help them identify and learn about the different kinds of animals they were most likely to encounter.

Also, in my experience with guided tours, I've found that guides vary tremendously in their knowledge of wildlife. Many, of course, are fantastic sources of information on the ecology and behavior of animals. Some, however, know only about certain kinds of animals, birds, for instance. And many others, I found, knew precious little about wildlife, and what information they did tell their groups was often incorrect. For example, many guides in Latin America, when asked the identity of any large lizard, respond that it is an "iguana." Well, there certainly are iguanas in the region, but there are also many other types of lizards, and people interested in wildlife need some way to identify more common ones. This book will help.

Last, like most ecotravellers, I am concerned about the threats to many species as their natural habitats are damaged or destroyed by people; when I travelled, I wanted current information on the conservation statuses of animals that

I encountered. This book provides the traveller with conservation information on southern Mexico in general, and on many of the animal family groups pictured or discussed in the book.

A few administrative notes: because this book has an international audience, I present measurements in both metric and English system units. By now, you might think, the scientific classification of common animals would be pretty much established and unchanging; but you would be wrong. These days, what with new molecular methods to compare species, classifications of various animal groups that were first worked out during the 1800s and early 1900s are undergoing sometimes radical changes. Many bird groups, for instance, are being rearranged after comparative studies of their DNA. The research is so new that many biologists still argue about the results. I cannot guarantee that all the classifications that I use in the book are absolutely the last word on the subject, or that I have been wholly consistent in my classifications. However, for most users of this book, such minor transgressions are probably irrelevant.

Finally, let me say that I tried, in several sections of the book, to make the information I present at least mildly entertaining. So many books of this type are written in a dry, terse style. I thought a lighter touch was called for – after all, many of the book's readers will be on holiday, and should not have to plod through heavy material. When I anthropomorphize – provide plants and animals with human characteristics – I do so for fun; plants and insects, I feel fairly safe in saying, do not actually think and reason. Readers who decide they do not appreciate my sense of humor may simply ignore those sections; remaining still should be a solid natural history guide to Mexican animals.

I must acknowledge the help of a large number of people in producing this book. First, most of the information here comes from published sources, so I owe the authors of those books and scientific papers a great deal of credit. The source I probably consulted most often during this book's preparation was *Costa Rican Natural History*, edited by Daniel Janzen (see reference page for complete citations); a great deal of the general information I present on tropical mammals, for instance, comes from this source. I freely acknowledge my debt to Janzen and the large numbers of contributors to that great compendium of information on the plants and animals of Central America; without it, my job would have been much harder. Other good sources of information were *A Neotropical Companion*, by J. C. Kricher, *Neotropical Rainforest Mammals: A Field Guide*, by L. H. Emmons, *A Guide to the Birds of Mexico and Northern Central America*, by S. N. G. Howell and S. Webb, *The Amphibians and Reptiles of the Yucatán Peninsula*, by J. C. Lee (particularly for information on the uses of amphibians and reptiles by ancient Mayan peoples), *Biological Diversity of Mexico*, edited by T. P. Ramamoorthy *et al.*, and *Birds of the Mayas*, by A. L. Bowes, and I recommend all these books for those wanting to delve deeper into their particular subjects.

I would like to take this opportunity to thank the many people who provided information for or helped in the preparation of this book, including Armando Sastre, Joanne Andrews, Ricardo Hernández Sánchez, Carlos Guichard, Julian Lee, Jon Campbell, Eugene Hunn, Gordon Orians, and Jay Savage. Special thanks to Richard Francis for writing the section on marine life, Jon Lyon for contributing the section on Yucatán habitats, and to Irma Trejo and Rodolfo Dirzo for the section on Oaxaca and Chiapas habitats; also to the artists who drew the wonderful illustrations, Priscilla Barrett (mammals), David Beadle (birds), David Dennis (amphibians and reptiles), Colin Newman (marine life), and John Myers (plants), to the Burke Museum at the University of Washington for kindly allowing me access to its facilities, and to my editor at Academic Press, Andrew Richford. Glen Murphy and Mark Peck of the Royal Ontario Museum permitted generous use of materials for producing the bird plates.

Please let me know of any errors you find in this book. I am also interested in hearing your opinions on the book, suggestions for future editions, and of your experiences with wildlife during your travels. Write care of the publisher or e-mail: Ecotravel8@aol.com

Les Beletsky

Chapter 1

Ecotourism: Travel for the Environmentally Concerned

- What Ecotourism Is and Why It's Important
- The History of Ecotourism
- How Ecotourism Helps; Ecotravel Ethics

What Ecotourism Is and Why It's Important

In many remote corners of the world, local people and sometimes entire villages are presently engaged in activities that even a few years ago might have seemed eccentric. In out-of-the-way spots in Costa Rica and Belize and Madagascar, people operate butterfly breeding farms. In Belize and Mexico's Yucatán Peninsula, poor communities have established nature reserves dedicated to monkeys. On the Pacific Coast of Oaxaca, in southern Mexico, local villagers run a facility devoted to sea turtles – not to process them for sale or consumption, but to explain and protect them. In rugged regions of the Mexican state of Chiapas, members of native groups pile visitors into ancient mini-buses and drive them to where the road ends and beyond. And rising in the middle of many tropical forests are aerial walkways, rope ladders and bridges being threaded by local entrepreneurs through and above the trees. But these seemingly weird projects are not the result of mass mental failings. Rather, they are highly relevant community or private projects created with a laudable goal in mind: to convince you, the ecotraveller, to visit.

Ecotravel is a new kind of travelling. Of course, people have always travelled and probably they always will. Historical reasons for travelling are many and varied: to find food, to avoid seasonally harsh conditions, to emigrate to new regions in search of more or better farming or hunting lands, to explore, and even, with the advent of leisure time, just for the heck of it (travel for leisure's sake is the definition of tourism). For many people, travelling fulfills some deep need; there's something irreplaceably satisfying about journeying to a new place: the sense of being in completely novel situations and surroundings, seeing things never before encountered, engaging in new and different activities.

During the 1970s and 1980s there arose a new reason to travel, perhaps the first wholly new reason in hundreds of years: with a certain urgency, to see natural habitats and their harbored wildlife before they forever vanish from the surface of the Earth. Ecotourism or ecotravel is travel to destinations specifically to admire and enjoy wildlife and undeveloped, relatively undisturbed natural areas, as well as indigenous cultures. The development and increasing popularity of eco-tourism is a clear outgrowth of escalating concern for conservation of the world's natural resources and biodiversity (the different types of animals, plants, and other life forms found within a region). Owing mainly to peoples' actions, animal species, plant species, and wild habitats are disappearing or deteriorating at an alarming rate. Because of the increasing emphasis on the importance of the natural environment by schools at all levels and the media's continuing exposure of environmental issues, people have enhanced appreciation of the natural world and increased awareness of environmental problems globally. They also have the very human desire to want to see undisturbed habitats and wild animals before they are gone, and those with the time and resources increasingly are doing so.

But that's not the entire story. The purpose of ecotravel is twofold. Yes, peo-ple want to undertake exciting, challenging, educational trips to exotic locales – wet tropical forests, wind-blown deserts, high mountain passes, mid-ocean coral reefs – to enjoy the scenery, the animals, the nearby local cultures. But the second major goal of ecotourism is often as important: the travellers want to help con-serve the very places – habitats and wildlife – that they visit. That is, through a portion of their tour cost and spending into the local economy of destination countries – paying for park admissions, engaging local guides, staying at local hotels, eating at local restaurants, using local transportation services, etc. – eco-tourists help to preserve natural areas. Ecotourism helps because local people ben-efit economically as much or more by preserving habitats and wildlife for continuing use by ecotravellers than they could by "harvesting" the habitats for short-term gain. Put another way, local people can sustain themselves better eco-nomically by participating in ecotourism than by, for instance, cutting down rainforests for lumber or hunting animals for meat or the illicit pet trade.

Preservation of some of the world's remaining wild areas is important for a host of reasons. Aside from moral arguments – the acknowledgment that we share the Earth with millions of other species and have some obligation not to be the continuing agent of their decline and extinction – increasingly we understand that conservation is in our own best interests. The example most often cited is that botanists and pharmaceutical researchers each year discover another wonder drug or two whose base chemicals come from plants that live, for instance, only in tropical rainforest. Fully one-fourth of all drugs sold in the USA come from nat-ural sources – plants and animals. About 50 important drugs now manufactured come from flowering plants found in rainforests, and, based on the number of plants that have yet to be cataloged and screened for their drug potential, researchers estimate that at least 300 more major drugs remain to be discovered. The implication is that if the globe's rainforests are soon destroyed, we will never discover these future wonder drugs, and so will never enjoy their benefits. Also, the developing concept of *biophilia*, if true, dictates that, for our own mental health, we need to preserve some of the wildness that remains in the world. Bio-philia, the word recently coined by Harvard biologist E. O. Wilson, suggests that because people evolved amid rich and constant interactions with other species and in natural habitats, we have deeply ingrained, innate tendencies to affiliate with other species and actual physical need to experience, at some level, natural habitats. This instinctive, emotional attachment to wildness means that if we eliminate species and habitats, we will harm ourselves because we will lose things essential to our mental well-being.

The History of Ecotourism

Tourism is arguably now the world's largest industry, and ecotourism among its fastest growing segments. But mass ecotourism is a relatively new phenomenon, the word being coined only during the 1980s. In fact, as recently as the 1970s, tourism and the preservation of natural habitats were viewed largely as incompatible pursuits. One of the first and best examples of ecotourism lies in Africa. Some adventurers, of course, have always travelled to wild areas of the Earth, but the contemporary history of popular ecotourism probably traces to the East African country of Kenya. Ecotourism, by one name or another, has traditionally been a mainstay industry in Kenya, land of African savanna and of large charismatic mammals such as elephants and giraffes, leopards and lions – species upon which to base an entire ecotourism industry.

During most of the European colonial period in East Africa, wildlife was plentiful. However, by the end of colonial rule, in the middle part of the 20th century, continued hunting pressures had severely reduced animal populations. Wildlife was killed with abandon for sport, for trade (e.g. elephant ivory, rhinoceros horn), and simply to clear land to pave way for agriculture and development. By the 1970s it was widely believed in newly independent Kenya that if hunting and poaching were not halted, many species of large mammals would soon be eliminated. The country outlawed hunting and trade in wildlife products, and many people engaged in such pursuits turned, instead, to ecotourism. Today, more than a half million people per year travel to Kenya to view its tremendous wildlife and spectacular scenery. Local people and businesses profit more by charging ecotourists to see live elephants and rhinoceroses in natural settings than they could by killing the animals for the ivory and horns they provide. Estimates were made in the 1970s that, based on the number of tourist arrivals each year in Kenya and the average amount of money they spent, each lion in one of Kenya's national parks was worth $27,000 annually (much more than the amount it would be worth to a poacher who killed it for its skin or organs), and each elephant herd was worth a stunning $610,000 (in today's dollars, they would be worth much more). Also, whereas some of Kenya's other industries, such as coffee production, vary considerably from year to year in their profitability, ecotourism has been a steady and growing source of revenue (and should continue to be so, as long as political stability is maintained). Thus, the local people have strong economic incentive to preserve and protect their natural resources.

Current popular ecotourist destinations include South Africa, Botswana, Kenya, and Tanzania in Africa; Nepal, Thailand, and China in Asia; Australia; and, in the Western Hemisphere, Belize, Costa Rica, Puerto Rico, Ecuador, Chile, and the Amazon Basin. More than five million foreign tourists arrive each year in Mexico, the majority of them sun-worshippers from the USA, Canada, and Europe, heading for West Coast beach resorts or visiting the Cancún area of the Yucatán Peninsula. Increasingly, particularly in tropical southern Mexico, these visitors are interested in wild habitats and wildlife, and are leaving the beaches behind for all or part of their trip to explore ecotourism – taking river trips through towering river canyons, hiking on trails through tropical forests, boating along the coast to view flocks of pink flamingos and mysterious mangrove forests.

How Ecotourism Helps; Ecotravel Ethics

To the traveller, the benefits of ecotourism are substantial (exciting, adventurous trips to stunning wild areas; viewing never-before-seen wildlife); the disadvantages are minor (sometimes, less-than-deluxe transportation and accommodations that, to many ecotravellers, are actually an essential part of the experience). But what are the actual benefits of ecotourism to local economies and to helping preserve habitats and wildlife?

The pluses of ecotourism, in theory, are considerable:

(1) Ecotourism benefits visited sites in a number of ways. Most importantly from the visitor's point of view, through park admission fees, guide fees, etc., ecotourism generates money locally that can be used directly to manage and protect wild areas. Ecotourism allows local people to earn livings from areas they live in or near that have been set aside for ecological protection. Providing jobs and allowing local participation is important because people will not want to protect the sites, and may even be hostile toward them, if they formerly used the now-protected site (for farming or hunting, for instance) to support themselves but are no longer allowed such use. Finally, most ecotour destinations are in rural areas, regions that ordinarily would not warrant much attention, much less development money, from central governments for services such as road building and maintenance. But all governments realize that a popular tourist site is a valuable commodity, one that it is smart to cater to and protect.

(2) Ecotourism benefits education and research. As people, both local and foreign, visit wild areas, they learn more about the sites – from books, from guides, from exhibits, and from their own observations. They come away with an enhanced appreciation of nature and ecology, an increased understanding of the need for preservation, and perhaps a greater likelihood to support conservation measures. Also, a percentage of ecotourist dollars are usually funnelled into research in ecology and conservation, work that will in the future lead to more and better conservation solutions.

(3) Ecotourism can also be an attractive development option for developing countries. Investment costs to develop small, relatively rustic ecotourist facilities are minor compared with the costs involved in trying to develop traditional tourist facilities, such as modern beach resorts. Also, it has been estimated that, at least in many regions, ecotourists spend more per person in the destination countries than any other kind of tourists.

A conscientious ecotraveller can take several steps to maximize his or her positive impact on visited areas. First and foremost, if travelling with a tour group, is to select an ecologically committed tour company. Basic guidelines for ecotourism have been established by various international conservation organizations. These are a set of ethics that tour operators should follow if they are truly concerned with conservation. Travellers wishing to adhere to ecotour ethics, before committing to a tour, should ascertain whether tour operators conform to the guidelines (or at least to some of them), and choose a company accordingly. Some tour operators in their brochures and sales pitches conspicuously trumpet their ecotour credentials and commitments. A large, glossy brochure that fails to mention how a company fulfills some of the ecotour ethics may indicate an operator that is not especially environmentally concerned. Resorts, lodges, and travel agencies that specialize in ecotourism likewise can be evaluated for their dedication to eco-ethics. Some travel guide books that list tour companies provide such ratings. The Ecotourism Society, an organization of ecotourism professionals, may also provide helpful information (USA tel: 802-447-2121; e-mail: ecomail@ecotourism.org; www.ecotourism.org).

Basic ecotour guidelines, as put forth by the United Nations Environmental Programme (UNEP), the World Conservation Union (IUCN), and the World Resources Institute (WRI), are that tours and tour operators should:

(1) Provide significant benefits for local residents; involve local communities in tour planning and implementation.
(2) Contribute to the sustainable management of natural resources.
(3) Incorporate environmental education for tourists and residents.
(4) Manage tours to minimize negative impacts on the environment and local culture.

For example, tour companies could:

(1) Make contributions to the parks or areas visited; support or sponsor small, local environmental projects.
(2) Provide employment to local residents as tour assistants, local guides, or local naturalists.
(3) Whenever possible, use local products, transportation, food, and locally owned lodging and other services.
(4) Keep tour groups small to minimize negative impacts on visited sites; educate ecotourists about local cultures as well as habitats and wildlife.
(5) When possible, cooperate with researchers; for instance, Costa Rican researchers are now making good use of the elevated forest canopy walkways

Map 1 Tropical Mexico and surrounding regions. Shaded areas show the Mexican states covered in the book.

in tropical forests that several ecotourism facility operators have erected on their properties for the enjoyment and education of their guests.

Committed ecotravellers can also adhere to the ecotourism ethic by patronizing lodges and tours operated by local people, by disturbing habitats and wildlife as little as possible (including fish and other coral reef wildlife, not to mention the coral reef itself!), by staying on trails, by being informed about the historical and present conservation concerns of destination countries, by respecting local cultures and rules, by declining to buy souvenirs made from threatened plants or animals, and even by actions as simple as picking up litter on trails.

Now, with some information on ecotourism in hand, we can move on to discuss tropical Mexico. The area covered in this book is shown in Map 1.

Chapter 2

Southern Mexico: Ecotourism, Geography, and Habitats

A Brief Eco-history of Southern Mexico

It's safe to say that Mexico is not the first place that leaps to mind when you think of nature travel. Costa Rica, Belize, yes, the Galápagos, yes, East Africa, yes, but Mexico – land of mega beach resorts and sunny, hot deserts, massive environmental problems and huge Mexico City – with 20+ million people and probably the world's unhealthiest air? While it's true that Mexico has its share of problems, environmental and otherwise, ecotourism there, particularly in the country's southern, tropical reaches, is growing. And, actually, there are several excellent reasons, both practical and ecological, that you should place southern Mexico high on your list of travel priorities. First, Mexico has astounding biodiversity (the different types of animals, plants, and other life forms found within a region). It's home to fully 10% of the world's non-fish vertebrate animals (7% of amphibian species, 12% of reptiles, 11% of birds, 10% of mammals) and about 12% of all plant species. In one survey of a 1000-hectare (2500 acre) site in a tropical rainforest in southern Mexico, researchers estimated that there were over 1000 plant species, 300 bird species, and about 150 amphibian and reptile species. Moreover, Mexico is a center of biological *endemism* – meaning that much of its biodiversity occurs nowhere else (see Close-Up, p. 236). For instance, about 56% of the country's amphibian and reptile species (that is, of its *herpetofauna*) is endemic; to see these species in the wild, you must visit Mexico.

Second, in addition to high biodiversity, Mexico has some important wild areas that need protection, and ecotourism is one of the ways in which that protection will likely come about. In some ways, travelling to Mexico to visit parks and reserves and to stare at birds and trees is a political statement and act of conservation – you want those habitats and animals preserved, and you are willing to visit them and spend into the local economy to support that preservation. The habitats are worthy of your support. Take, for example, one habitat type that most people are aware is fast disappearing – forests. Recently it has been estimated that fully half of the world's forest cover is gone – cut and burned away. The main forests that need to be preserved now are increasingly called *frontier forests* – large expanses of forest still relatively wild and undisturbed; big and wild enough that, if protected now, they will maintain the rich biodiversity they hold. Mexico's record of forest destruction is dismal (as is that of many other countries, of course, including the USA and most Central American nations). About 92% of Mexico's original forest cover is gone and 77% of what's left is threatened. However, there are still some large, virgin expanses – about 87,000 sq km (33,500 sq miles) of frontier forest, much of it tropical forest in Oaxaca, Chiapas, and the Yucatán Peninsula.

Therefore, from biodiversity and conservation perspectives, Mexico is richly deserving as a high-priority ecotravel destination. But there are good, practical reasons to visit as well. Perhaps foremost is that once away from a few urban areas (Mérida, Oaxaca City, Tuxtla Gutiérrez) and mass tourism developments (Cancún and the coastal tourist corridor to its south), the countryside of southern Mexico is mostly uncrowded, with many charming sites and some spectacular scenery. Tourists over much of this area are few and far between. For instance, while recently visiting several smaller Mayan ruin sites scattered across the Yucatán Peninsula, some located not far from the world-famous pre-Columbian site of Chichén Itzá, I found myself to be the sole visitor. Added to the luxury of often travelling in splendid isolation is the fact that ecotourism is a nascent development over most of southern Mexico – new areas, some previously inaccessible, are just opening up, tours just now being created. Many of the sites described in this book have not yet been visited by great numbers of travellers; they are truly off the beaten path. Take, for example, new ecotourism development at the remote El Triunfo Biosphere Reserve in Chiapas (see p. 46), where, on average, only about 80 people per month visit. Go there for a week, experience the lush, wet cloud forest, and I can guarantee you won't get many "been-there-done-that" responses when you return home to show your slides and tell your travel stories.

Another positive feature: travelling to Mexico, from elsewhere in North America but from other places as well, is fairly inexpensive, especially when compared with trying to reach some other ecotravel destinations; travelling within the country, likewise, can be done quite cheaply. Gasoline is relatively cheap, the price akin to that in the USA. Local people are usually very friendly in Mexico, and the countryside in the southern states, and especially around ecotravel destinations, is fairly safe. Finally, southern Mexico has a lot to offer a traveller. Indeed, for a varied trip during which you might want to combine serious ecotravel to a remote rainforest with a few days of luxurious malingering at a world-class beach resort, and snorkeling in warm blue waters along one of the world's longest barrier reefs, it's probably one of the best places anywhere.

Are there problems associated with ecotravel to southern Mexico? Sure, but there are problems almost anywhere you go. It's a matter of balance, and the region's pluses easily outweigh the minuses. The Yucatán Peninsula is a hot, bright place – hot enough to be uncomfortable. Much of it has been deforested and much of the forest you do see is young *secondary* growth – the forest regenerating after being cut or burned away. Wildlife over the peninsula is not overly abundant – aside from some birds, lizards along trails and at archaeological ruin sites, and occasional monkeys, wildlife is difficult to see. Certain regions of Oaxaca and Chiapas are wilder places, with wildlife perhaps a bit easier to spot in protected areas – but here, too, habitat destruction and hunting have taken a great toll on wildlife. Ecotourism, as I've already mentioned, is only lightly developed in southern Mexico, which is nice from the standpoint of keeping ecotraveller numbers low, but it also means that some areas that you might like to see are relatively inaccessible – the Calakmul Biosphere Reserve (p. 48), for instance, is still difficult to access. Last, owing to political instability in Chiapas that began in 1994, and continued tension over land ownership and resource allocation issues among poor Indian groups, rebel organizations, and the government, there is undeniably some danger in the area; even if no tourists have been hurt, there is still a whiff of threat and revolution in the cool mountain air around Chiapas' San Cristóbal.

This book covers the entire Yucatán Peninsula (states of Campeche, Yucatán, and Quintana Roo), plus the large southern states of Oaxaca and Chiapas. Tabasco state is treated briefly because it connects the Yucatán Peninsula to Chiapas, but ecotourism there is little developed.

Yucatán Peninsula

Perhaps three million sun-worshippers per year from the USA, Canada and Europe flock to the Yucatán Peninsula, but this migration is a fairly new phenomenon. Very few foreigners visited the Yucatán Peninsula before the creation of Cancún as a beach resort. Historically the peninsula was a very sparsely settled region of Mexico, especially the most eastern state, Quintana Roo, where Cancún now draws tourists and locals alike. The reasons for sparse settlement and visitation were many. Much of the peninsula was covered with hard-to-penetrate forest and good roads were absent through the region's interior. A main problem faced by potential settlers and residents of the hot, humid peninsula was the difficulty of obtaining fresh water; much of the land area lacks lakes and rivers. Some agriculture was possible, mainly sugar cane and *henequén*, an agave cactus from which hemp rope is made, and for a time during the 1800s and early 1900s, the Yucatán Peninsula, especially Yucatán state, enjoyed a brief economic heyday by providing the raw material for much of the world's rope. The advent of synthetic fibers mostly killed off the natural fiber rope business, leaving the Yucatán Peninsula as recently as the 1950s and 1960s a bright, hot, lightly visited region of Mexico where a trickle of tourists came to view the stone ruins of ancient civilizations.

Then in the early 1970s the Mexican government, seeking to create a beach resort attraction closer than Acapulco and Puerto Vallarta to the large population centers on the USA's East Coast, selected for mega-development a tiny fishing village near the northeast corner of the peninsula – Cancún. Over the next 25 years, scores of resort hotels were built along a 20-km (12-mile) sand spit that arches off the mainland into the Caribbean, and the village of a few hundred souls grew to a service and support city of 300,000. So successful was Cancún that the resorts and crowds spilled out of the hotel zone there and cascaded southwards, down the Caribbean Coast, overrunning Isla Mujeres and the island of Cozumel, until, by now, large resort hotels and other tourist attractions dot the coast for 130 km (80 miles), between Cancún and Tulum, creating the *Corredor Turístico*. The Cancún area now generates about 25% of Mexico's tourist revenues. The Mexican government, seeking to duplicate this economic success elsewhere, is considering even more tourism/resort development farther south on the Yucatán's Caribbean Coast, between the Sian Ka'an Biosphere Reserve's southern tip and Xcalak (a small beach resort area near Belize popular with scuba divers).

Cancún, Cozumel, and the associated tourist corridor hold little of interest for the true ecotraveller, but they do serve admirably as gateways to the rest of the peninusula and as service providers for those interested in coastal underwater exploration. Ecotourism is only slowly getting started in this area. Some hotel owners recognize the growing interest of their guests in nature and are starting to act. Unfortunately, but perhaps not surprisingly, the first corporate inclinations to cash in on this interest was to create artificial natural attractions – nature amusement parks such as the heavily advertised Xcaret, south of Cancún. The almost total absence of ecotourism promotion in Cancún is conspicuous – you are bombarded with offers of cruises, water sports, visits to Tulum and Chichén Itzá – but no one at the airport or your hotel hands you a brochure about the nearby Sian Ka'an or Ría Lagartos Biosphere Reserves, and no tour agencies on street corners offer to take you to a rainforest or protected wetland. The Mexican government, and tour operators in the region (in all of Mexico, for that matter) have been very slow on the uptake in regard to ecotourism, but it's starting to change, propelled by that bedrock feature of capitalism, consumer interest. Increasingly, visitors want to escape the planned beach resorts and see something of the countryside, perhaps visit, even if only briefly, wild areas. For instance, after its guests repeatedly asked for tours to wilderness areas, one coastal resort that caters to European visitors told a conservation agency operating tours of a large biosphere reserve (Sian Ka'an) that it wanted to reserve permanent places on its tourboats one day a week for its guests.

So ecotourism is growing in the region. That's good, because the peninsula has much to recommend it from wildlife and wild habitat perspectives, being home to about 40 species of amphibians, 140 reptiles, and 140 mammals. More than 450 bird species have been spotted there, many of them migratory – the peninsula serves a vital function as a resting and feeding site for birds migrating southwards in October and November and northwards in March and April (see Close-Up, p. 117). Habitats ecotravellers will want to see include coastal mangroves and other wetland habitats at Ría Celestún, Ría Lagartos, and Sian Ka'an Biosphere Reserves, virgin tropical jungle at Calakmul Biosphere Reserve, and coral reefs along the peninsula's eastern and northern coasts.

Ecotourism's roots on the peninsula are fairly recent, traceable to a few local conservation organizations which, with financial and technical backing from international agencies, started the ball rolling. For instance, take the enthusiastic work of Amigos de Sian Ka'an ("Sian Ka'an" is Mayan for "where the sky is born"). (Work of other local conservation agencies is detailed in Chapter 4.) This private, non-profit organization, headquartered in Cancún, is involved with raising funds for the Sian Ka'an Biosphere Reserve (the reserve itself, a federal entity, cannot seek donations), working with local people who live in and near the reserve to minimize harmful ecological effects of their fishing and forest utilization, conducting ecological research within the reserve and, in the interest of enhancing conservation of the reserve, developing and promoting ecotourism there and actually conducting tours. They have obtained financial backing and other support from such organizations as The Nature Conservancy (with which they have a working partnership), World Wildlife Fund, and the MacArthur and Rockefeller Foundations.

The reserve, one of Mexico's largest, contains many pristine areas, and is located down the coast from Cancún, just south of Tulum. It's so close to the tourist corridor, in fact, that many worry that should tours to the reserve ever really catch on, the place would be flooded with tourists, thus ruining it as a remote, uncrowded eco-attraction. Right now, though, it's a quiet wilderness area of about 617,000 hectares, or 1.6 million acres, including underwater sites in two large bays. The reserve is about a third forest, a third wetlands, and a third coastal and marine areas. Major habitats within the reserve are deciduous tropical forest, wet savanna with sawgrass, mangroves, marshes, coastal dunes, and many miles of offshore coral reef. Flora in the reserve includes more than 1200 plant species, and fauna includes 350 bird species (particularly common are water-associated birds and birds-of-prey), big cats, tapir, monkeys, manatee, sea turtles, and considerable numbers of crocodiles. Also included within the reserve are more than 20 small archaeological ruin sites and two small fishing villages, whose several thousand inhabitants still seek to make their livings in and around the reserve, mainly by gathering clawed crustaceans (lobsters) from the shallow bays.

Much of Sian Ka'an is still relatively pristine for the simple reason that human occupation of the area was always sparse and, importantly, much of the reserve consists of inaccessible wetlands. Like the Everglades in the USA's Florida, it is not possible to walk through much of the reserve; there are few trails. The area first received protected status in 1982, was designated a biosphere reserve in 1986 (as the Amigos group was formed), and a UNESCO World Heritage Site in 1987. Like most biosphere reserves (see p. 57), the park is divided into *core zones*, large wilderness areas with intact biodiversity in which human activity is severely restricted to scientific research, and buffer zones, in which some human activities such as ecotourism, consistent with preserving the wild nature of the area, are allowed. The villagers who live within the reserve are in buffer/transition zones. The reserve is administered by SEMARNAP, the Mexican Secretariat of the Environment, through the National Institute of Ecology, and in cooperation with Amigos de Sian Ka'an.

Amigos de Sian Ka'an works with local villages in and around the reserve to try to develop community projects or cottage industries so that these people can make their livings in or near the reserve as they have in the past, but not significantly harm its habitats or wildlife. In other words, Amigos promotes *sustainable* use of the reserve's buffer zones. For instance, they are working with villagers to establish a small chicle industry (harvesting *chicle*, chewing gum base, from sapodilla trees in the reserve, which does not kill the trees) and an ornamental plant nursery program, and educating local people about sustainable hunting. Amigos, with its own biologists, also conducts ecological studies at the reserve, some practical, for example, trying to determine potential negative impacts of ecotourism there, and some more basic, such as population studies of the local crocodiles and research into replacing invading alien plant species with native ones. Amigos also created the ecotourism program that now exists at the reserve. Seeking to combine tourism with fundraising for conservation and management purposes, they bought some boats, hired biologist-guides, and started giving tours. Small groups, usually six to eight people, are taken on five-hour tours of the reserve, mostly by boat. About 25% of the reserve's operating budget now comes from tour revenues. Therefore, it's a highly successful ecotourism program, one that, should you be in the area, is very much worth your participation. (Contact: Amigos de Sian Ka'an, Ave. Coba #5, Plaza América, easily found in downtown Cancún; tel: (98) 84-9583; fax: (98) 87-3080; e-mail: sian@cancun.rce.com.mx)

Oaxaca and Chiapas

Oaxaca (wah-HAH-kah) and Chiapas are big, mostly poor Mexican states with large Indian populations. Both are primarily agricultural, although Chiapas has some significant industry. Both are remote and rugged, with extensive highland regions dominated by beautiful pine and pine–oak forests, and separated from the rest of Mexico by high mountains. Both states are heavily deforested. Much of the forest in Oaxaca has been cut and burned to clear the land for agriculture, and by the need for firewood. In Chiapas, the problem is the same, exacerbated by a fairly recent increase in cattle ranching. Forests are cut to provide grazing land for cattle; but the land quickly deteriorates, necessitating further deforestation. However, the remoteness of these states and the rugged mountains covering much of them have prevented deforestation and development over certain sections, and some significant wilderness areas remain. Oaxaca and Chiapas, in fact, are key to Mexico's biodiversity and are centers of endemic species, particularly among reptiles and amphibians (see Close-Up, p. 236). About 680 bird species occur in Oaxaca (about the same number that occur in all of North America north of the Rio Grande), and 465 of them breed there. More than 600 species of birds occur in Chiapas, as do about 185 mammal species. Fully 30% of Mexico's considerable number of animal species and 20% of its plants occur in Chiapas' Selva Lacandona region alone.

Chiapas has a recent history of conservation awareness, attributable mainly, many residents say, to the efforts of one man, Miguel Alvarez del Toro (1917–1996). Alvarez del Toro, at one time a Mexico City taxidermist, moved to Chiapas in the early 1940s to take a position with the new Institute of Natural History (Instituto de Historia Natural, IHN), a state museum and ecology research center. He eventually became the institute's director and rose to a position of prominence in the zoological world as the foremost expert on many aspects of Mexican wildlife. During the 1970s he wrote books on the birds and mammals of Chiapas that summarized knowledge of those groups. Later in life Alvarez del Toro was a leader of Mexico's environmental movement, and was instrumental in the establishment of the Sumidero Canyon area outside Tuxtla as a protected national park (p. 43). In recognition of his efforts, and in his honor, a plaque has been placed along the Grijalva River in Sumidero, and the Tuxtla Zoo (p. 42) bears his name. Today Chiapas' IHN is at the forefront of Mexican conservation efforts. In association with international organizations such as The Nature Conservancy, which provide supplementary funding, IHN seeks to protect Chiapas' biodiversity in several ways: they work with local residents in important conservation priority areas to develop sustainable uses of natural resources so as to minimize negative ecological effects; they undertake environmental education and wildlife research; and they conduct hands-on protection of parks and reserves. IHN currently manages several major protected areas of Chiapas, including El Triunfo Biosphere Reserve, Selva del Ocote Special Biosphere Reserve (or El Ocote Ecological Reserve), and La Encrucijada Ecological Reserve.

Traditionally, foreign tourists have come to Oaxaca and Chiapas to visit pre-Columbian ruin sites, particularly Monte Albán near Oaxaca City and Palenque in northern Chiapas; to view colonial-era architecture; to experience local Indian cultures and see and purchase Indian crafts and artworks; and more recently, to visit beach resorts such as the new developments around Huatulco, on Oaxaca's Pacific Coast. The Oaxaca City area has always been a big draw as a cultural attraction. Chiapas historically has been a less-visited state, perhaps because many of its scenic areas are difficult to reach and its main city, Tuxtla Gutiérrez, is a less-appealing, modern, more commercial capital. In early 1994 a rebel group based in the high mountains of central Chiapas, the Zapatista National Liberation Army (EZLN), with grievances against the federal government concerning resource allocation and, especially, land ownership in the region, seized the sizable town of San Cristóbal de las Casas and several smaller towns and villages. Government soldiers quickly pushed the revolutionaries out, but, as of the date of this writing (mid-1998), although there is a long truce in effect, political tension and occasional violence in the area continues. Tourism to Chiapas was at first negatively affected by the brief uprising, but it has been recovering. In fact, a new kind of tourism, *political tourism* – travelling to San Cristóbal to see what's going on or even to show support for the rebels – was born in the area and has become popular. Tourism is now by far the greatest single contributor to the economy in and around San Cristóbal.

As for ecotourism, I would say that if it has been slow off the mark in the Yucatán Peninsula, it's even less developed in Oaxaca and Chiapas. But as in the Yucatán, it's showing signs of steady growth, seemingly with more remote, exciting sites opened to visitation each year. Increasingly, local conservation organizations and even, to some extent, the Mexican government, are deciding that one of the prime ways to preserve wilderness areas in these states is through the development of ecotourism. For example, local conservation organizations, with the assistance of international agencies, are experimenting with ecotourism in the Lacandon rainforest, in eastern Chiapas, which is one of Mexico's top priority sites for conservation. The Mexican government itself, in the guise of the National Institute of Ecology, has developed and is promoting ecotourism at El Triunfo Biosphere Reserve (p. 46). Currently these operations are small-scale, but more development is planned (see Chapter 4).

Geography and Climate

Southern Mexico has complex geography, topography, and climate, but we can simplify things a bit by considering first the Yucatán Peninsula and then looking separately at Oaxaca, Chiapas, and Tabasco.

Yucatán Peninsula

The Yucatán Peninsula basically is a large, flat, thumb-shaped limestone rock, with a thin covering of topsoil, that projects northwards from the southern tip of the North American continent into the Caribbean Sea. Politically it consists primarily of the Mexican states of Campeche, Yucatán, and Quintana Roo (Map 2; if we wanted to get technical, physically the peninsula also includes small bits of western Tabasco and Chiapas states, northern Belize, and northern Guatemala). The northern third of the peninsula is extremely flat, the terrain occasionally rising to no more than about 50 m (165 ft) above sea level. There are no rivers and few lakes in this region, surface water being restricted to *cenotes* and *aguadas*. The former are deep holes in the limestone in which subterranean rivers flow, and the latter are shallow depressions in the limestone in which rainwater collects, producing semi-permanent ponds. In the central region of the peninsula there are some hilly areas (for instance, the Puuc Hills region near Uxmal) where elevations approach 250 m (800 ft). There are some small lakes and a few rivers in the southern part of the peninsula, in addition to cenotes and aguadas.

Quintana Roo (Map 2) occupies the eastern third of the peninsula, bordering the Caribbean. It supports a resident (non-tourist) human population of about 700,000 over its 50,212 sq km (19,387 sq miles). Most of the population lives along the coast, much of the inland area consisting of hot, humid forests with clay-containing soils that are unsuitable for many types of agriculture. Beaches, bays, reefs, and islands along the coast support the state's main industries, tourism and seafood harvesting. Yucatán state, population about 1.6 million, with an area of 38,400 sq km (14,827 sq miles), occupies the middle of the peninsula and includes most of the northern coast. Fishing villages dot the coast, and fishing and agriculture (corn, sugar cane, oranges, henequén, cattle) dominate the economy of this flat, brushy state. Campeche, mostly a hilly, scrubby region, but also one that contains the large rainforest expanse of the Calakmul Biosphere Reserve, occupies the western third of the peninsula. Its economy, in addition to fishing and agriculture (sugar cane, rice, corn), is supported by several small resort areas that line the narrow beaches along the Gulf of Mexico. Campeche encompasses about 50,800 sq km (19,619 sq miles) and has a population of 650,000. The three peninsula states have a combined area similar to that of the USA's Arkansas or to England. Main cities are Cancún and the state capitals, Campeche (Campeche state), Mérida (the peninsula's largest city, with 600,000 people; Yucatán state), and Chetumal (Quintana Roo).

It's not surprising that the Yucatán Peninsula, in its tropical low-elevation setting, is fairly uniformly hot and humid. Travelling here means experiencing day after day of unrelenting sun, which is fine for Cancún area beaches but less pleasant for ecotravel. In the rainy season, May through October, late afternoon showers and thunderstorms cool things off a bit, and often right along the coast, especially the Gulf Coast, temperatures are a bit lower. Average daily temperature on the peninsula all year is about 25 or 26 °C (77 to 79 °F), but daytime highs range often into the mid-30s (°C, or mid-90s, °F). The actual amount of rainfall per year varies from as little as 50 cm (20 in) in some coastal areas of Campeche to as much as 140 cm (55 in) in northern Quintana Roo. Generally there is more rain as you move southwards on the peninsula.

Tourism in the Yucatán peaks (along with hotel prices) during the drier winter months, December through March, when Americans and Canadians descend

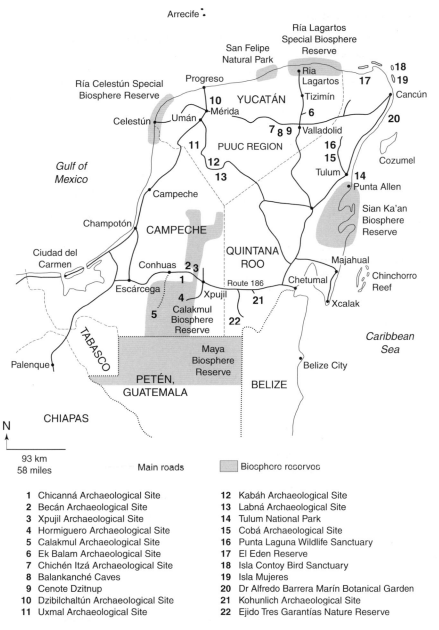

Map 2 Mexico's Yucatán Peninsula, including the states of Campeche, Yucatán, and Quintana Roo. Shown are main roads and approximate locations of biosphere reserves and other eco-attractions.

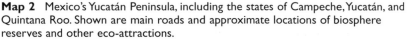

1 Chicanná Archaeological Site	12 Kabáh Archaeological Site
2 Becán Archaeological Site	13 Labná Archaeological Site
3 Xpujil Archaeological Site	14 Tulum National Park
4 Hormiguero Archaeological Site	15 Cobá Archaeological Site
5 Calakmul Archaeological Site	16 Punta Laguna Wildlife Sanctuary
6 Ek Balam Archaeological Site	17 El Eden Reserve
7 Chichén Itzá Archaeological Site	18 Isla Contoy Bird Sanctuary
8 Balankanché Caves	19 Isla Mujeres
9 Cenote Dzitnup	20 Dr Alfredo Barrera Marín Botanical Garden
10 Dzibilchaltún Archaeological Site	21 Kohunlich Archaeological Site
11 Uxmal Archaeological Site	22 Ejido Tres Garantías Nature Reserve

in large numbers. During the summer months, Americans are scarce, but European and Mexican tourists crowd Cancún, Cozumel and cooler coastal areas. Good ecotravel times are spring (April and May) and fall (October and November), periods when traditional tourism ebbs.

Oaxaca, Chiapas, and Tabasco

Oaxaca is a large Mexican state (94,200 sq km, or 36,370 sq miles) the size of Indiana or Portugal that borders 480 km (300 miles) of the Pacific Ocean. Its capital, and only real city, is also called Oaxaca. The state population is close to 3.5 million, concentrated in the central valleys around the capital, coastal areas, and in the western part of the state. Agriculture includes especially sugar cane, corn and green alfalfa (lucerne). Very warm, lowland regions include the Pacific coastal plain, the Oaxacan contribution to the Isthmus of Tehuantepec (the narrowing of the Mexican land mass where North America appears to connect to Central America, and the Gulf of Mexico is separated from the Pacific Ocean by only about 220 km, or 135 miles; see Map 3), and the north-central section bordering the state of Veracruz. The remainder of Oaxaca, two-thirds of it, is mountainous. Most of the state is comprised of the Mesa del Sur, or southern mesa, which consists of highland areas that include eight mountain ranges and the three central valleys. Oaxaca City, at an elevation of 1550 m (5085 ft), sits in the middle of the state, where the central valleys converge, surrounded by high mountains that average 2000 to 2500 m (6560 to 8200 ft) in elevation. The two highest Oaxacan peaks are Zempoaltepec (3395 m, or 11,138 ft) in the Zempoaltepec Mountains east of Oaxaca City, and Yucuyacua (3375 m, or 11,074 ft) in the Yucuyacua Mountains west of Oaxaca. In its Sierra Juárez Mountains, Oaxaca holds one of Mexico's deepest cave

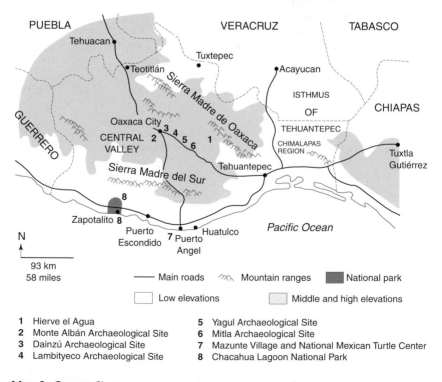

1	Hierve el Agua	
2	Monte Albán Archaeological Site	
3	Dainzú Archaeological Site	
4	Lambityeco Archaeological Site	

5	Yagul Archaeological Site
6	Mitla Archaeological Site
7	Mazunte Village and National Mexican Turtle Center
8	Chacahua Lagoon National Park

Map 3 Oaxaca. Shown are main roads and approximate locations of eco-attractions. Shaded areas show middle and high elevations (in this map, those at 600 m, 1968 ft, or above).

systems, the Sistema Cheve. The caves are near the village of Concepción Pápalo at an elevation of about 2800 m (9100 ft); they extend for at least 24 km (15 miles), to a depth of 1380 m (4500 ft).

Temperature in Oaxaca varies considerably depending on elevation and other factors. Coastal and lowland areas in the north are perpetually very warm or hot; high-elevation sites can be quite cool, with winter frosts and local people always wearing heavy clothing. The central valleys and Oaxaca City are usually warm, high 20s to low 30s °C (80s to mid 90s, °F) during the day, dropping to between 10 and 15 °C (50 to 60 °F) on cooler nights. Average rainfall varies extensively, from as little as 45 cm (18 in) per year in some dry sites to about 500 cm (200 in) in some high-elevation cloud forests. The Pacific lowland region, because it has a pronounced dry season in February through April, enjoys a fairly arid climate; but rain can be heavy during other seasons. Many areas visited by tourists, including Oaxaca City and the central valleys, get about 65 mm (26 in). Most rain falls in June through September, which is also the warmest period in lowland areas. It's also the busiest time for tourists, together with Christmas and Easter seasons.

Chiapas, like Oaxaca, is a large state (at 74,200 sq km, or 28,650 sq miles, the size of South Carolina or Scotland) with complex geography. The capital, Tuxtla Gutiérrez, with about 10% of the state's population of 3.6 million, is located at about 500 m (1700 ft) elevation in the state's western sector (Map 4). Major agricultural crops include corn, beans, coffee, and bananas. Much of central interior Chiapas consists of rugged mountains and other highland topography. Working northward from Chiapas' 265 km-long (165 mile) Pacific coast, there are a series of northwest-to-southeast running topographical features: first, just north of the lowland coastal plain, is the Sierra Madre de Chiapas mountains, mostly between 1000 and 2500 m (3300 to 8000 ft), but rising up to 3000+ m (10,000+ ft); the high point is the Tacaná volcanic peak right on the Guatemalan border, at 4093 m (13,428 ft). Next comes the large Central Valley, or Central Depression, formed mainly of the Río Grijalva valley, at elevations of between 600 and 1200 m (2000 to 4000 ft), with Tuxtla Gutiérrez sitting in its western reaches. To the north of the valley is another highland region called the Meseta Central, most of which is between 1200 and 2400 m (4000 and 8000 ft). This region includes the Sierra Los Altos de Chiapas mountains and the state's second city (but the favorite of most tourists), San Cristóbal de las Casas, at about 2100 m (6900 ft). The extreme northern part of Chiapas (where Palenque is) and the northeastern sector adjacent to Guatemala (the Selva Lacandona region) are at low elevations.

Temperatures in Chiapas vary with elevation and location. The Pacific Coast and low-elevation north, near Tabasco, and northeast, near Guatemala, are warm and humid, with daytime highs averaging in the low 30s °C (high 80s and low 90s, °F). It's cool in the highlands, such as in San Cristóbal and in the many mountain-side villages along the road between San Cristóbal and Ocosingo; daytime temperatures between 10 and 20 °C (50s and 60s, °F) are common, and winter nights are frosty. As in Oaxaca, rainfall varies considerably depending on location. Most areas get between 50 and 100 cm (20 and 40 in) per year, including Tuxtla and much of the Central Valley, but high-elevation cloud forests and rainforests can get much more. Regardless of site, most rain falls between May and October.

Tabasco is a fairly small state (25,268 sq km, or 9756 sq miles), the size of Maryland or Sicily, lying to the north of Chiapas, with a 190 km-long (120 mile) exposure to the Gulf of Mexico. The state has a population of about 1.8 million, 20% of which lives in and around the capital, Villahermosa. Main crops are sugar

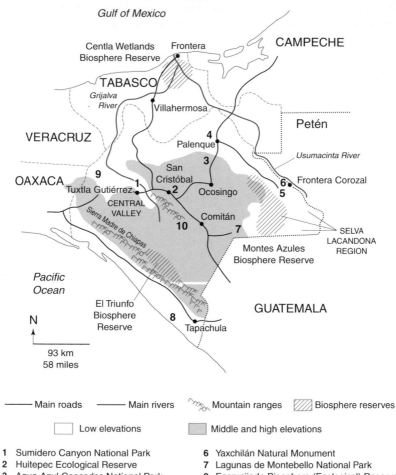

Map 4 Chiapas and Tabasco. Shown are main roads and approximate locations of biosphere reserves and other eco-attractions. Shaded areas show middle and high elevations (in this map, those at 600 m, 2000 ft, or above).

cane, bananas and corn; oil extraction near the coast is a major industry. Tabasco is very low and flat near the Gulf but more hilly in its southern reaches; the topography of the entire state, however, never exceeds 300 m (1000 ft) elevation. Tabasco is hot, humid, and rainy, receiving about 150 cm (60 in) of precipitation per year. The low altitude, the rain, and two major rivers (the Usumacinta and Grijalva converge in the state, form a huge delta, and empty into the Gulf) and their floodplains make much of Tabasco a land of swamps and small lakes. Rainy season is between May and October, but Tabasco is hot and humid all year.

Visitors to these tropical regions should bear in mind that, even during rainy parts of the year, seldom does it rain all day. A typical pattern, for instance, is sunny mornings but late afternoon showers. Also remember that, in contrast to temperate regions, where season largely determines temperature, in the tropics, elevation has the most important effect – the higher you are, the cooler you will be.

Habitats

Tropical Forest: General Characteristics

The most striking thing about many tropical forests is their high degree of species diversity. Temperate forests in Europe or North America often consist of only several tree species. The norm in many tropical forests is to find between 50 and 100 tree species (or more!) within the area of a few hectares or acres. In fact, sometimes after appreciating a specific tree and then looking around for another of the same species, it is not easy to locate one. Ecologists say tropical areas have a much higher *species richness* than temperate regions – for plantlife, as well as for many animals such as insects and birds. The reasons for geographic differences in species richness are not well understood, but are an area of current research interest.

During first visits to tropical forests, people from Europe, North America, and other temperate-zone areas are usually impressed with the richly varied plant forms, many of which are not found in temperate regions. Although not every kind of tropical forest includes all of them, a number of highly typical plant forms and shapes are usually seen.

Tree Shape and Forest Layering
Many tropical trees grow to great heights, straight trunks rising dramatically before branching. Tropical forests often appear layered, or *stratified*, and several more or less distinct layers of vegetation can sometimes be seen. A typical tropical forest has a surface herb layer (ground cover), a low shrub layer, one or more lower levels of shorter trees, and a higher, or *canopy*, tree layer (Figure 1). In reality, there are no formal layers – just various species of trees that grow to different, characteristic, maximum heights. Trees are sometimes referred to as *emergent*, lone, very tall trees that soar high above their neighbors (emergents are characteristic only of tropical forests); *canopy*, those present in the upper layer; *subcanopy*, in the next highest layers; or *understory*, short and baby trees (Figure 1). Many of the *crowns*, or high leafy sections, of tropical trees in the canopy are characteristically-shaped, being short and very broad, resembling umbrellas (Figure 1).

Large-leaved Understory Plants
Tropical forests often have dense concentrations of large-leaved understory shrubs and herbs (Figures 1 and 2). Several plant families are usually represented, including aroids, family Araceae, which include plants such as *Dieffenbachia*, or dumb cane (Figure 12), and climbers such as *Philodendron* (Figure 12); and *Heliconia*, family Heliconiaceae, which are large-leaved perennial herbs (Figure 12).

Tree Roots
Any northerner visiting a tropical forest for the first time quickly stops in his or her tracks and stares at the bottoms of trees. The trunks of temperate zone trees may widen a bit at the base but they more or less descend straight into the ground. Not so in the tropics, where many trees are *buttressed*: roots emerge and descend from the lower section of the trunk and spread out around the tree before entering the ground (Figure 2). The buttresses appear as ridges attached to the

Figure 2 Interior view of a typical tropical forest.

Figure 1 Exterior view of a typical tropical forest.

sides of a trunk, ridges that in larger, older trees, are big and deep enough to hide a person (or a coiled snake!). The function of buttresses is believed to be tree support and, indeed, buttressed trees are highly wind resistant and difficult to take down. But whether increased support is the primary reason that buttressing evolved, well, that's an open question, one that plant biologists study and argue over. Another unusual root structure associated with the tropics is *stilt*, or *prop, roots*. These are roots that seem to raise the trunk of a tree off the ground. They come off the tree some distance from the bottom of the trunk and grow out and down, entering the ground at various distances from the trunk (Figure 2). Stilt roots are characteristic of trees, such as mangroves, that occur in habitats that are covered with water during parts of the year, and many palms. Aside from anchoring a tree, functions of stilt roots are controversial.

Climbers and Vines

Tropical trees are often conspicuously loaded with hanging vines (Figures 1 and 2). Vines, also called *climbers, lianas,* and *bush-ropes,* are species from a number of plant families that spend their lives associated with trees. Some ascend or descend along a tree's trunk, perhaps loosely attached; others spread out within a tree's leafy canopy before descending toward the ground, free, from a branch. Vines are surprisingly strong and difficult to break; many older ones grow less flexible and more woody, sometimes reaching the diameter of small trees. Common vines that climb trees from the ground up are philodendrons and those of the genus *Desmoncus* (or bayal), rattan-like vines used for basket-making.

Epiphytes

These are plants that grow on other plants (usually trees) but do not harm their "hosts" (Figure 2). They are not parasites – they do not burrow into the trees to suck out nutrients; they simply take up space on trunks and branches. (Ecologically, we would call the relationship between a tree and its epiphytes *commensal*: one party of the arrangement, the epiphyte, benefits – it gains growing space – and the other party, the tree, is unaffected.) How do epiphytes grow if they are not rooted in the host tree or the ground? Roots that grow along the tree's surface capture nutrients from the air – bits of dust, soil, and plant parts that breeze by. Eventually, by collecting debris, each epiphyte develops its own bit of soil, into which it is rooted. Epiphytes are especially numerous and diverse in middle and higher-elevation rainforests, where persistent cloud cover and mist provide them ideal growing conditions. Orchids, with their striking flowers that attract bees and wasps for pollination, are among the most famous kinds of epiphytes. Bromeliads (Figure 11), restricted to the Americas, are common epiphytes with sharply pointed leaves that grow in a circular pattern, creating a central bucket, or *cistern*, in which collects rainwater, dust, soil, and plant materials. Recent studies of bromeliads show that these cisterns function as small aquatic ecosystems, with a number of different animals – insects, worms, snails, among others – making use of them. Several groups of amphibians are known to spend parts of their life cycles in these small pools (some salamanders, for example), and a number of species of tiny birds nest in bromeliads. (Not all bromeliads are epiphytes; some, such as pineapple, grow on the ground as largish, spiny plants.) Other plants that grow as epiphytes are mosses and ferns.

Palms

The trees most closely associated with the tropics worldwide are palms. Being greeted by palm trees upon exiting a jet is a sure sign that a warm climate has been reached. In fact, it is temperature that probably limits palms mainly to tropical and subtropical regions. They grow from a single point at the top of their stems, and so are very sensitive to frost; if that part of the plant freezes, the plant dies. Almost everyone recognizes palms because, for trees, they have unusual forms: they have no branches, but all leaves (which are quite large and called *fronds*) emerge from the top of the single trunk; and their trunks are usually of the same diameter from their bottoms to their tops. Many taller palms have stilt roots propping them up. Some palms have no trunks, but grow as small understory plants. Coconut palms, *Cocos nucifera*, found throughout the world's tropical beaches, occur along southern Mexico's coasts, and a number of other palm species are quite common.

Habitats and Common Vegetation

Some of the more common and conspicuous trees and other plants of the region are illustrated on pp. 26 to 34.

Yucatán Peninsula (by Jon Lyon)

The Yucatán Peninsula is linked to the Central American mainland but its geology and vegetation are unique. The peninsula is essentially a geologic platform underlain by gently dipping beds of limestone, chalk and other sedimentary sediments ranging in age from the Cretaceous Period (130 million years ago) to fairly recent (the Pleistocene epoch). This means that the soils in the region are typically alkaline, there is a great deal of *karst* topography (limestone-supported landscape), and there is an extensive underground circulation of water with minimal areas of surface water flow. The presence of rivers and lakes is restricted to the southernmost third of the peninsula. Rainfall is heaviest in the summer months and there is a distinct dry season from December to May.

The peninsula's vegetation is similar in some ways to both the Central American and Caribbean regions. However, the Yucatán does contain a considerable number of endemic plant species and contains more species linked to the Caribbean vegetation than any other region in Mexico. The major habitat types on the Yucatán Peninsula clearly are related to the various forest vegetation types found there. Much of the peninsula, however, has been extensively modified by human activities such as clearing and burning dating back to the Mayan period. Thus, many of the forests found today are secondary, disturbed forests.

A clear vegetation gradient runs from north to south along the peninsula, reflecting a concomitant change in climate; there is a distinct gradient of increasing moisture, precipitation, and humidity as you move from the north and northwest of the peninsula towards the south and southwest into the Petén of Guatemala and northern Belize. The forest vegetation, moving from north to south, follows this climatic gradient, resulting in the presence of distinct forest types: tropical scrub forest (sometimes referred to as *thorn forest*) in the northwest; a relatively thin band of tropical deciduous forest moving south; then a wide band of more tropical semi-deciduous forest extending in a wide swath east to west across the peninsula. There is a small region of scrubland and lowland marsh (*tulares*) in the eastern part of the peninsula, in the southern region of Quintana Roo. There are also small patches of grassland and savanna scattered across southern Campeche and Quintana Roo.

Forests on the peninsula have a history of being highly disturbed and continue to be disturbed through land-clearing and timber exploitation. Thus, there are many early successional plant communities in the region. The most prevalent early successional forest types include assemblages known locally as *guarumal* (dominated by *Cecropia*), *zarzosa* (dominated by plant types such as *Hamelia, Indigofera, Cornutia,* and *Cassia*) and *acahuales* (dominated by copal, *Bursera simaruba*; ciruelo, *Spondias mombin*; and *Alvaradoa amorphoides*). Given this disturbance history, the forest landscape across the Yucatán consists of a mosaic of these successional forests interspersed with patches of the other main forest vegetation types and extensive areas of forest clearings.

Forest and Habitat Types

Mangrove Forests (manglares). A fringe of mangrove forest at one time was found around virtually the entire peninsula. Red mangrove (mangle colorado; *Rhizophora mangle*) (Figure 4) is found in saline areas with white mangrove (mangle blanco; *Laguncularia racemosa*), and black mangrove (mangle negro; *Avicennia germinans*) (Figure 4) found in more brackish waters. The coastal strands also typically support bands of coastal (or *littoral*) forests located on higher areas along beaches. Many of the mangrove forests have been cleared or have been highly disturbed due to clearing and coastal development. The largest expanse of mangrove is in the extreme northwest corner of the peninsula.

Herbaceous Wetlands (tulares). This vegetation type is most prevalent in the southeastern corner of Quintana Roo. The vegetation is dominated by marsh vegetation that reaches from 1 to 3 m (3 to 10 ft) in height. These are poorly drained wetlands found in both fresh and brackish water and that contain a high organic matter content. Common herbaceous plants in these wetlands include cattail (*Typha*), bulrush (*Scirpus*), and *Cyperus*.

Tropical Scrub Forest. These forests have also been classified as thorn forest (bosque espinoso), deciduous seasonal forest or dry deciduous forest, owing to the low canopy of the forests and the stunted appearance of many of the tree species. Scrub forests are restricted to the northernmost regions of the Yucatán and in the southeastern reaches of Quintana Roo. The northern scrub forest region is often called the Steppe region in the Yucatán due to the low rainfall and the relatively high latitude. Canopies tend to be scrubby and broken and typically reach from 5 to 10 m (16 to 33 ft), although some may attain height up to 20 m (65 ft) if soil and moisture conditions are favorable. This forest tends to contain fewer palms and a high proportion of legumes. In addition, many tree species have spines and the majority shed their leaves during the dry season, giving the forests a strong deciduous flavor. The forest is dominated by legumes, including species of *Acacia* and *Albizza*, as well as *Mimosa albida* (Figure 5) and *Lysiloma sabicu* (Figure 5). Other common species include ebano (*Pithecellobium flexicaule, Pithecellobium ebano*), tenaza (*Pithecellobium pallens*), huizache (*Acacia farnesiana*) (Figure 5), and tecomate (*Crescentia cujete*) (Figure 5). Understories in these forests range from sparse and sandy to impenetrable thickets. The thorn forests are often peppered with poorly drained patches called *bajos.*

Tropical Deciduous Forest. This forest forms a somewhat narrow band across the northern Yucatán, sandwiched between the scrubby forests to the north and the more evergreen forests to the south. These forests are transitional and made of mosaics of patches that contain species found in deciduous, thorn, and evergreen forests. Canopies range from closed to slightly open and are typically 8 to 10 m (26 to 33 ft) in height but can achieve heights of 20 m (65 ft). In the dry season, the majority of the tree species shed their leaves, giving the forest its deciduous characteristics. The most characteristic tree in the deciduous forest is yaxnik (*Vitex gaumeri*) (Figure 6). Yaxnik often forms associations with other species including ramón (*Brosimum alicastrum*) (Figure 6), copal (*Bursera simaruba*) (Figure 6), *Sideroxylon gaumeri*, citamche (*Caesalpina gaumeri*) (Figure 6), *Guettarda combsii*, tzalam (*Lysiloma bahamensis*) (Figure 7), and brasil (*Haematoxylon brasiletto*). On deeper and richer soils, guanacaste (*Enterolobium cyclocarpum*) (Figure 12), cedro (*Cedrela mexicana*), ceiba (*Ceiba pentandra*) (Figure 10), and jocotillo (*Astronium graveolens*) are commonly found.

Tropical Semi-deciduous Forest. This forest type is restricted to southern Campeche and Quintana Roo (as well as in northern Belize and across the northern portion of the Petén). These forests have canopies that can reach from 15 to 40 m (50 to 130 ft), but typically are from 20 to 30 m (65 to 100 ft). The canopy may either be closed or partially open. In the dry season from 20% to 40% of the trees shed their leaves. Because of the limestone soils, many of the common tree species are lime-loving, including ramón (*Brosimum alicastrum*) (Figure 6), copal (*Protium copal*) (Figure 8), cedro (*Cedrela mexicana*), zapote (*Manilkara zapota*), tzalam (*Lysiloma bahamensis*) (Figure 7), plantanillo (*Alseis yucatánensis*) (Figure 7), zapote borracho (*Pouteria campechiana*) (Figure 8), and pucte (*Bucida buceras*) (Figure 7). Where ramón is dominant, the forests are often referred to locally as *ramónals* and where zapote is dominant the forests are called *zapotal*. Some palm species may also be common, including cohune (*Orbigyna cohune*) (Figure 7), escoba (*Chrysophila argentea*), and sabal (*Sabal morrisiana*). In these forests, there is also an association known as *caobal* (dominated by mahogany, *Swietenia macrophylla* (Figure 9) and zapote, *Manilkara zapota*).

Oaxaca, Chiapas, and Tabasco (by Irma Trejo Vazquez; translated by Rodolfo Dirzo)

Oaxaca and Chiapas are among the most diverse of the Mexican states in terms of their ecosystems and vegetation. Chiapas is ranked first among the states in number of plant species, with an estimated 10,000; about 8000 species occur in Oaxaca. Tabasco has much lower plant diversity but it contains several major vegetation types and contains Mexico's most extensive wetlands. Throughout the region the most conspicuous types of vegetation include several kinds of tropical forest, ranging from seasonally dry forests to very wet rainforests, as well as temperate coniferous forests, cloud forests, wetlands, and some semi-arid vegetation types. Coniferous and broadleaf temperate forests, or a combination of both, are found in the higher elevation portions of the various mountain ranges of Oaxaca and Chiapas, while Tabasco harbors only lowland vegetation.

Oaxaca

In cooler, high-elevation mountainous regions, above 2750 m (9000 ft), it's possible to find fir forests with trees 20 to 30 m (65 to 100 ft) tall. Frequently these forests are dominated by firs such as abeto (*Abies oaxacana*) and *Abies guatemalensis*, together with cipres (*Cupressus lindleyi*), as well as several species of pine.

Pine forests are found at elevations between 800 and 2200 m (2600 to 7200 ft) in the mountains of the Sierra Madre de Oaxaca and Sierra Madre del Sur. On slopes facing the Pacific Ocean, trees in these forests tend to be taller and have abundant epiphytes; on slopes facing inland, trees are smaller, sparser, and carry fewer epiphytes. The Sierra Madre de Oaxaca region contains about half of all the pine species that occur in Mexico, including pino real (or pino blanco; *Pinus ayacahuite*) (Figure 8), pino lacio (*Pinus michoacana*), pino moctezuma (*Pinus mon-*

tezumae), *Pinus douglasiana*, ocote colorado (*Pinus patula*), and pino hortiguillo (*Pinus lawsoni*).

There are several varieties of broadleaf forests, the type in a given area dependent on the abilities of dominant tree species to make use of available moisture. In drier areas, oaks (*encinos*) predominate; the more moisture that is available, the more pines predominate at the expense of oaks. Intermediate situations are common, with pine–oak forests (Figure 3) covering large areas. The most common trees in these forests are encino amarillo (*Quercus castanea*) (Figure 8), encino colorado (*Quercus sororia*), roble (or encino prieto; *Quercus crassifolia*), encino blanco (*Quercus peduncularis*), encino napis (*Quercus magnolifolia*), encino laurelillo (*Quercus laurina*), and madroño (*Arbutus xalapensis*) (Figure 9).

Cloud forests are found in cool, wet mountain areas on Pacific slopes where fog is frequent, commonly between 1400 and 2250 m (4600 to 7400 ft). For instance, cloud forests occur in northern Oaxaca, along the road south from Teotitlán. Trees usually reach a maximum height of about 30 m (100 ft), and carry heavy loads of epiphytes. Tree ferns are common. Some of the more common trees are liquidambar (*Liquidambar styraciflua*) (Figure 9), flor de la manita (*Chirantodendron pentadactylon*) (Figure 9), *Ulmus mexicana*, *Wenmmania pinnata*, and cucharilla (*Clethra* sp.).

Below 1400 m (4600 ft) in elevation, a warmer climate prevails and tropical vegetation predominates (for example, around Tuxtepec in northeastern Oaxaca and in the Chimalapas region in eastern Oaxaca). Tropical rainforests here reach above 30 m (100 ft) in height. Trees are mostly evergreen, with straight trunks and sometimes with buttresses. Common trees include sombrerete (*Terminalia amazonia*) (Figure 10), palo de agua (*Vochysia hondurensis*), zapotillo (*Pouteria unilocularis*), bálsamo amarillo (*Sweetia panamensis*), guapaque (*Dialium guianense*), ceiba (*Ceiba pentandra*), and palms such as palma real (*Scheelea liebmannii*).

Semi-evergreen Forest. Another variety of tropical rainforest, occupying low humidity lowlands, for example, around the Pacific coastal Chacahua Lagoon National Park. Trees here reach 25 to 30 m (80 to 100 ft), and some drop their leaves during the dry season. The most common trees are ramón (*Brosimum alicastrum*), chicozapote (*Manilkara zapota*) (Figure 10), caimito (*Chrysophyllum mexicanum*), gateado (*Astronium graveolens*), matapalo (*Ficus involuta*), palo mulato (or chaca; *Bursera simaruba*) (Figure 6), guarumbo (*Cecropia obtusifolia*) (Figure 10), and póporo (*Crataeva tapia*).

Sub-deciduous Forest. Another variety of lowland tropical forest, with trees to 20 to 30 m (65 to 100 ft), found, for example, near the Oaxacan coast near the state of Guerrero stretching to Puerto Angel. A larger proportion of trees in these forests drop leaves during the dry season. Common trees: ramón (*Brosimum alicastrum*) (Figure 6), guanacaste (*Enterolobium cyclocarpum*) (Figure 12), ceiba (*Ceiba pentandra*) (Figure 10), *Pterocarpus acapulcensis*, cocuite (*Gliricidia sepium*), and rosa amarilla (*Cochlospermum vitifolium*).

Medium Height Deciduous Forest. Another variety of lowland tropical vegetation, found around the coastal Huatulco area, with trees up to 25 m (80 ft); many trees here drop their foliage during the dry season. Common trees: *Pterocarpus rohrii*, camarón (*Calycophyllum candidissimum*), *Albizzia huachape*, *Jacaratia mexicana*, and cuajiote (*Bursera morelensis*) (Figure 11).

Seasonally Dry Forest. This forest type, which occurs on the coast, from Huatulco to Tehuantepec, has trees around 8 m (25 ft) high, often with highly branched trunks; trees usually lose their leaves during the dry season. Large cacti

Figure 3 Pine–oak woodlands at higher elevations, Oaxaca and Chiapas.

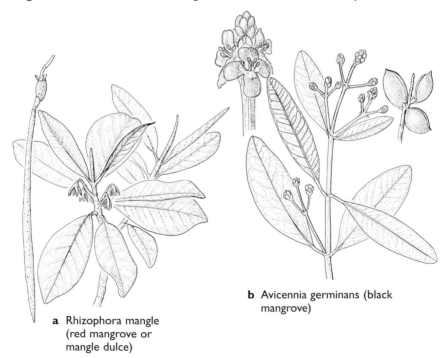

a Rhizophora mangle (red mangrove or mangle dulce)

b Avicennia germinans (black mangrove)

Figure 4

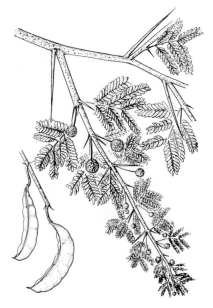

a Acacia farnesiana (huizache)

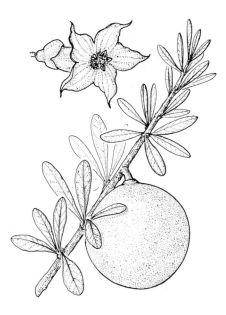

b Crescentia cujete (tecomate or jícara)

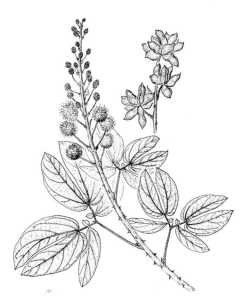

c Mimosa albida

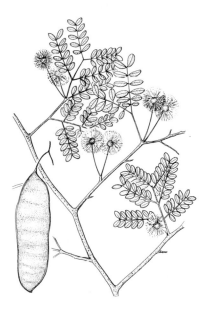

d Lysiloma sabicu

Figure 5

a Vitex gaumeri (yaxnik)

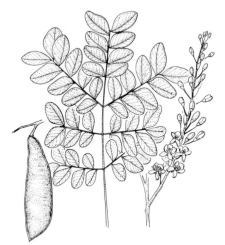

b Brosimum alicastrum (ramón)

c Bursera simaruba (copal or palo mulato)

d Caesalpina gaumeri (citamche)

Figure 6

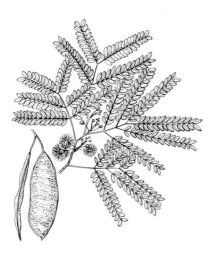

a Lysiloma bahamensis (tzalam)

b Orbigyna cohune (cohune palm)

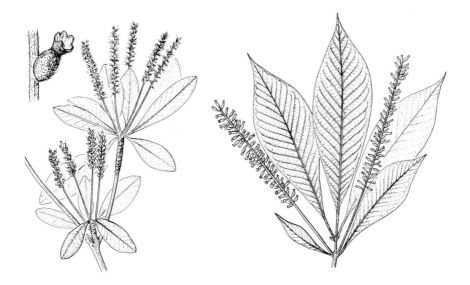

c Bucida buceras (pucte)

d Alseis yucatánensis (plantanillo)

Figure 7

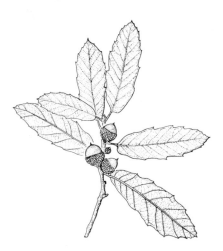

b Pouteria campechiana (zapote borracho)

a Protium copal (copal)

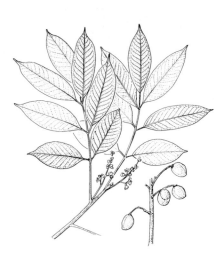

d Quercus castanea (encino amarillo)

c Pinus ayacahuite (pino real)

Figure 8

a Arbutus xalapensis (madroño)

b Liquidambar styraciflua (liquidambar)

c Chirantodendron pentadactylon
(flor de la manita)

d Swietenia macrophylla (caoba or
mahogany)

Figure 9

a Terminalia amazonia (sombrerete)

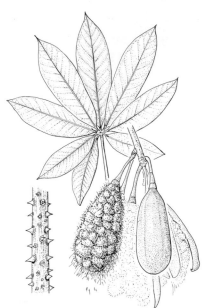

b Ceiba pentandra (ceiba or kapok)

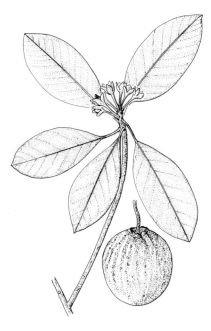

c Manilkara zapota (chicozapote)

d Cecropia obtusifolia (guarumbo)

Figure 10

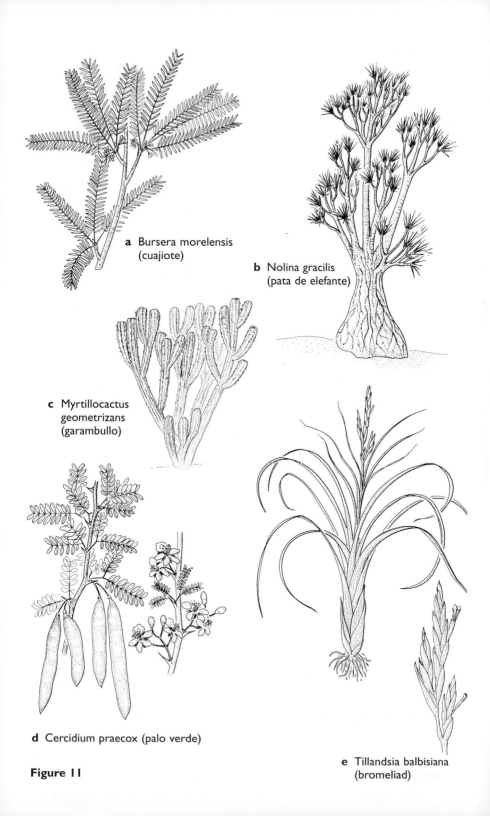

a Bursera morelensis
(cuajiote)

b Nolina gracilis
(pata de elefante)

c Myrtillocactus
geometrizans
(garambullo)

d Cercidium praecox (palo verde)

e Tillandsia balbisiana
(bromeliad)

Figure 11

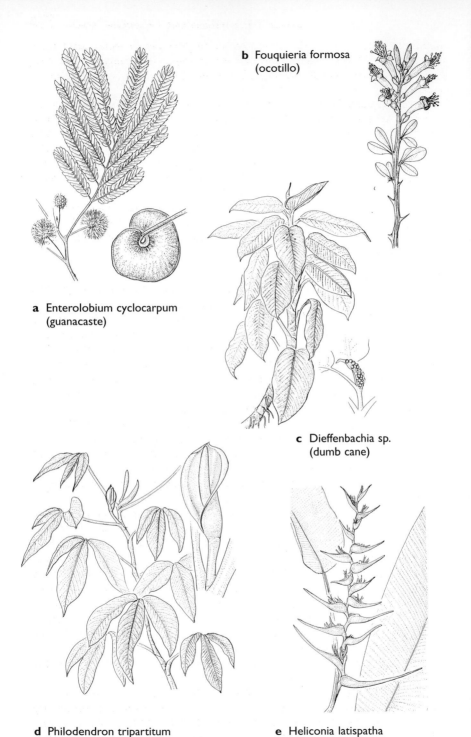

b Fouquieria formosa (ocotillo)

a Enterolobium cyclocarpum (guanacaste)

c Dieffenbachia sp. (dumb cane)

d Philodendron tripartitum

e Heliconia latispatha

Figure 12

intermingle with the trees. Common trees are cuajiote (*Bursera morelensis*) (Figure 11), rosamorada (*Tabebuia rosea*), *Bucida machrostachya*, *Lonchocarpus emarginatus*, pochotl (*Ceiba aesculifolia*), *Stenocereus pecten-arboriginum*, and *Cnidosculus megacanthus*.

Scrub Vegetation. Low scrub vegetation (*matorrales*) can be found in the driest parts of Oaxaca, such as in the Cuicatlan Valley between Oaxaca City and Tehuacan. Common species here are pata de elefante (*Nolina gracilis*) (Figure 11), ocotillo (*Fouquieria formosa*) (Figure 12), palo verde (*Cercidium praecox*) (Figure 11), and cacti such as pitayo viejo (*Cephalocereus chrysacanthus*), candelabro (*Pachycereus weberii*), quiotilla (*Escontria chiotilla*), organo (*Neobuxbaumia macrocephala*), and garambullo (*Myrtillocactus geometrizans*) (Figure 11).

Mangroves. Mangroves, with plants such as mangle dulce (or mangle colorado; *Rhizophora mangle*) (Figure 4), mangle blanco (*Laguncularia racemosa*), and mangle negro (*Avicennia germinans*) (Figure 4), are associated with coastal areas, such as around Chacahua Lagoon National Park.

Chiapas

The most notable aspect of Chiapas' vegetation is its extensive areas of tropical rainforest, particularly the Lacandon rainforest (Selva Lacandona). However, temperate forests, both coniferous and broadleaf, as well as cloud forests, are also prominent. Large areas of savannas and mangroves also occur. (Mangrove forests have the same species as those in Oaxaca, with mangle cenizo, *Conocarpus erectus*, also being common.) Near the Pacific coast, seasonally dry forests with many thorn-bearing plants are common. Savanna-like areas also occur in this region, with spectacular trees such as jícara (*Crescentia cujete*) (Figure 5). In the Central Valley, around Tuxtla Gutiérrez, other types of savanna occur, with treelets such as nanche (*Byrsonima crassifolia*) and cacaito (*Curatella americana*). It is still possible to find sites here covered by seasonally dry forest, with trees reaching to 12 to 15 m (40 to 50 ft), especially surrounding the city of Tuxtla and Sumidero Canyon. Some of the common trees here are *Lysiloma divaricata*, *Cordia alliodora*, *Bucida machrostachya*, *Ceiba acuminata*, *Cedrela salvadorensis*, *Gyrocarpus mocinoii*, *Plumeria rubra*, and *Bursera excelsa*.

The most extensive and best preserved of Mexico's tropical rainforests are found in Chiapas. The tallest tropical trees in Mexico are in the Montes Azules Biosphere Reserve area; some, such as mahogany (or caoba), reach up to 70 m (230 ft). In more accessible areas, such as in Bonampak and Yaxchilán Natural Monuments, and Palenque and Agua Azul National Parks, it is possible to see some of this habitat's typical trees, including caoba (*Swietenia macrophylla*) (Figure 9), ceiba (*Ceiba pentandra*) (Figure 10), sombrerete (*Terminalia amazonia*) (Figure 10), guapaque (*Dialum guianense*), chicozapote (*Manilkara zapota*), mamba (*Pseudolmedia oxyphyllaria*), *Bernoullia flammea*, and guarumbo (*Cecropia obtusifolia*). Palms such as camedora (*Chamaedorea tepejilote*), chocho (*Astrocaryum mexicanum*), and palma real (*Scheelea liebmannii*), as well as ferns, are abundant in the understory, and vines, lianas, and epiphytes are also common.

On the Pacific slopes of the Sierra Madre de Chiapas, a host of different forest habitats, an entire vegetational gradient, can be observed, particularly on the trip to the El Triunfo Biosphere Reserve. Seen are medium-height tropical forest, tropical tall evergreen forest, and cloud forest. Common cloud forest trees include liquidambar (*Liquidambar styraciflua*) (Figure 9), encino (*Quercus oocarpa*), *Matudea trinervia*, *Dendropanax populifolius*, *Ocotea chiapensis*, *Trophis cuspidata*, *Oreopanax* sp., and magnolia. Bromeliads, orchids, and cycads are very common. These cloud forest species also occur in the Lagunas de Montebello National Park region.

The tropical lowland vegetation of Chiapas is quite similar to that found in Oaxaca (p. 25), sharing many of the same characteristics and tree species.

Near the city of San Cristóbal, in the area of the Sierra Los Altos de Chiapas, several types of temperate, coniferous, and broadleaf forests are found:

Oak Forests. Oak forests often cover large portions of Chiapas' mountains. Some of the most common oaks are encino (*Quercus brachystachys*) in the vicinity of San Cristóbal; encino blanco (*Quercus peduncularis*) in the Lagunas de Montebello area; encino amarillo (*Quercus glaucescens*) in the Agua Azul area; and *Quercus sebifera*, only about 4 m (13 ft) tall, in the Comitán area.

Pine Forests. In higher, cooler sites, pine forest predominates, with pino real, (*Pinus ayacahuite*) (Figure 8), the tallest Chiapas pine (to 40 m, 130 ft), as well as pino moctezuma (*Pinus montezumae*), *Pinus pseudostrobus*, *Pinus teocote*, and pino prieto (*Pinus oocarpa*).

Fir Forests. These forests are found in some regions with steep slopes and high humidity.

Tabasco

Most of this state consists of Gulf of Mexico coastal plains, where several major rivers meet the sea, forming huge wetland areas. Northeast of the city of Villa-hermosa, on the way to Frontera, are the extensive Pantanos de Centla (Centla swamps). This is a major area of mangrove forests, as well as several semi-aquatic habitats given various local names: *mucales*, dominated by mucal (*Dalbergia brownei*), majagua (*Hibiscus tilaceus*), and mangle negro (or mangle botoncillo; *Conocarpus erectus*); extensive areas of *tulares*, with cattail (*Typha latifolia*); *popales*, with popal (*Thalia geniculata*) up to 2 m (6.5 ft) high and heliconia (*Heliconia latispatha*) (Figure 12); and, in zones subject to temporary flooding, *tintales*, with tinto (*Haematoxylon campechianum*), pucte (*Bucida buceras*) (Figure 7), and palma real (*Sabal mexicana*). Small fragments of tropical lowland forests can be found throughout the state, particularly medium-height and some tall evergreen forests, but deforestation has been extensive. The predominant landscape is that of remnant forests, isolated trees left standing, and wet grasslands used for cattle ranching.

Chapter 3

Parks, Reserves, and Getting Around

Getting Around Southern Mexico

Driving Here and There

Most of the main roads in southern Mexico, you may be surprised to learn, are in quite good repair. Driving to most of the sites listed in this chapter is fairly easy, at least during drier parts of the year. Local buses will get you to some but not all of them. Most people visit these places with tour groups, including tours put together by local travel agencies in Oaxaca City, Tuxtla, Mérida, Cancún, etc. Driving in the Yucatán Peninsula is usually relaxed – most terrain is flat and roads tend to be nice and straight. Oaxaca and Chiapas, in contrast, hold an impressive amount of mountainous country with twisting roads that make for adventurous driving (see p. 43). Roadsigns are OK along main highways, but terrible within cities and towns. In fact, it's impossible to exit Cancún's hotel zone and find the main road to Chichén Itzá without getting lost at least twice; and everyone will make wrong turns in Tizimín, near Ría Lagartos, and in Umán on the way to Ría Celestún; but, of course, getting lost sometimes adds charm to a trip. Gasoline is readily available from the ever-ready, strategically spaced Pemex stations. If you rent a car in the Yucatán Peninsula, get a good spare tire; also, pay the extra money for air-conditioning or face certain heat-stroke.

Notes Concerning Interactions with Humans

Let me say first that most of the people you meet in Mexico are really nice. But a few groups deserve special mention: cops and kids. Cops are very rarely stationed along main roads, waiting for speeders and other criminals. Rather, you run into them mainly in the cities and towns – traffic police, city police, civil militia. They are thick along downtown streets, and will whistle and pull you over for many a small infraction. You don't want to deal with this situation – so drive extremely carefully in towns and cities. If pulled over and you're very lucky, you might be able to play the dumb tourist, have the cop feel sorry for you, and get away without handing over any money – this has actually worked for me. The Mexican Army, apparently always in the form of boys in their late teens who keep their machine guns trained on your legs, is another story entirely. In Chiapas, due to political problems, and anywhere in Mexico near the Guatemala or Belize borders, expect frequent army checkpoints at which passport and tourist card may need to be produced. They are looking in particular for drugs and will search baggage. If there's a lot of vehicle traffic, such checkpoints can slow down your trip, but usually it's just a minor inconvenience.

As for kids at parking areas at some of the archaeological ruins, at cenotes, at Agua Azul waterfalls, etc., who want to wash your car, either let them, but agree first on the price, or tell them, no, but if they carefully watch your car ("se lo cuido") while you're away you'll hand over some pesos. Kids at some of the main highway entrances to cities run over and wash your windshield while you're stopped at a light – often before you can say not to. Your options are screaming at them, ignoring them and driving away, or giving them a few pesos. If windows are open, watch valuables on car seats, etc.

Topes and Topes Culture

There are few stoplights in southern Mexico. Rather, traffic is slowed around human habitations by the presence of topes (TOE-pace, plural of tope), or speed-bumps. If you drive in this region you will get quite tired of shifting gears between topes, which sometimes are spaced every 100 m (300 ft) along the main highway as it cuts through a village or town. Some topes are merely small rumbles in the road, others are monsters that scrape the oilpan of your car no matter how slowly you cross. Then there are the seemingly newer, high-tech "vibradores" – which I leave to your imagination. Tope construction is interesting. Some are concrete, some are blacktop, some are hard-packed dirt, some, near the coast, are old nautical ropes. The most important thing to remember about topes is that they can be highly cryptic – they can hide and surprise you. Most are warned about with signs: "Topes, 100 m," "Topes, 50 m," and "Topes" with an arrow down indicating that right here, at last, are the aforementioned topes. Unfortunately, some topes, the cunning ones, are not marked – and they have a sneaky way of melding into the road's surface so that they are virtually invisible as you approach. To hit topes at 60 or 70 kph – well, it's not pretty. Drive carefully and concentrate on the road. If you see traffic ahead slowing and stopping for no apparent reason, slow down quickly – it's topes. When the sun is low and slim tree shadows fall across the road, driving gets tricky – every shadow appears as possible topes.

Another interesting facet of topes: because cars and trucks must reduce speed for topes, actually slow to a crawl, some village people, mostly youngsters, have come to spend much of their free time hanging around topes – what I call, "topes culture." They stand or sit or bike-ride by the roadside at the topes, talking in small groups, waiting for vehicles – this is their entertainment. They check out cars and occupants as they pass. More enterprising kids are topes vendors – they sell food and trinkets to the slowed traffic. In the evening, villagers tend to gather at the main highway that splits their village and at topes – often, it seems, the only paved spot in town.

Descriptions of Parks, Reserves, and Other Eco-sites

The parks and reserves described below were selected because they are the ones most often visited by ecotravellers or because they have a lot to offer. Some of the sites are obvious places for hiking along trails and wildlife watching, but in southern Mexico, most remote wild areas are still undeveloped for ecotourism and, consequently, very few people reach those sites. On the other hand, the attractions many people do visit are archaeological ones, and these are as good a place as any for wildlife sightings – they are often highly protected forested areas near water. Table 1 rates many frequently visited archaeo-sites as wildlife areas. The animals profiled in the color plates are keyed to parks and reserves in the following way: the profiles list the Mexican states (OAX = Oaxaca, CHI = Chiapas, TAB = Tabasco, CAM = Campeche, YUC = Yucatán, QRO = Quintana Roo) in which each species is likely to be found, and the parks listed below are arranged by state. State and park locations are shown in Maps 2, 3, and 4 (pp. 15, 16, and 18). If no special information on reaching parks and on wildlife viewing is provided for a particular park, you may assume the site is fairly easy to reach and that wildlife can be seen simply by walking along trails, roads, beaches, etc. Tips on increasing the likelihood of seeing mammals, birds, reptiles, or amphibians are given in the introductions to each of those chapters.

Oaxaca

Central Valley, Within Easy Day-trip Distance of Oaxaca City

Hierve el Agua

This is a commonly visited naturalistic tourist site within easy driving distance of Oaxaca City. Located about two hours south and east of the city, past the town of Mitla, it is only a short drive off the main road. The attractions are interesting mineral springs and the rock formations created over the centuries by the mineral deposits from the water flow. The springs bubble out of the rocky ground (which, in some places, has been reshaped by government workers to create terraced pools) and produce some large, spectacular mineral deposits. The shallow water spreads out over a rock surface and finally oozes over the water-smoothed edge of a cliff. From the cliff-top there are great views of the agricultural region on the valley floor far below. A short trail, the following of which requires shimmying down some slippery boulders where a slip could cause serious injury, leads through a beautiful cactus and agave forest to a secondary site of cliff-side

Table I Ecotravellers' Quick Guide to Archaeological Sites

Commonly visited archaeological sites: sure, they're good spots to look at heaps of ancient rock in sweltering heat, but how do they rate as places to see wildlife?

Size indicates the area that contains most of the major, excavated temples, pyramids, etc, that is, the area most visitors walk around; it does not indicate that actual area of the reserve in which the major ruins are located, which is often much larger.

Wildlife rating is based on the relative amounts of wildlife sighted on my visits (mostly birds, lizards and occasional mammals such as monkeys) and conversations with tour-guides. (1 = awful, 10 = super)

Site	Size	Crowds?	Views?	Forested?	Trails?	Wildlife rating
Oaxaca						
Monte Albán	medium	at times	amazing	no	no	5
Mitla	small	at times	no	no	no	1
Yagul	smallish	no	great	no	some	5
Dainzú	small	no	from top of hill behind ruins	no	no	3
Chiapas						
Bonampak	smallish	no	no	yes	some	7
Yaxchilán	medium	no	from top of ruins	yes	some	8
Palenque	large	at times	from top of ruins	yes	yes	7
Campeche						
Chicanná	small	no	no	yes	limited	6
Becán	medium	no	from top of ruins	yes	yes	8
Xpujil	small	no	no	yes	no	6
Hormiguero	smallish	no	no	yes	limited	8
Yucatán						
Ek Balam	largish	no	no	yes	no	7
Chichén Itzá	large	often	from top of ruins	no	no	4
Dzibilchaltún	medium	yes	no	yes	limited	6
Uxmal	large	at times	yes	yes	some	7
Kabáh	medium	at times	from top of ruins	yes	limited	6
Labná	smallish	no	from top of ruins	yes	no	5
Quintana Roo						
Tulum	smallish	often	oceanside	some	limited	2
Cobá	large	possible	from top of ruins	yes	yes	7
Kohunlich	large	no	from top of ruins	yes	yes	8

mineral deposits with sweeping views. Hierve el Agua, inevitably, has vendors selling crafts and refreshments, but they remain in a limited area near the parking lot. There are picnic tables, often used by local families for weekend outings. In addition to the springs, also worth seeing is the green mountain scenery and pine forests along winding mountain roads from Mitla to the Hierve el Agua turnoff, and beyond, onwards to Ayutla (a rustic, high-elevation town).

Monte Albán

A medium-sized archaeological site, the most heavily visited one in Oaxaca. Located only a short distance southwest of Oaxaca City, on a mountain-top plateau easily reached by bus or taxi. The ancient Zapotec ruins are set in a very open, almost treeless site, fully exposed to the sun, and crowds are possible. But it is a stunning, unique site, with clear, high mountain views of the Central Valley in all directions; even if not high on a nature-lover's list, the place is worth a visit. Behind some of the ruins are paths that meander around the periphery of the site and start down the mountain toward settlements below. A good many species of lizards can be spotted on the ruins and along these paths. In the morning, these peripheral, scrubby parts of the site make for decent birding habitat – flycatchers, including Vermilion Flycatcher, robins, swallows, sparrows, grosbeaks, and various oriole species are commonly seen.

Mitla, Yagul, Lambityeco, Dainzú

Four frequently visited Mixtec and Zapotec archaeological sites southeast of Oaxaca City, all lying along and very near to the road to Mitla. All four can be seen easily during a single day-trip. Mitla is located within a town, so is not a wildlife site. Lambityeco and Dainzú are small, open sites with little to recommend them from wildlife or ecological perspectives, although a gravel track, good for birding, leads through and beyond the Dainzú ruins to rocky, brushy pastures beyond. Easily the best of the four sites, from archaeological, scenic, and wildlife points of view, is Yagul. It is set on a hilltop with 360-degree panoramic vistas of the agricultural valley below. Beyond and above the stone ruins are some beautiful, cactus-laden trails with occasional breathtaking views of the surrounding countryside. Along the trails are lizards and many cactus habitat birds, among them woodpeckers, flycatchers, wrens, and sparrows.

Pacific Coast

Mazunte Village and National Mexican Turtle Center

Located 12 km (7 miles) west of the coastal town of Puerto Angel. Before sea turtle harvesting was made illegal in Mexico in the early 1990s, Mazunte was a center of the turtle industry. Now many in the area try to make a living through ecotourism. With help from environmental organizations and some government funds, local people have established an ecological reserve for sea turtles and a turtle lab and museum meant to attract visitors. The turtle center conducts research on sea turtles and has a small aquarium to show visitors several species of sea and freshwater turtles (www.eden.com/~tomzap/turtle.html). Sea turtles still use the beaches extensively in this area and westward for egg-laying. Fishermen in the area offer boat trips to look for ocean-going turtles, dolphins, seabirds, etc. Some visitors enjoy hiking in this area along the beach and its associated lagoons. One hike begins west of Mazunte at the tiny coastal settlement of Ventanilla and proceeds west along the beach about 3.5 km (2 miles) to a lake.

Chacahua (or Chacagua) Lagoon National Park

Located along the coast about 60 km (37 miles) west of the small resort town of Puerto Escondido, 130 km (80 miles) west of Puerto Angel. The park consists of several mangrove-fringed lagoons, river outflows, and surrounding coastal areas.

Local people from the coastal village of Zapotalito take visitors on one- to three-hour boat trips through the lagoons, along the mangroves and small islands. Wildlife frequently seen includes crocodiles and turtles and a good number of water-associated birds, including herons and egrets, spoonbills, storks, seabirds and marsh birds. Sea turtles nest extensively in beach areas of the park. Most people visit via tours from Puerto Escondido, but you can also drive to Zapotalito, which is not far off the main highway. (Area: about 14,200 hectares, 35,000 acres. Elevation: near sea level. Habitats: mangroves, coastal dunes and beaches, semi-deciduous tropical forest, savanna.)

Isthmus of Tehuantepec

As of this writing, there were no regular ecotourism sites in the isthmus, although one was in the planning stages. Local isthmus people in the villages of San Miguel Chimalapa and Santa Maria Chimalapa pushed for creation of a protected federal reserve, the Los Chimalapas Campesino Reserve, and conservation organizations are helping to establish ecotourism in the area. Also, in extreme western Chiapas lies the Selva del Ocote Special Biosphere Reserve. Although not far from Tuxtla Gutiérrez, it is fairly difficult to access. The reserve seeks to protect large, relatively undisturbed patches of evergreen and semi-deciduous tropical forest at elevations of between 180 and 1450 m (600 to 4800 ft).

Chiapas

Miguel Alvarez del Toro Zoo (Tuxtla Zoo, or Zoomat)

The Tuxtla Zoo, on the outskirts of the city and an easy taxi ride from the city center, is widely considered one of Mexico's best. It's a fine place that exclusively exhibits Chiapas wildlife and, because it's fairly extensive and has plenty of shade trees, its a great place to spend an entire afternoon. If I remember correctly, it's also free. More than 200 species of animals, including many mammals, birds, and reptiles (not to mention tarantulas, scorpions and cockroaches!) are housed mainly in largish, naturalistic enclosures. In addition to caged exhibits, free-ranging zoo-roamers include agoutis (p. 222), a host of chachalacas (p. 139), Great Currasow (Plate 27), and Highland Guan (Plate 27) (at least, they all roamed free when I was there – I assume it was intentional and that they were not all escapees.) My favorites at the zoo were a bevy of playful baby coatis (Plate 79) climbing all over the trees in their enclosure, the nocturnal house with incredibly active night mammals, the good collection of snakes, including many venomous ones, and a huge enclosure with a dozen or so frisky, bickering Scarlet Macaws (Plate 34) – their screeching and squawking could be heard throughout the zoo. This may be the only zoo to exhibit the Resplendent Quetzal (Plate 43).

Faustino Miranda Botanical Garden

This garden is in the middle of the city of Tuxtla, part of the downtown cultural center (Parque Madero Complex), adjacent to theaters, museums, and the children's amusement park. It exhibits mostly local plant species, but it's smallish with little wildlife – a squirrel here, a city bird there. It opens at 9:00 am, so late in the morning for birding. The paths are heavily sprayed in mornings to kill mosquitos, and the place reeks of the stuff when it opens – to the point of making you queasy. Skip it.

Sumidero Canyon National Park

A great ecotourism bargain, easily reached by bus or taxi from Tuxtla. For a few dollars, you get a two-hour, 35-km (22-mile) river boat trip through a breathtakingly beautiful, towering, winding canyon. The canyon walls, sometimes of sheer stone, sometimes with trees growing from small ledges, rise up to 1000 m (3300 ft) from the water. The word "spectacular" is overused in travel books, including mine, but Sumidero truly is so – photos do not do it justice. On my visit I expected a pleasant river trip, but what I got was a crick in my neck from constantly scanning upwards at the towering walls and a whack in the ribs from a companion who told me to stop saying "wow" – the European tourists were becoming nervous of me. The guide took the small boat into a shallow bat cave at the base of the canyon (the sleepy bats a mere meter (3 ft) or so above our heads), into other small cave-like rock formations with stalactites, and into the fine spray mist that fell into the river from a small waterfall that arose from a seep from the canyon wall about midway up; the perpetual spray had created below it a small, mossy green patch of tropical rainforest – different from the surrounding, dryer habitat. Wildlife commonly seen in Sumidero includes some crocodiles and many birds: seabirds, water birds, herons and egrets, marsh birds, parrots in the trees high above, vultures, kingfishers, and, should you bring binoculars, songbirds. If you go, remember to look up and not down, because the river itself, the Grijalva, is fairly dirty because it carries much of the big city of Tuxtla's sewage. It's now mostly slow-moving owing to the Chicoasén dam, one of the world's tallest; the dam, finished in 1980, is the turn-around point for all river trips. Sumidero is a popular tourist destination. Boats leave frequently from the town of Chiapa de Corzo, about 15 km (9 miles) from Tuxtla. Don't let the water pollution and possibility of encountering small gangs of tourists discourage you – it's a worthwhile few hours. Taxis, for mucho pesos, will also take you to high lookouts at the canyon rim. (Area: 21,790 hectares, 53,800 acres. Elevation: 500 to 1500 m, 1600 to 4900 ft. Habitats: aside from the river, mostly deciduous forest, with some conifers.)

The Road from Tuxtla Gutiérrez to San Cristóbal de Las Casas and on to Palenque

If you like to drive, try doing so in Chiapas. It's challenging, fun, and the scenery can't be beat, but it's not for the faint-hearted. Driving here requires *total* concentration. Roads are fairly good so you can speed along, but constant S-curves and hair-pin turns tend to slow you down. Roads, through beautiful pine forests, cling to the sides of mountains, often above the clouds, with no shoulders or guard rails. Heavy trucks and buses appear ahead of you moving uphill at a snail's pace, and close behind you, speeding, going downhill. Around each bend, new obstacles appear on the road, just like in a video game – people, dogs, donkeys, cattle, major depressions in the road surface (one of my favorite road signs apparently translated as "danger – shifting geology"). Add to this assorted army and police checkpoints and fantastic views that tend to divert attention, and you get one heck of a driving experience. There are relatively few good places to pull off the road if you want to nose around, birdwatch, or hike a bit, and most of the land is privately held by individuals or village groups. If the countryside really intrigues you, try stopping in some of the high mountain villages through which the main road passes and asking about walking along village trails. If that doesn't work, there's a small nature reserve (135 hectares, 330 acres) about 4 km (2.5 miles) northwest of San Cristóbal (on the road to Zinacantán and San Juan Chamula), the Huitepec Ecological Reserve (also called the Pronatura Reserve), in which you can walk for an hour or so along a steep trail in cloud forest and oak woodlands (at about 2400 m, 8000 ft).

Agua Azul Cascades National Park

About 75 minutes south of Palenque, a few kilometers off the main road to San Cristóbal, lie the famous blue waters of the Agua Azul cascades. Photos in tour-books show wonderful pastel blue and blue-green terraced falls, but the striking blue colors are apparent only in some seasons (usually spring) at some water levels. What most visitors see is a series of mist-producing waterfalls, none individually very high, but, in sum, a nice site. Trails from the parking area lead up-river through forest to view the higher falls. (For the adventurous, there's a trail to more spectacular falls – three of them, actually – about 3 km (2 miles) downstream. The trail passes through ejido-owned agricultural lands, and so it's smart to take along as a guide one of the kids who hangs around the parking area.) Local people crowd the site on weekends and holidays for swimming and to picnic or to sit at the plastic tables of the many small restaurants that line the trail, sipping cool drinks. The cascades are pretty and some tourbooks gush about them, but from a wildlife viewpoint it's disappointing and the site is now too commercialized and deforested; some would place it in the tourist trap category. *Caution*: It's usually really sunny, hot, and humid in the area and you will be tempted to take a swim in the river. Be careful and swim only where the locals do; the swirling waters have claimed a few ecotravellers. (Area: 2580 hectares, 6400 acres. Elevation: 200 to 500 m, 650 to 1600 ft.)

Palenque National Park

This park contains what many experienced archaeo-travellers consider the best, indeed, their favorite, ancient ruins site in southern Mexico. If you were disappointed by the too-open, park-like settings of Tulum and Chichén Itzá, Palenque is the remedy. It's also a good place for wildlife, and opens at 8:00 am, so morning birding is possible. The extensive ruins, some excavated and partially restored, some still melded with jungle, are set amongst towering trees of dense lowland tropical forest. Huge, epiphyte-encrusted trees are everywhere; strangler figs (a kind of fig that starts out as a vine going down a tree's trunk, but eventually surrounds the tree and kills it) are common. Many of the ruins are connected by jungle trails – wet, leaf-and-vine-strewn, sometimes rocky, sometimes muddy – sufficiently wild and unkempt to be good places to keep a lookout for snakes. Trails connect the museum (about 1 km away) to the main ruins, and other trails lead through parts of the national park not associated with the ruins (one begins behind the Temple of Inscriptions and runs for several miles; ask about such trails at the park entrance or at the museum). You can also follow the river at the ruins some distance and birdwatch along it. Small and mid-sized lizards, including Jesus Christ lizards, are commonly seen, as are many birds, particularly parrots, motmots, flycatchers, and many small songbirds; howler monkeys are fairly common. Visits to Palenque declined dramatically during the mid-1990s due to political troubles in Chiapas, but tourism in the region is recovering nicely; still, most days Palenque is relatively uncrowded. (Area: 1770 hectares, 4400 acres. Elevation: sea level to 300+ m, 1000+ ft. Habitats: the park preserves one of the last patches of intact evergreen tropical rainforest in the region.)

Bonampak and Yaxchilán Natural Monuments

Located very nearly at the end of the Earth (or as close as you can get within 1600 km, or 1000 miles, of Miami), the splendid Bonampak and Yaxchilán archaeological zones, in their remote jungle settings, are worth going out of your way to reach. The sites are situated in the Lacandon rainforest (Selva Lacandona) near the Guatemalan border. Conservation organizations and Mexico's National Institute of Ecology have worked with the local Lacandon community to establish tours to these sites (see p. 61). The two-day tours originate in the town of Palenque, where hotels or travel agencies can put you in touch with the tour operators. Bonampak is the smaller site, in terms of excavated ruins and things to do. An hour or two there is probably suffcient time to look around. There is one main group of ruins, but they contain an extraordinary sight – colorful wall and ceiling frescoes painted, supposedly, more than 1000 years ago. There's not much in the way of trails, but there is a long, grassy runway along which you can walk and birdwatch. Common birds at the ruins and at the village of Lacanja Chansayab, where tours stay overnight, are woodpeckers, doves, anis, flycatchers, wrens, orioles, blackbirds, and tanagers. The basic tour includes a more than two-hour walk to Bonampak from the village, the first 45 minutes of which are along a nice forest trail. A new road to Bonampak also permits tours to drive directly to the site, and is open to bicycles. (Area: 2620 hectares, 6470 acres. Elevation: 200 to 500+ m, 650 to 1600+ ft. Habitats: evergreen tropical rainforest.)

Yaxchilán is the more extensive, more impressive site, and the one better for wildlife viewing. It's also more fun to get to, as it involves an hour boat trip on the Usumacinta River (the border between Mexico and Guatemala). River access is from the town of Frontera Corozal, where conservation organizations have assisted local people in setting up ecotourism businesses, including river transportation to Yaxchilán (see p. 61). The actual ruins enjoy a magnificent setting among huge, buttressed trees crowded with epiphytes and hanging vines. The ruins themselves are interesting, with access to most of them requiring walking through an ancient, totally dark stone tunnel with wet stone steps. Some mysterious narrow trails lead off from the periphery of the ruins, presumably made by the archaeologists who worked at the recently opened site. Although better than Bonampak, animal sightings often are only so-so. Wildlife along the river is limited because most of the way small farms line the river's edge, not virgin jungle. Also, because it's usually late morning or early afternoon before you arrive at Yaxchilán, birding is limited. Still, it's a great trip; lizards, some birds, and howler monkeys are commonly seen, and the main point, I think, is the fun and adventure just getting there – the remoteness of the place and the scenery. (Area: 4360 hectares, 10,800 acres. Elevation: sea level to 200 m, 650 ft. Habitats: river, evergreen tropical rainforest.)

Montes Azules Biosphere Reserve

This huge reserve seeks to protect a significant portion of the Lacandon rainforest (Selva Lacandona), which is part of the largest expanse of tropical rainforest in Mexico (and so in North America), and perhaps Mexico's top conservation priority. Together with the adjacent Petén region of Guatemala and the Yucatán's Calakmul Biosphere Reserve, the area contains the largest stand of virgin rainforest north of the Amazon region of South America. Although it includes much less than 1% of Mexico's total area, preliminary surveys indicate that Montes Azules boasts approximately 20% of the country's biodiversity (its total number of plant and animal species), including 3000+ plant species, 320 orchids, 112 mammals, 300+ birds, 84 reptiles, 25 amphibians, and 800+ butterflies. Access to Montes Azules is not easy – it's a remote area. Most people wanting to visit the Selva Lacandona do so by trips to Bonampak and Yaxchilán (technically outside the borders of Montes Azules). However, conservation organizations have been work-ing with one ejido (land-owning community association – see p. 62), Emiliano Zapata, near Laguna Miramar (Lake Miramar) to establish an ecotourism site. Laguna Miramar is the largest lake in Selva Lacandona. It's located in a wilderness area at about 200 m (650 ft) elevation, surrounded by mountains. Visitors to the site stay in a communal hut, and there are trails, caves, and small sets of archae-ological ruins to explore, in addition to lake-associated activities such as canoe-ing. Main access is via the road from Ocosingo to San Quintin, on the reserve's western border. Air and boat access are also possible. For information, contact the Dana Association in San Cristóbal; tel: 967-80-468; e-mail: danamex@mail.inter-net.com.mx (Area: 331,200 hectares, 818,000 acres. Elevation: 120 to 1300 m, 400 to 4250 ft. Habitats: evergreen and semi-deciduous tropical rainforest, palm forest, pine and oak forest, swamp forests, lakes, rivers.)

Lagunas de Montebello National Park

This scenic spot is located about 3.5 hours from Tuxtla, or 2.5 hours from San Cristóbal, near the Guatemalan border. The main attraction is a string of 50+ pretty lakes, some nicer than others, ringed closely by mixed coniferous and coniferous–deciduous woodlands. The lakes formed when the ceilings of under-ground caverns collapsed, exposing underground rivers. Many of the lakes take on various shades of green and blue, depending on water and soil chemistry. There are scenic overlooks from roadside pullovers, trails around the lakes, and many dirt tracks and logging roads leading off from the main road along which you can walk and birdwatch. Usually the place is fairly deserted, although it's a popular picnic spot for local families on weekends and holidays. Several of the more popular lakes have local villagers encamped there, operating restaurants and souvenir stands; many of the more remote lakes tend to be deserted. A new paved road within the park makes getting around very easy. Some of the gravel turn-offs to individual lakes are quite steep. Lagunas de Montebello is a bit out of the way, but it makes a nice day trip and is a pretty drive from San Cristóbal. (Area: 6022 hectares, 14,900 acres. Elevation: 1300 m, 4250 ft. Habitats: lakes with freshwater marshes, pine forest, oak forest, cloud forest.)

El Triunfo Biosphere Reserve

Protecting one of the region's last untouched patches of perpetually misty high-land cloud forest, El Triunfo is an ecotravel gem, recently opened and fairly easy to visit. The area, in southwest Chiapas in the southern Sierra Madre de Chiapas mountains, is very wet, as warm and humid Pacific air masses rise up to cross the mountains, cool, and pour down rain. With help from Mexico's National Insti-tute of Ecology and the state Institute of Natural History, local villagers have established a small ecotourism business in the high cloud forest. In dryer months, from November to May, small groups take four-, seven-, or nine-day trips to a cen-tral bunkhouse in the reserve at about 2000 m (6600 ft) elevation. Reaching it requires a four- to six-hour climbing hike, with mules carrying supplies. Once at base camp, days are filled with hiking the more than 60 km (37 miles) of trails, searching through the towering wet forest and treeferns for quetzal (Plate 43; this is one of the better places to spot this gorgeous bird), manakins, hummingbirds, and such rarities as the Horned Guan and Azure-rumped (Cabanis') Tanager (both endemic to southern Chiapas and adjacent Guatemala), not to mention tapir (Plate 80) and Jaguar (Plate 76). Access is via a village to the north of the reserve, Angel Albino Corzo, or from Tapachula, to the southeast. To go, contact the eco-tourism program at the Institute of Natural History at phone/fax: 011-52-961-42334 or 40779; e-mail: ihnreservas@laneta.apc.org (Area: 119,000 hectares, 294,000 acres. Elevation: 500 to 2500 m, 1600 to 8200 ft. Habitats: highland cloud forest, pine and oak forest, evergreen forest, highland scrub.)

Tabasco

Centla Wetlands Biosphere Reserve (Pantanos de Centla)

North and east of Villahermosa, the capital of Tabasco, lies an extensive area of wetlands formed by the floodplains and combined deltas of the Usumacinta and Grijalva Rivers. About 30% of Mexico's surface freshwater flows through this region. Access is difficult, but day tours to marsh areas can be arranged through travel agencies in Villahermosa. The route is usually north on Highway 180 to the town of Frontera, then on a turn-off south toward the marshes. It's mostly a bird-watching area good for birds-of-prey and water-associated birds (marsh birds, herons and egrets, ducks, storks), but mammals such as howler monkeys, large cats, and manatees are seen occasionally, and crocodiles are common. (Area: 302,700 hectares, 748,000 acres. Elevation: sea level to 100+ m, 330+ ft. Habitats: semi-deciduous tropical rainforest, palm forest, mangrove swamps, freshwater marshes.)

Campeche

Chicanná, Becán, and Xpujil Archaeological Sites

These are three nice but smallish Mayan ruin sites located within the same area of southern Campeche. All have very easy access from the main road crossing the base of the Yucatán Peninsula, Route 186, and all three can be visited within a few hours, *no es problema*. One nice aspect of the area: it's often a bit cooler (just a bit) than the western and northern sections of the peninsula. Of the three sites, Becán is probably the best for wildlife. It's fairly extensive: there are several sets of massive ruins connected by paths and forested, leaf-strewn trails. Some of the trails seem to head off to nowhere, perhaps used by archaeologists to explore for other ruins. On a recent trip in June, I was the only visitor at all three sites. Mammals are scarce, but small lizards are commonly seen, as are a good selection of birds (parrots, woodpeckers, flycatchers, orioles, tanagers, and a host of others); it's also a good snake spot. There's a fairly new hotel in the area just off the main road that, while the most expensive place for miles around, at least is trying to employ ecologically benign practices. It uses solar power and rainwater, recycles water, and has a waste treatment plant. It's the Ramada Chicanná Ecovillage Resort (Campeche tel: 52-981-622-33).

Hormiguero Archaeological Site

A fairly small Mayan site with two main excavated structures, located in a great forest setting. Reaching the site is somewhat adventurous. It's located down unpaved roads south and then west of the town of Xpujil, within the Calakmul Biosphere Reserve. Most visitors go with organized tours. It's possible to drive to the site in a regular sedan, although the rutted, rocky, grassy road can be quite bad in any season and impassable during rainy season (June–October); 4-wheel drive is recommended. Another problem is that, because the road continues south toward the Guatemalan border, army anti-drug checkpoints and patrols are frequent. There are some trails around the ruins; wildlife watching also can be done by hiking along the road to the site. Because other trips into the Calakmul rainforest are often even more difficult, a trip to Hormiguero can be a nice way to see some of this area.

Calakmul Biosphere Reserve
A huge expanse of tropical forest and other habitats, located both north and south of Route 186, the main road through the base of the Yucatán Peninsula. The reserve, continuous with the Maya Biosphere Reserve of northern Guatemala's Petén Province, seeks to protect some of the last stands of tropical forest in the region that have not been cut or burned for farming or ranching. Wildlife here includes Jaguar, Ocelot, Margay, Jaguarundi, Puma, tapir, brocket deer, howler and spider monkeys, and more than 350 species of resident and migratory birds, including abundant parrots and the Ocellated Turkey (Plate 27). With the help of conservation organizations, some of the 15,000 people who live in the reserve are learning about ways to support themselves that reduce impacts on the forest (see p. 58), and ecotourism development eventually will be included. As of this writing, the best way to see some of the reserve is via tours to the Hormiguero ruins (see above) or the Calakmul ruins (more than an hour's drive south from the Route 186 turnoff at Conhuas on a fairly good paved, gravel, and rock road). Signs generally are few or lacking, so going with a guide is recommended. If you're not with an organized tour, ask around the town of Xpujil for guides with vehicles. Some visitors with deep pockets helicopter to the Calakmul ruins (there's a nearby landing pad that locals swear is a great place to see deer, peccaries, and birds). (Area: 723,000 hectares, 1,785,000 acres. Elevation: sea level to 200+ m, 650+ ft. Habitats: evergreen and semi-deciduous tropical rainforest, savanna, freshwater lakes and wetlands.)

Yucatán

Ek Balam Archaeological Site
Located only 10 km (6 miles) off the main road between Valladolid and Ría Lagartos, Ek Balam is a wonderful, fairly large, very lightly visited Mayan ruin site quite near to Chichén Itzá. A day after visiting Chichén Itzá along with several hundred other people, I took a tourbook's advice, made the short trip to Ek Balam, and found that aside from the lonely caretaker, I was the only human at the 10 sq km (4 sq mile) site. I spent several hours hiking around and, for wildlife, I found it to be much better than Chichén Itzá, where the forest has been cleared. At Ek Balam, most ruins are still buried in jungle and unexcavated – in fact, only one largish group of ruins has been excavated and partially restored. Because the ruins sit in a forest setting, there's lots of cool shade. Lizards abound at the site (especially Spiny-tail Iguana and whiptails), as do lots of good birds – anis, woodpeckers, flycatchers, jays, blackbirds, tanagers. If you are at all interested in birds and want to get a look at one of the world's most stunning examples, bring binoculars to Ek Balam (or to lots of other archaeological sites, where they seem to thrive) and train them on the common and conspicuous Turquoise-browed Motmots. Pictures, including the fine one in Plate 45, simply do not do the bird justice; the

bright turquoise head glinting in the sun and the beautiful racket-shaped tail ends have to be seen to be really appreciated.

Chichén Itzá Archaeological Site

The world-famous, heavily visited Mayan ruin site, located in the center of the northern portion of the Yucatán Peninsula, west of Valladolid. This is the Big One, where almost all tourists to the peninsula, save those who refuse to leave Cancún's beaches, eventually find themselves. Tour buses disgorge hundreds of people per hour; however, the site is so large that it rarely seems mobbed. From ancient history and archaeological perspectives I guess the place is a must-see, but the ruins are in an open, park-like setting, most of the forest is gone, and even birdwatching is difficult with lots of people around. Wildlife frequently seen includes many small lizards and large Spiny-tailed Iguanas, and a good many birds that enjoy open or forest-edge habitats: vultures, doves, woodpeckers, fly-catchers, swallows, motmots, blackbirds, orioles, tanagers. My favorite aspects of Chichén Itzá: birdwatching at the Sacred Cenote – a huge waterhole at the bottom of a steep pit – and the shiny ambulance the authorities have standing-by adjacent to the ruins, ever-ready to whisk tourists off to the hospital after their heat-stroke-induced falls from the steep steps of the high temples.

Balankanché Caves

Caves, essentially big holes in the ground, are, to my mind, usually not worth visiting, especially for the mild claustrophobe. When we want to be in small, dark, enclosed places, we can stay home and watch TV. I list this cave here because it's on the main road near the Chichén Itzá ruins and many travellers stop to take a peek. Also, I like the local Spanish word for caves, *grutas*. For a fee, small groups are led through an impressive, fairly extensive cave system with several larger caverns where Mayan artifacts were found and are preserved – presumably in the same positions in which the ancient Mayans left them. Small bats (at least five species) are plentiful and close to your nose, and the cave ends at a chamber that recedes into an underground lake (cenote). Scary!

Cenote Dzitnup

Right off the main road just outside of the town of Valladolid, on the way to Chichén Itzá. Cenotes (seh-NOE-tays) are water holes, sinkholes filled with water, created when a section of rock ceiling covering an underground river collapses. Over the Yucatán Peninsula, much of which lacks surface water (rivers and lakes), cenotes were in ancient times (and still are, in many areas) the only source of large amounts of water for drinking, bathing, agriculture, etc. In addition to service as local swimming holes, cenotes have lately gained ecotourism prominence as scuba diving sites for those tired of the wonders of Cancún's barrier reef. Cenotes can be very deep and those proximate to ancient settlements sometimes hold archaeological treasures in their depths. Tourbooks commonly point out several cenotes in the region. Some are above ground, small lakes, really; some are at the bottom of steep stone pits (such as Chichén Itzá's Sacred Cenote); and others (such as Dzitnup) are wholly underground, somewhat spooky, reached by descending into a cave. From a wildlife viewpoint, remote cenotes are often good places to visit because, as the only source of open water over wide areas, they tend to attract thirsty animals. Some birds – swallows, motmots, and others – are reliably found around these waterholes. In addition, many subterranean cenotes harbor endemic species of blind fish and shrimp.

Dzibilchaltún Archaeological Site

Easily reached, the site is just 10 km (6 miles) north of the city of Mérida. This is a nice, mid-sized Mayan ruin site, totalling some 539 hectares (1330 acres). It's usually fairly uncrowded, but on weekends and holidays local families come to stroll, picnic, and cool off in its Sacred Cenote. A few large groupings of massive ruins are widely spaced in a very open setting. The place's best feature may be the long, rock-lined dirt path through the forest that connects some of the clusters of ruins. Lizards abound along the path, and the open forest with small cacti and epiphyte-laden trees is a good birding spot. Bring mosquito repellent.

Uxmal, Kabáh, and Labná Archaeological Sites

These are the more commonly visited Mayan ruin sites south of the city of Mérida, along what tourbooks and road signs refer to as *La Ruta Puuc*. "Puuc" is Mayan for "hills" and the Puuc region is the hilly, dry area in the west-central portion of the Yucatán Peninsula. Uxmal is the largest and best of the sites, but also the one most visited by large groups, tour buses, etc. There are extensive ruins at Uxmal and, if you can take the heat, lots to see. Small lizards are common and there's a good population of large Spiny-tailed Iguanas. Frequently seen are beautiful motmots and jays. All three sites, which can be visited together in a full morning or afternoon, are more or less forested, have much the same kinds of wildlife, and Uxmal at least, has some nice views of the surrounding countryside (views can be had at other sites by climbing the high ruins). Bring mosquito repellent, because the bugs hang out in the shady areas – which is where you'll want to be.

Ría Lagartos Special Biosphere Reserve

A picturesque set of mangrove-fringed lagoons, canals, and coastal waters in which to look for water-associated birds and reptiles, located directly north of the town of Valladolid, about 3.5 hours from Cancún. The primary attraction is pink American Flamingos (Plate 22), which breed in the reserve. Village fishermen take small groups out in their boats for two-hour trips in search of flamingos, and usually find them. Commonly seen are small crocodiles and all sorts of birds – seabirds, water birds, marsh birds, birds-of-prey. Sea turtles nest on the coastal beaches, and Jaguar, so they say, roam the area. The boat captains know that tourists want to photograph the flamingos close up and in flight (which, admittedly, is quite a sight and makes for spectacular photos), so against the rules they move slowly toward them until the birds fly. The constant harassment has caused the local population to thin. (In fact, San Felipe Natural Park, on the coast to the west of Ría Lagartos, was recently established when a sizable fraction of the flamingo population began breeding there – moving perhaps in response to harassment from Ría Lagartos tour boats.) If you go, tell your boat driver you do not want to get too close (100 m, 330 ft, is the recommended minimum distance). Most of the reserve is wetlands – mangroves and marshes – and, as such, inaccessible to people. In addition to protecting flamingo breeding areas, the reserve is also important as resting and feeding habitat for migratory birds. When you drive into Ría Lagartos, you will be flagged down by fishermen wanting to take you in their boats; if not, proceed to the waterfront. Prices can be steep; despite what some tourbooks say, some negotiation is possible. (Area: 48,000 hectares, 118,000 acres. Elevation: sea level to 10 m, 33 ft. Habitats: mangroves, marshes, savanna, coastal dunes, deciduous forest.)

Ría Celestún Special Biosphere Reserve

This reserve, located an easy drive about 1.5 hours west of the city of Mérida, protects one of the only nesting colonies of American Flamingos (Plate 22) in Mexico (others are in and around the Ría Lagartos reserve), and is perhaps the flamingos' main wintering area. Most visitors take boat rides in the main, mangrove-fringed lagoon, in search of flamingos which, as at Ría Lagartos, seem to be getting harder to locate. The big wading birds apparently move to areas where they are not harassed by tour boats. (In fact, recent research at Ría Celestún shows the harmful effects on the birds: some groups are disturbed an average of 13 times a day for a total of about three hours and, as a result, they feed about 40% less time than do undisturbed birds; it takes the birds about 20 minutes to return to their "normal" behavior after a tour boat disturbs them. Remember to tell your boat driver you want to see the flamingos but not disturb them.) This is a bigger, more successful ecotourism operation than that at Ría Lagartos. Many tour boats operate from below the highway bridge you cross just before entering the town of Celestún. Most tours last between one and two hours. In addition to looking for flamingos, the tour boats can take you through some nice water trails through narrow mangrove passages and to a swimming area set in a mangrove clearing where a cool freshwater spring bubbles up into the lagoon. Aside from flamingos, commonly seen are birds-of-prey, pelicans, anhingas, cormorants, herons and egrets, ibises, and lots of frigatebirds. Visitors wanting to walk or hike among lagoons or along the beach start at the main plaza in town and head northwards on the sandy lanes until the town peters out. The reserve is especially important for protecting resting and feeding habitat for migratory birds. More than 300 bird species have been spotted there. (Area: 59,000 hectares, 146,000 acres. Elevation: sea level to 10 m, 33 ft. Habitats: mangroves, marshes, savanna, coastal dunes, mudflats, deciduous forest.)

Quintana Roo

Tulum National Park

This is a famous Mayan archaeological site located 1.5 hours south of Cancún, along the main highway. Tulum's good point is that the ruins are in a stunning, oceanside setting; were it in a remote site, it would be magnificent. Unfortunately, Tulum is at the south end of the *Corredor Turístico*; every foreign visitor to the Yucatán Peninsula ends up there. The relatively small site is usually crowded – and I mean large, Disneyland-type crowds. In fact, completing the picture of a real American-style tourist trap, there are motorized trams to haul visitors between the parking area and the ruins. I visited recently at 4:00 pm, an hour before closing, hoping for fewer people, but the place was mobbed. Try visiting when the place opens, at 9:00 am. It's not what I call a wildlife site, but some lizards, common songbirds, and seabirds are often seen. (Area: 664 hectares, 1640 acres, but the main ruins take up only a small fraction of this. Elevation: sea level to 20 m, 65 ft. Habitats: stone ruins in park-like setting, tourists, semi-deciduous forest, coastal scrub.)

Cobá Archaeological Site

The Cobá ruins, although only an hour or so inland from the Tulum ruins, couldn't be more different. Cobá is a huge forested site, covering many square kilometers. Three main sets of excavated ruins are located some distance from each other, separated by wide forest trails. There are also many narrow side trails to explore;

some lead to Lake Cobá and other lakes. Best of all, Cobá is often semi-deserted. It's a good place to come for birdwatching (opens at 8:00 am), and wildlife sightings of all kinds are possible. Common birds include marsh and wading birds in the lakes, vultures, doves, flycatchers, motmots, becards, jays, orioles, and tanagers. There's also quite a view of the surrounding forest from the top of a 42-m (138 ft) pyramid. This is a splendid eco-attraction not far from Cancún – if you're in the area, don't miss it. (Area: 20,000 hectares, 49,400 acres. Elevation: near sea level. Habitats: semi-deciduous tropical forest, lakes and marshes.)

Punta Laguna Wildlife Sanctuary (also called Spider Monkey Reserve)
Located perhaps 20 km (12 miles) north of the Cobá ruins on a fairly good paved road, the tiny village of Punta Laguna, with some help from conservation agencies, has begun a small ecotourism project. About 200 hectares (490 acres) of tropical semi-deciduous forest bordering a small lake have been set aside as an ecological reserve. It's a very nice, but still lightly visited, small-time operation. On a recent weekend morning visit, I was the only ecotourist in the area. (But at the time of this writing, tourism operators were negotiating with the village to build lodges around the lake and to bring in good numbers of day-tourists for picnicking, swimming, etc. – so the site may or may not remain small-time and rustic.) For a small entrance and guide fee, local boys show you around the reserve and point out various tree and animal species. The trail system is fairly extensive, many trails following ancient Mayan stone paths. Spider monkeys and some other mammals such as squirrels are seen fairly regularly, and it's also a good spot for lizards and lots of birds – parrots, hawks, trogons, woodpeckers, woodcreepers, antbirds, motmots. (Elevation: near sea level.)

Sian Ka'an Biosphere Reserve
This is a large protected area on the east coast of the Yucatán Peninsula, starting just south of Tulum. Included in the reserve are large tracts of tropical wet savanna and other types of wetlands – coastal lagoons, marshes, flooded forest. About 350 species of birds have been spotted at Sian Ka'an (of the 475 species that occur on the Yucatán Peninsula). Most visitors see water-associated birds: herons and egrets, ibis and spoonbills, ducks, storks, pelicans, frigatebirds, gulls and terns, occasional flamingos. However, even a brief foray on land near the entrance site can quickly yield sightings of birds-of-prey, parrots, doves, toucans, woodpeckers, flycatchers, mockingbirds, jays, orioles, and tanagers. Jaguar, Margay, Ocelot, and Jaguarundi occur here, as do spider and howler monkeys, tapirs, peccaries, and deer. Manatees are frequently sighted in the bays and coastal waters. Four species of sea turtles use Sian Ka'an beaches for nesting, and crocodiles prowl the coast and inland waterways. Due to the large wetlands content, most of the reserve is inaccessible; there are few trails. Most visitors take tours of the reserve with Amigos de Sian Ka'an, a non-profit conservation organization (see p. 11). The five-hour tours start near Tulum and proceed to the reserve by minibus. Biologist-guides give excellent introductions to local ecology, flora and fauna, then load the small groups into boats for wildlife viewing, cruising narrow canals through the wetlands, and even for a swim. If you are in the region, try not to miss this trip. Contact Amigos de Sian Ka'an in Cancún: tel: 98-84-9583; fax: 98-87-3080; e-mail: sian@cancun.rce.com.mx. You can also access the reserve by driving the unpaved road south from Tulum to the village of Punta Allen, or on unpaved roads that head east into the reserve from Route 307, south of Tulum. (Area: 617,000 hectares, 1.6 million acres. Elevation: sea level to 20 m, 65 ft.

Habitats: deciduous tropical forest, seasonally flooded forest, savanna, coastal lagoons, mangrove swamps, marshes, freshwater canals, 110-km long, 68 mile, offshore coral reef.)

El Edén Reserve

This is a remote, private ecological reserve located about two hours (50 km, 30 miles) northwest of Cancún. It was established in 1990 for research and conservation purposes and, with reservations, ecotravellers can visit and stay overnight in rustic cabins. Habitats include semi-deciduous tropical forest and wetlands. There are trails for hiking, and it's a great birdwatching site (more than 200 species have been spotted there). Crocodiles are plentiful. Contact the reserve at phone/fax: Cancún 98-80-5032; e-mail: mlazcano@cancun.rce.com.mx; www.ucr.edu/pril/peten/images/el_eden/Home.html   (Area: 1500 hectares, 3700 acres. Elevation: near sea level. Habitats: deciduous tropical forest, swamp forest, savanna, cenotes, many types of wetlands.)

Isla Contoy Special Biosphere Reserve (also called Isla Contoy Bird Sanctuary)

Reached via a pleasant one- to two-hour boat trip from Cancún or Isla Mujeres, usually with a commercial tour company (a few are authorized to take people to the restricted-access island; some make a day of it, including lunch, free-flowing booze, and a mid-ocean stop for snorkeling the reef near Isla Contoy). The island, made a reserve to protect colonies of seabirds that breed there, not to mention four species of sea turtles that use its beaches for nesting, is long and very narrow. There's a visitors' center, a high tower for birdwatching, and a few trails – but visitors are allowed access to only about 20% of the island. One trail leads to a lagoon where you can get quite close to breeding seabirds such as frigatebirds. Pelicans, boobies, gulls and terns, and perhaps spoonbills, all breed on the island and about 70 bird species have been spotted there. Some reptiles frequently seen on the island are two snakes, Roadguard (Plate 10) and Boa Constrictor (Plate 13), large Spiny-tailed Iguana (Plate 16), and Barred Whiptail (Plate 19). (Area: 176 hectares, 435 acres. Habitats: sandy and rocky beach, coastal dunes, mangroves, low thorny shrub and cacti.)

Dr Alfredo Barrera Marín Botanical Garden

Located along the main highway maybe 35 km (22 miles) south of Cancún, this is a very nice, very shady eco-attraction, definitely worth an hour or two of exploration. For an entrance fee of only a few pesos, you get 60 hectares (150 acres) of broad, shady paths and trails through a good collection of native and non-native plant species. It's a good birding spot (although it does not open until 9:00 am) particularly because small bird baths have been placed strategically at intervals along some of the paths, drawing birds down for ease of viewing – trogons, woodpeckers, woodcreepers, antbirds, becards, wrens, cowbirds, blackbirds, orioles, and tanagers are common. The Garden has special displays of medicinal and succulent plants. At the rear fringe of the Garden are trails lined with palms, bromeliads, mangroves, and trees with heavy loads of epiphytes.

Kohunlich Archaeological Site

This large site is located in a remote, forested setting, about 60 km (37 miles) west of the town of Chetumal. It's a few kilometers off the main road, Route 186, that cuts across the base of the peninsula. Many Mayan ruins buffs consider the Kohunlich site one of their favorites – it's in a beautiful setting amongst jungle trees and cohune palms (p. 29), and there are rarely more than a few visitors in attendance. This is a good place for seeing birds that prefer forest edges and open, scrubby habitats. Other wildlife is often plentiful and diverse; the area has a good population of tapirs.

Chapter 4

Environmental Threats and Conservation

- Major Threats and Conservation Record
- Conservation Programs and Ecotourism
  Biosphere Reserves
  Ecotourism
  Ejidos, Conservation, and Ecotourism

Major Threats and Conservation Record

Mexico is a huge country (one of the world's 15 largest) with a big population (96 million in 1996) growing at a fast pace. There are major environmental threats, chiefly destruction of natural habitats, and the country has had, until very recently, a poor environmental record and outlook. Suffering from widespread poverty, governmental neglect and corruption, and little organized local interest in conservation, Mexico for most of its history has been a state where business and agricultural interests held nearly absolute power over development decisions and the uses of natural resources. Economic growth, at almost any cost, was the mantra, everyone's chief concern; conservation was a very low priority for business, government, and most citizens alike. One indication: Mexico did not join the Convention on International Trade in Endangered Species (CITES), the chief international agreement to stop trade in threatened and endangered plants and animals, in effect since 1975, until 1991, the last Latin American nation to do so. During the past few years, some big changes have taken place. Triggering growing environmental awareness in the country have been increasing education, local and international publicity over Mexico's pollution levels and poor environmental record, the gradual weakening of the formerly omnipotent Institutional Revolutionary Party (PRI), and the emergence of a sizable middle-class who vacation in the countryside and take strong interest in living healthy lives. The dramatic decline in environmental quality has been noticed, and people and government have begun steps to reverse the decline.

Mexico has a host of environmental problems, including significant chemical pollution from factory discharges and waste dumping, but the main threat to its biodiversity probably stems from deforestation and other natural habitat losses. Forest habitat is lost for a number of reasons. The major factor is land use – land is cleared for crop agriculture, cattle grazing, human colonization, and for business development. A rapidly multiplying human population and economic growth propels and constantly increases these uses. (For instance, currently there is explosive population growth among the Yucatán Mayan people, owing to high birth rate, low infant mortality rate, and high immigration rate – mostly young people arriving to seek jobs in Cancún.) Other causes of forest loss are over-exploitation for timber and fuelwood, and natural agents such as fire and disease. The use of trees as fuel for heating and cooking takes an especially heavy toll on forests. In Oaxaca, for instance, it's estimated that each family burns on average about 12 kg (26 lb) of wood per day. Very few forested areas of Mexico are free from human disturbance; in fact, most forests contain scattered settlements whose residents are usually very poor and who still practice age-old slash-and-burn agriculture. The rate of forest loss, one of the highest in the world, is officially estimated at anywhere between 300,000 and 400,000 hectares (740,000 to one million acres) per year, but is actually higher – perhaps double these estimates. (Reforestation rates average less than 100,000 hectares, or 250,000 acres, per year.) Forests in some regions are destroyed faster than others, for instance, over half the forest cover in the Selva Lacandona region has been lost since 1960 (see below).

Compounding the ongoing threat to its forests and other natural habitats is the dawning realization that Mexico is a treasure trove of biodiversity and a center of endemic species – those that occur only in one place (see Close-Up, p. 236); destroy them in Mexico and they will be extinct. During the 1980s and early 1990s, as an initial, necessary step for conservation, Mexico underwent surveys and censuses of much of its biodiversity. Some results are in, and they are stunning. Mexico, it turns out, probably holds more species of plants and animals than any other country on Earth but two. For instance, there are some 30,000 plant species, of which between 50% and 60% are endemic, 49 species of pines (more than half the world's total), 450 mammals (Brazil, which is more than twice Mexico's size has only 394 mammals), about 1000 birds, 693 reptiles, 285 amphibians, and more than 2000 fish. As of the mid-1990s, many species were known to be already threatened: 64 mammals, 36 birds, 18 reptiles, three amphibians, and about 85 fish.

Wildlife surveys, in addition to identifying and counting species, note where they are located and in which habitat types they occur, so that biologists can target and prioritize geographic regions and habitats for conservation attention. That is, they can determine which habitat types are most threatened and where, and which ones, based on their biodiversity, are most worth trying to save. For instance, a major part of the last large expanse of virgin tropical forest in North America, Chiapas' Selva Lacandona (the Lacandon Jungle), near Mexico's border with Guatemala, with its high level of biodiversity, has now become a top conservation priority for Mexico and for international conservation organizations. Likewise, wetland areas along the Yucatán Peninsula's coasts were recognized for their biodiversity holdings and identified as critical migratory bird stopover habitat and, as a result, recently were given protection as biosphere reserves. Overall, based on its degree and rate of habitat loss and the amount of biodiversity it holds, Mexico is now considered one of the 15 most environmentally threatened places in the world.

As I said above, environmental concern is growing in Mexico and corrective actions increasing. International publicity about Mexico's pollution problems has helped. Drawing most interest probably is the USA/Mexico border region, where for decades USA companies and others set up industrial and manufacturing plants on the Mexico side to take advantage of lax Mexican environmental regulations and cheap labor costs, while at the same time maintaining close access to USA markets. The practice took on special significance to international media when it was realized that dumping wastes into northern Mexican rivers and bays, for instance, polluted not just Mexico but adjacent regions of the USA as well. Recent trade deals among North American nations (NAFTA, etc.) have associated environmental provisions – side agreements – that obligate signatory nations to regulate relevant industry to reduce pollution. Grassroots support in Mexico for conservation and against rampant development is also growing. A recent golf resort development within easy driving distance of Mexico City, for use of the city's elite, which in the recent past would have been built easily and quickly, was held up because local townspeople did not want their wild countryside marred by development. This activism is a new way of doing things for Mexico and speaks well for its future conservation efforts. There is a new Green Party, but it garnered less than 1% of the vote in mid-1990s national elections.

In 1996, the new government of President Ernesto Zedillo consolidated several ecology and wildlife services to form a new Secretariat of the Environment, Natural Resources, and Fisheries (SEMARNAP) and staffed it with respected environmentalists. Most observers believe this reorganization represents a sincere effort to begin addressing the nation's need for strong environmental protection and conservation. However, as in most other nations of Latin America, even when public atmosphere and governmental goodwill exist for conservation, necessary funds are usually lacking. In relatively poor countries (Mexico's per capita gross domestic product in 1994 was US$7900), where there is always critical need for increased spending for roads, public health, and education, money for conservation is always hard to find. Parks and reserves can be established with the stroke of a pen, but often they are protected reserves in name only. There are no funds to hire the managers and wardens to patrol and protect the habitats and wildlife. Sometimes a single guard or caretaker is responsible for overseeing vast tracts of land, and can do little to prevent habitat destruction and wildlife poaching.

Mexico has been slow to realize that natural habitats attract capital, and that wild habitats have, to some degree, the capacity to help save themselves, through ecotourism. But efforts are underway now by international and local non-profit conservation organizations and by state and federal governments to use ecotourism to help preserve some of Mexico's ecological treasures.

Conservation Programs and Ecotourism

Biosphere Reserves

Chief among Mexico's conservation efforts must be its establishment of reserves to protect areas with threatened or fragile habitats and the plants and animals they contain. The country has long had several protected areas, but the pace of declaring reserves, especially larger ones with critical, high-biodiversity habitats, significantly quickened during the 1980s and 1990s, as the deteriorating nature of Mexico's last wild areas became known. As of this writing, Mexico has more than 70 protected areas that, in total, make up more than 5% of the national territory. Many of these protected lands are smaller parcels in the form of *national parks* (defined by Mexico as areas with special historical, scientific, ecological, or aesthetic value) and *natural monuments* (areas with unique natural beauty or scientific value), but most of the total area is included within large wild zones known internationally as *biosphere reserves*.

The first biosphere reserves were designated in 1976 and, as of 1997, there were 337 of them in 85 countries, covering 200 million hectares (500 million acres). Mexico now has about 10 biosphere reserves, defined by the United Nations scientific arm (UNESCO) as protected areas, generally larger than 10,000 hectares (24,700 acres), that contain one or more important biological zones and that include significant pristine, or wilderness, areas, untouched by people. Smaller protected areas (less than 10,000 hectares, or 24,700 acres) that meet most of the requirements of biosphere reserves, are known as *special biosphere reserves*, and Mexico also has more than 10 of these (for example, Ría Celestún and Ría Lagartos reserves). Mexico's biosphere reserves are generally acknowledged to be its best-protected lands. The purpose of biosphere reserves, after 20 years of trial and error learning, has evolved to be three-fold: to conserve biological and cultural diversity; to develop and serve as models for sustainable land use; and to provide areas for environmental research, monitoring, training, education, and tourism.

Biosphere reserves are structured in a special way to facilitate these goals. They contain one or more *core* or *nuclear* zones, true wilderness areas designated for the strongest degree of long-term conservation protection; only research relevant to conservation and perhaps limited, low-impact tourism are allowed in these zones. Surrounding the core zones are *buffer* zones, meant to protect the core from human intrusion and in which limited activities relevant to conservation are permitted, such as education, research, ecotourism, non-destructive recreation, and low-impact uses of natural resouces. Surrounding or adjacent to buffer zones are *transition* zones, where local communities may live and, usually with training from conservation agencies, engage in sustainable uses of natural resources. (*Sustainable* means using plants and animals in ways that are economically profitable for the local economy yet not ecologically harmful; use, in other words, that will not lead to significant ecosystem damage or to decline in biodiversity.)

A good example of how one of these reserves works and how local and international conservation organizations play major roles, can be found in the Yucatán Peninsula's Calakmul Biosphere Reserve. This large reserve on Mexico's southern border is continuous with Guatemala's Maya Biosphere Reserve; its primary mission is to preserve the biodiversity contained within this largest patch of remaining tropical forest north of the Amazon region (the "patch" also includes the forests of the Maya Biosphere Reserve and of Chiapas' Selva Lacandona). At least 350 bird species occur here, as do about 100 mammals and more than 50 reptiles and amphibians. Calakmul, named for the ancient city of the same name whose ruins still exist in the southern part of the reserve, consists of two large core zones surrounded by buffer/transition zones made up of ejidos (lands held communally by local villages; see below) and their associated forest reserves.

The most important conservation organization working today in the Yucatán Peninsula and, in particular, to protect Calakmul and conserve its biodiversity is Pronatura Yucatán, a Mexican group with headquarters in Mérida. (Pronatura does the hands-on work but receives financial and other assistance from organizations such as The Nature Conservancy, World Wildlife Fund, and the US Agency for International Development.) About 15,000 people live in and around Calakmul and make their livings there, often by means that are environmentally harmful. The biggest problem is deforestation, primarily caused by traditional slash-and-burn farming and logging for timber. Protecting these critical forests lies not in simply establishing a reserve and then telling residents they can no longer use the natural resources they depend on. That kind of treatment, experience shows, leads only to local resentment and lack of compliance with conservation rules. Rather, most modern conservation organizations, such as Pronatura, believe that the way to protect remaining forests and biodiversity in core areas is to work with local people in surrounding zones so that they, too, have strong incentives to preserve wild habitats. In other words, if local residents can be taught to manage their forests in sustainable ways, it will be better in the long run for the local people – it will improve their lives – and also better for conservation.

So what, exactly, does Pronatura do in Calakmul? In a phrase, lots of things – some of which, at first, you might not imagine are strongly related to conservation of trees, birds, and bugs. But most are aimed at improving the lives of local residents who live in or near the reserve so that their dependence on harmful environmental practices is reduced. For instance, Pronatura conducts workshops in many villages in environmental education (composting, gardening, hog farming, etc.), reproductive health and family planning, and traditional medicine (with help from community experts); they train villagers for and help with rainwater catchment projects (lack of water during dry season is a major community problem) and infrastructure improvements (putting up signs, etc.). Their major emphasis, however, is on training local people in *agro-ecology* projects – teaching them to support themselves in environmentally friendly ways. One of the main projects at Calakmul is teaching people to plant crops of legumes – plants such as beans that, as they grow, tend to put nutrients back into soils. These crops actually improve soils so that the same fields can be used year after year. Repeated use means that new fields do not have to be cleared every few years from the forest when soils become nutrient-poor (the traditional slash-and-burn method). Another major agro-ecology project pursued now in Calakmul is honey production. More than 120 individual producers in about 14 villages are keeping bee hives and harvesting honey, which is marketed as *Jungle Honey*. For the time and effort investment, honey production is quite profitable (after all, the insects making the stuff work for below minimum-wage). It has been quite successful of late, keeps people working all year (whereas much local employment is only seasonal) and, from a conservation perspective, is truly an eco-friendly industry. Pronatura helps in the training aspects of honey production (finding local experts to give workshops on bee-keeping and on working with aggressive Africanized honeybees, helping to obtain materials, and getting people started), but not in the commercial aspects – local people market their own product.

In the end, it will probably be huge biosphere reserves in Mexico and other countries – where both core wilderness areas and large buffer zones enjoy high degrees of protection – that will offer many species their best chances of survival. As in any new discipline, the scientific field of conservation biology is rife with controversies, but there is general agreement that preserving large wild areas is probably the only way to save many species from rapid extinction. For instance, viable populations of larger mammals, such as Jaguar, White-lipped Peccaries, and tapirs (Plates 76, 80) cannot possibly be successful in small patches of forest.

Ecotourism

Aside from moving all people from the Earth, no conservation method is likely to be perfect – they all have pluses and minuses. Ecotourism is no exception, and I mention some negatives at the end of the chapter. However, on balance, most conservationists agree that ecotourism can be used successfully to help preserve many wild habitats. Indeed, conservation organizations in southern Mexico are increasingly fostering ecotourism with the hope that visits to remote sites by tourists, and the economic surge they bring, can help preserve those sites. One of the Yucatán Peninsula's early efforts at ecotourism, visits to the Sian Ka'an Biosphere Reserve, led by the Amigos de Sian Ka'an organization, is detailed on p. 11. Ecotourism in the remote Calakmul Biosphere Reserve is beginning – the conservation agency Pronatura Yucatán, in fact, recently started training local guides. Pronatura, again with technical advice and financial assistance from organizations such as The Nature Conservancy, also has been instrumental in helping two communities on the peninsula's coast establish ecotourism businesses. In both Ría Lagartos and Celestún, Pronatura worked with the local communities to establish boat tours for visitors to see flamingos and other wildlife inhabiting the biosphere reserves that surround these two towns. In addition to businesses relating to boat tours, the organization works with other sections of the communities to promote eco-friendly cottage industries. For instance, in the town of Celestún, a women's group was organized that now runs a successful small business gathering ocean shells, making from them household objects such as napkin holders, and selling them in the town plaza. Ría Lagartos and Ría Celestún Special Biosphere Reserves protect important wetlands, habitat for many water-associated birds and other kinds of animals. In fact, owing to its critical function as bird nesting habitat and also wintering habitat and stopover habitat for many migratory birds, and to the high number of threatened species it contains, Ría Lagartos in 1986 was placed on the international "RAMSAR" list of essential wetlands (as Mexico's only entry). The list is the result of the Convention on Wetlands of International Importance, signed in Ramsar, Iran in 1971; it prioritizes the world's wetlands for conservation purposes.

So ecotourism on the Yucatán Peninsula started only recently and is growing slowly. Its biggest obstacle may be that local governments, especially in Quintana Roo and Yucatán states, are, so far, doing little in the way of promotion (Campeche is a bit more on the ball, and has even collaborated with Pronatura in training nature guides). Also, SEMARNAP, the federal agency now charged with protecting natural resources, is cautious about ecotourism, probably because they know that if it is not handled correctly, it can lead to over-visitation of sites and environmental damage.

Chiapas is perhaps a bit behind even the Yucatán Peninsula in ecotourism development because of the remoteness of some of its attractions and its recent political unrest. Nonetheless, Chiapas now has several fantastic eco-attractions and is beginning to promote them and render more of them reachable by ordinary people. The Chiapas state government appears to support ecotourism strongly, and conservation organizations are increasingly using ecotourism in the state to try to save wilderness areas. Conservation International (CI), for instance, is one agency heavily involved with trying to preserve the Selva Lacandona (Lacandon jungle) ecosystem that lies in northeast Chiapas near the Guatemalan border. The region is irreplaceable from a biodiversity perspective. At about 1.8 million hectares (4.5 million acres), the Selva Lacandona contains about 30% of all animal and 20% of all plant species found in Mexico, and represents a good part of the last large remaining tract of North American tropical rainforest. But this wonderful wild region is under attack by an increasing human population and associated deforestation; it's estimated that half the forest has been cleared since 1960. Looking at a satellite image of the area is sobering and slightly eerie: on the Guatemalan side of the Usumacinta River is the unbroken greenery of virgin tropical rainforest, a tribute to the remoteness of the region and lack of roads on the Guatemalan side of the border that prohibits access to settlers; on the Mexican side however, where the Selva Lacandona is situated, are huge patches of deforestation, the result of human penetration into the area after the Mexican government pushed roads into the region for their military to patrol the border.

On paper, some of the Selva Lacandona is now protected, with the Montes Azules Biosphere Reserve probably the best protected and most important part. But protection on paper is not sufficient. Local residents, most often poor and poorly educated, must be convinced that wild areas are worth saving, and ecotourism development is one way to do that. CI and other organizations work with these local communities to help them create infrastructure for ecotourism. They hold workshops to train local people in potential ecotourism activities such as offering guide services, lodging, and transportation, and, fundamentally, describing what tourists want from the experience of visiting the area. The most successful outcome to date of these training activities are the wonderful tours that now depart daily from Palenque for the remote Mayan ruin sites of Bonampak and Yaxchilán (see p. 45). CI helped set up the organization for these tours, but local people from the indigenous communities located around the archaeological sites now run and profit from the tours. The success of the operation is evident: other indigenous communities have begun asking how they, too, might participate in ecotourism, and CI has been working to expand their success to other, even more remote regions of the Selva Lacandona (for example, in the Laguna Miramar region).

Ejidos, Conservation, and Ecotourism

One of the few acknowledged truths of ecotourism is that if local people and communities of modest means are to accept and support conservation measures and new nature reserves, they should be informed and consulted during all phases of the development process. For their continued support, local people should benefit economically from a park, reserve, or ecotourism facility, but local communities should also have a hand in the decision-making concerning the development and maintenance of ecotourism sites. In southern Mexico, complicating the ecotourism and conservation picture is the fact that impoverished local people own or occupy most of the land. These people can be divided into four main groups, and conservation agencies must sometimes work with all of them to achieve conservation success. A *community* is a large group of indigenous people that hold, as a group, huge chunks of tribal lands. The Lacandon Community, for instance, with which CI worked to establish tours to the Bonampak and Yaxchilán Mayan sites, is composed of the Lacandones, Choles, and Tzeltales peoples.

An *ejido* (eh-HEE-doe) is a smaller group, usually about 400 to 600 people, which owns a smaller parcel of communal land, often, in total, about 50 hectares (125 acres) per person, but sometimes much less. About 70% of Mexico's remaining forests are now owned by communities and ejidos. A third kind of people who often must figure prominently in conservation strategies are *new immigrants* – the many people who, for example, recently fled into the Selva Lacandona region of Chiapas to escape the political unrest in other parts of the state. New immigrants immediately begin cutting trees for houses and fuelwood, and need jobs, and therefore can have strong impacts on forests. The fourth category of local people important for conservation are *private landowners*. Land containing perhaps 20% of Mexico's remaining forests is held by small landowners and, in larger pieces, by the remnants of the *hacienda* system – the large pieces of Mexican real estate handed out by the Spanish crown during colonial times as rewards for services rendered. The attitudes of these different groups toward conservation depends on their use of forest resources. The Lacandon people, for instance, actually live in the jungle – it is their home – so they quickly understand the need for conservation. They see the logging and burning of forests as threats to their very way of life. But other groups, not so intimately tied to the forests, are not clear on the need for conservation. Some ejidos, for example, may understand that there is some money to be made through ecotourism, but really don't understand or care about the conservation implications of such activities. Conservation agencies work with these communities to educate them about the need for conservation and the relationship between ecotourism and conservation.

Ejidos themselves are increasingly offering ecotourism services. Inequitable land ownership was a main cause of the Mexican Revolution early in the 20th century; 90% or more of rural families at the time owned no land. Land redistribution was enshrined in the post-revolution 1917 Constitution. Land was confiscated mainly from the colonial haciendas and by 1981, nearly 100 million hectares (250 million acres) had been distributed to more than 1.5 million people. (Unfortunately, much of this land was of poor quality for agriculture.) As a way to prevent lands from ever again becoming concentrated in the hands of a few wealthy individuals, the ejido system was created. Each person was given land to work in a collective agricultural unit. In addition, each ejido (made up of several hundred people) was provided with large tracts of common land for use of all people in the ejido for grazing livestock and for collecting forest products – fuelwood and timber. (It is these associated forests lands held by ejidos that are often important for conservation purposes.) Until recently, all ejido lands were owned by the Mexican nation in perpetuity; they could not be sold, and had to be returned if not worked (the Mexican constitution was changed a few years ago to permit ejidos to be sold). The ejido is a self-governing land unit, a cooperative with elected officials; it decides how best to use its lands and what to do with profits from ejido agriculture and commerce (V. Halhead 1984). And some ejidos, at the urging of conservation organizations, are turning to ecotourism as a new, eco-friendly way to generate income.

Some of these ejidos are already operating rustic ecotourism lodges, and others are in the planning stages. The ejido Tres Garantías in southern Quintana Roo (located about 30 km, or 18 miles, south of Route 186, the main road that cuts across the base of the Yucatán Peninsula) has set aside half its forested lands as a nature reserve, built an enclosed camping hut with mosquito netting in the middle of the jungle, and opened its doors to ecotravellers. For very reasonable rates you can sleep there, get fed, and wander the lands and trails. CI has been working with the ejido Emiliano Zapata in the Laguna Miramar region of the Selva Lacandona to initiate ecotourism. Laguna Miramar is a large lake in a wilderness area in eastern Chiapas. Visitors camp in a communal hut and spend their days exploring the lake, nature trails, and nearby archaeological ruins. Eventually, CI would like to extend this kind of ejido-based ecotourism to other sites in the Selva Lacandona, such as the Laguna Ocotal area in the northern reaches of the Montes Azules Biosphere Reserve. Another good example of ejidos participating in ecotourism is the cloud forest trips to the El Triunfo Biosphere Reserve (p. 46) in southeastern Chiapas. Chiapas' Institute of Natural History worked closely with local ejidos (about 25 ejidos are located within the reserve's buffer zones) to establish the tours, and members of ejidos work as guides, cooks, etc. Similarly, the Institute of Natural History hopes to work with ejidos near the new La Encrucijada Biosphere (or Ecological) Reserve on the Pacific in southeastern Chiapas to open that site also to ecotourists.

Ecotourism is a kind of sustainable development and can certainly contribute to conservation. But does ecotourism always help local economies and significantly preserve visited habitats and wildlife? This question is important because increasingly, the fostering of ecotourism is suggested by indigenous people in developing nations, by the nations themselves, and by international conservation organizations, as one of the best methods to preserve natural resources and biodiversity almost anywhere that they are threatened. Many people who monitor tourism – researchers and government officials – believe that in the rush to make money from ecotourism, benefits are often overstated and problems ignored.

Many private companies purporting to be "ecotour" operators are "eco-" in name only; they are interested solely in profits, and are not concerned about local economies or the wild areas into which they take tourists. There is increasing concern about monetary "leakage": despite attempts to keep most of the ecotourist revenues in local destination economies, many of those dollars, more than 50% by recent estimates, leak back to large urban areas of destination countries and even to developed nations; relatively little actually is spent on conservation. (In southern Mexico, ecotravellers can visit ejidos or take tours operated by indigenous groups, thereby assuring that 100% of the fees charged go to local people.) In some countries, ecotourism is an unstable source of local employment and economic well-being. Tour bookings are heavily dependent on seasonal trends, on the weather, on a country's political situation, and on worldwide currency fluctuations. Finally, popularity as an ecotourism site may inevitably lead to its failing. As ecotourism expands dramatically, sites that are over-used and under-managed are damaged. Trails in forests gradually enlarge and deepen, erosion occurs, crowds of people are incompatible with natural animal behavior. A good example: the number of visitors to Ría Lagartos Biosphere Reserve on the northern coast of the Yucatán Peninsula has been declining, probably because the flamingos that are the primary attraction are increasingly moving to other areas to escape tour boat harassment. Also, ecotourism's success harms itself in another way: when any area becomes too popular, many travellers wanting to experience truly wild areas and quiet solitude no longer want to go there; that is, with increasing popularity, there is an inexorable deterioration of the experience.

Thus, ecotourism is not a miracle cure-all for conservation; these days it is understood to be a double-edged sword. Clearly, large numbers of people visiting sites cannot help but have adverse impacts on those sites. But as long as operators of the facilities are aware of negative impacts, careful management practices can reduce damage. Leakage of ecotourist revenue away from the habitats the money was meant to conserve is difficult to control, but some proportion of the money does go for what it is intended and, with increased awareness of the problem, perhaps that proportion can be made to grow. Travellers themselves can take steps to ensure that their trips help rather than hurt visited sites (see p. 4).

Chapter 5

How to Use This Book: Ecology and Natural History

- What is Natural History?
- What is Ecology and What Are Ecological Interactions?
- How to Use This Book
  Information in the Family Profiles
  Information in the Color Plate Sections

What is Natural History?

The purpose of this book is to provide ecotravellers with sufficient information to identify many common animal species and to learn about them and the families of animals to which they belong. Information on the lives of animals is generally known as *natural history*, which is usually defined as the study of animals' natural habits, including especially their ecology, distribution, classification, and behavior. This kind of information is of importance for a variety of reasons: animal researchers need to know natural history as background on the species they study, and wildlife managers and conservationists need natural history information because their decisions about managing animal populations must be partially based on it. More relevant for the ecotraveller, natural history is simply interesting. People who appreciate animals typically like to watch them, touch them when appropriate, and know as much about them as they can.

What is Ecology and What Are Ecological Interactions?

Ecology is the branch of the biological sciences that deals with the interactions between living things and their physical environment and with each other. *Animal ecology* is the study of the interactions of animals with each other, with plants, and with the physical environment. Broadly interpreted, these interactions take into account most everything we find fascinating about animals – what they eat, how they forage, how and when they breed, how they survive the rigors of extreme climates, why they are large or small, or dully or brightly colored, and many other facets of their lives.

An animal's life, in some ways, is the sum of its interactions with other animals – members of its own species and others – and with its environment. Of particular interest are the numerous and diverse ecological interactions that occur between different species. Most can be placed into one of several general categories, based on how two species affect each other when they interact; they can have positive, negative, or neutral (that is, no) effects on each other. The relationship terms below are used in the book to describe the natural history of various animals.

Competition is an ecological relationship in which neither of the interacting species benefit. Competition occurs when individuals of two species use the same resource – a certain type of food, nesting holes in trees, etc. – and that resource is in insufficient supply to meet all their needs. As a result, both species are less successful than they could be in the absence of the interaction (that is, if the other species were not present).

Predation is an ecological interaction in which one species, the *predator*, benefits, and the other species, the *prey*, is harmed. Most people think that a good example of a predator eating prey would be a mountain lion eating a deer, and they are correct; but predation also includes interactions in which the predator eats only part of its prey and the prey individual often survives. Thus, deer eat tree leaves and branches, and so, in a way, they can be considered predators on plant prey.

Parasitism, like predation, is a relationship between two species in which one benefits and one is harmed. The difference is that in a predatory relationship, one animal kills and eats the other, but in a parasitic one, the parasite feeds slowly on the "host" species and usually does not kill it. There are internal parasites, like protozoans and many kinds of worms, and external parasites, such as leeches, ticks, and mites.

Some of the most compelling of ecological relationships are *mutualisms* – interactions in which both participants benefit. Plants and their pollinators engage in mutualistic interactions. A bee species, for instance, obtains a food resource, nectar or pollen, from a plant's flower; the plant it visits benefits because it is able to complete its reproductive cycle when the bee transports pollen to another plant. In Central America, a famous case of mutualism involves several species of acacia plants and the ants that live in them: the ants obtain food (the acacias produce nectar for them) and shelter from the acacias and, in return, the ants defend the plants from plant-eating insects. Sometimes the species have interacted so long that they now cannot live without each other; theirs is an *obligate* mutualism. For instance, termites cannot by themselves digest wood. Rather, it is the single-celled animals, protozoans, that live in their gut that produce the digestive enzymes that digest wood. At this point in their evolutionary histories, neither the termites nor their internal helpers can live alone.

Commensalism is a relationship in which one species benefits but the other is not affected in any way. For example, epiphytes (p. 21), such as orchids and bromeliads, that grow on tree trunks and branches obtain from trees some shelf space to grow on, but, as far as anyone knows, neither hurt nor help the trees. A classic example of a commensal animal is the remora, a fish that attaches itself with a suction cup on its head to a shark, then feeds on scraps of food the shark leaves behind. Remora are commensals, not parasites – they neither harm nor help sharks, but they benefit greatly by associating with sharks. Cattle Egrets (Plate 23) are commensals – these birds follow cattle, eating insects and other small animals that flush from cover as the cattle move about their pastures; the cattle, as far as we know, couldn't care one way or the other (unless they are concerned about that certain loss of dignity that occurs when the egrets perch not only near them, but on them as well.)

A term many people know that covers some of these ecological interactions is *symbiosis*, which means living together. Usually this term suggests that the two interacting species do not harm one another; therefore, mutualisms and commensalisms are the symbiotic relationships discussed here.

How to Use This Book

The information here on animals is divided into two sections: the *plates*, which include artists' color renderings of various species together with brief identifying and location information; and the *family profiles*, with natural history information on the families to which the pictured animals belong. The best way to identify and learn about Mexican animals may be to scan the illustrations before a trip to become familiar with the kinds of animals you are likely to encounter. Then when you spot an animal, you may recognize its type or family, and can find the appropriate pictures and profiles quickly. In other words, it is more efficient, upon spotting a bird, to be thinking, "Gee, that looks like a flycatcher," and be able to flip to that part of the book, than to be thinking, "Gee, that bird is yellowish" and then, to identify it, flipping through all the animal pictures, searching for yellow birds.

Information in the Family Profiles

Classification, Distribution, Morphology. The first paragraphs of each profile generally provide information on the family's classification (or *taxonomy*), geographic distribution, and *morphology* (shape, size, and coloring of the animals). Classification information is provided because it is how scientists separate animals into related groups and often it enhances our appreciation of animals to know these relationships. You may have been exposed to classification levels sometime during your education but if you are a bit rusty, a quick review may help: *Kingdom* Animalia: aside from plant information, all the species detailed in the book are members of the animal kingdom. *Phylum* Chordata, *Subphylum* Vertebrata: most of the species in the book are vertebrates, animals with backbones (exceptions are the corals and other marine *invertebrate* animals discussed in Chapter 10). *Class*: the book covers several vertebrate classes such as Amphibia (amphibians), Reptilia (reptiles), Aves (birds), and Mammalia (mammals). *Order*: each class is divided into several orders, the animals in each order sharing many characteristics. For example, one of the mammal orders is Carnivora, the carnivores, which includes mammals with teeth specialized for meat-eating – dogs, cats, bears, raccoons, weasels. *Family*: families of animals are subdivisions of each order that contain closely related species that are very similar in form, ecology, and behavior. The family Canidae, for instance, contains all the dog-like mammals – coyote, wolf, fox, dog. Animal family names end in "-dae;" subfamilies, subdivisions of families, end in "-nae." *Genus*: further subdivisions; within each genus are grouped species that are very closely related – they are all considered to have evolved from a common ancestor. *Species*: the lowest classification level; all members of a species are similar enough to be able to breed and produce living, fertile offspring.

Example: Classification of the Keel-billed Toucan (Plate 48):

Kingdom: Animalia, with more than a million species
Phylum: Chordata, Subphylum Vertebrata, with about 40,000 species
Class: Aves (Birds), with about 9000 species
Order: Piciformes, with about 350 species; includes honeyguides, woodpeckers, barbets, and toucans
Family: Ramphastidae, with 55 species; includes barbets and toucans
Genus: *Ramphastos*, with 11 species; toucans
Species: *Ramphastos sulfuratus*; Keel-billed Toucan

Some of the family profiles in the book actually cover animal orders; others describe families or subfamilies.

Species' distributions vary tremendously. Some species are found only in very limited areas, whereas others range over several continents. Distributions can be described in a number of ways. An animal or group can be said to be *Old World* or *New World*; the former refers to the regions of the globe that Europeans knew of before Columbus – Europe, Asia, Africa; and the latter refers to the Western Hemisphere – North, Central, and South America. Southern Mexico falls within the part of the world called the *Neotropics* by biogeographers – scientists who study the geographic distributions of living things. A Neotropical species is one that occurs within southern Mexico, Central America, South America, and/or the Caribbean Islands. The terms *tropical*, *temperate*, and *arctic* refer to climate regions of the Earth; the boundaries of these zones are determined by lines of latitude (and ultimately, by the position of the sun with respect to the Earth's surface). The tropics, always warm, are the regions of the world that fall within the belt from latitude 23.5 degrees North (the Tropic of Cancer) to latitude 23.5 degrees South (the Tropic of Capricorn). The world's temperate zones, with more seasonal climates, extend from 23.5 degrees North and South to the Arctic and Antarctic Circles, at 66.5 degrees North and South. Arctic regions, more or less always cold, extend from 66.5 degrees North and South to the poles. The position of southern, or tropical, Mexico with respect to these zones is shown in Map 5.

Several terms help define a species' distribution and describe how it attained its distribution.

Range. The particular geographic area occupied by a species.

Native or Indigenous. Occurring naturally in a particular place.

Introduced. Occurring in a particular place owing to peoples' intentional or unintentional assistance with transportation, usually from one continent to another; the opposite of native. For instance, pheasants were initially brought to North America from Europe/Asia for hunting, Europeans brought rabbits and foxes to Australia for sport, and the British brought European Starlings and House Sparrows to North America.

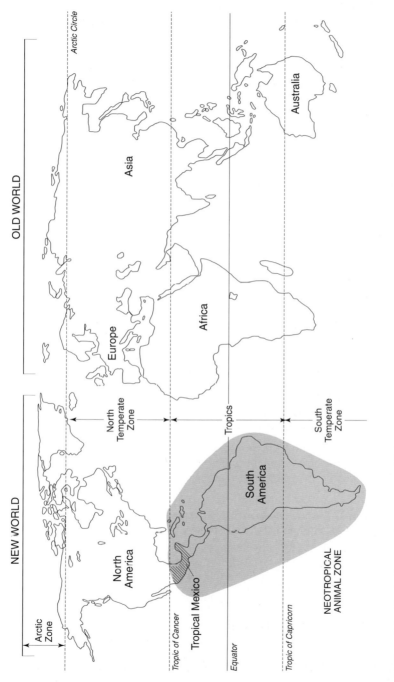

Map 5 Map of the Earth showing the position of Tropical Mexico, the area covered in the book; Old World and New World zones; tropical, temperate, and arctic regions; and the Neotropical animal life zone.

Endemic. A species, a genus, an entire family, etc., that is found in a particular place and nowhere else (see Close-Up, p. 236). Galápagos finches are endemic to the Galápagos Islands; nearly all the reptile and mammal species of Madagascar are endemics; all species are endemic to Earth (as far as we know).

Cosmopolitan. A species that is widely distributed throughout the world.

Ecology and Behavior. In these sections, I describe some of what is known about the basic activities pursued by each group. Much of the information relates to when and where animals are usually active, what they eat, and how they forage.

Activity Location – *Terrestrial* animals pursue life and food on the ground. *Arboreal* animals pursue life and food in trees or shrubs. Many arboreal animals have *prehensile* tails, long and muscular, which they can wrap around tree branches to support themselves as they hang to feed or to move about more efficiently. *Cursorial* refers to animals that are adapted for running along the ground.

Activity Time – *Nocturnal* means active at night. *Diurnal* means active during the day. *Crepuscular* refers to animals that are active at dusk and/or dawn.

Food Preferences – Although animal species can usually be assigned to one of the feeding categories below, most eat more than one type of food. Most frugivorous birds, for instance, also nibble on the occasional insect, and carnivorous mammals occasionally eat plant materials.

> *Herbivores* are predators that prey on plants.
> *Carnivores* are predators that prey on animals.
> *Insectivores* eat insects.
> *Granivores* eat seeds.
> *Frugivores* eat fruit.
> *Nectarivores* eat nectar.
> *Piscivores* eat fish.
> *Omnivores* eat a variety of things.
> *Detritivores*, such as vultures, eat dead stuff.

Breeding. In these sections, I present basics on each group's breeding particulars, including type of mating system, special breeding behaviors, durations of egg incubation or gestation (pregnancy), as well as information on nests, eggs, and young.

Mating Systems – A *monogamous* mating system is one in which one male and one female establish a pair-bond and contribute fairly evenly to each breeding effort. In polygamous systems, individuals of one of the sexes have more than one mate (that is, they have harems): in *polygynous* systems, one male mates with several females, and in *polyandrous* systems, one female mates with several males.

Condition of young at birth – *Altricial* young are born in a relatively undeveloped state, usually naked of fur or feathers, eyes closed, and unable to feed themselves, walk, or run from predators. *Precocial* young are born in a more developed state, eyes open, and soon able to walk and perhaps feed themselves.

Ecological Interactions. These sections describe what I think are intriguing ecological relationships.

Lore and Notes. These sections provide brief accounts of folklore associated with the profiled groups, and any other interesting bits of information about the profiled animals that do not fit elsewhere in the account.

Status. These sections comment on the conservation status of each group, including information on relative rarity or abundance, factors contributing to population declines, and special conservation measures that have been implemented. Because this book concentrates on animals that ecotravellers are most likely to see – that is, on more common ones – few of the profiled species are immediately threatened with extinction. The definitions of the terms that I use to describe degrees of threat to various species are these: *Endangered* species are known to be in imminent danger of extinction throughout their range, and are highly unlikely to survive unless strong conservation measures are taken; populations of endangered species generally are very small, so they are rarely seen. *Threatened* species are known to be undergoing rapid declines in the sizes of their populations; unless conservation measures are enacted, and the causes of the population declines identified and halted, these species are likely to move to endangered status in the near future. *Vulnerable to threat*, or *Near-threatened*, are species that, owing to their habitat requirements or limited distributions, and based on known patterns of habitat destruction, are highly likely to be threatened in the near future. For instance, a fairly common bird species that breeds mainly in coastal mangrove swamps might be considered vulnerable to threat if it is known that people are cutting mangroves at a high rate. Several organizations publish lists of threatened and endangered species.

Where appropriate, I also include threat classifications from the Convention on International Trade in Endangered Species (CITES) and the United States Endangered Species Act (USA ESA). CITES is a global cooperative agreement to protect threatened species on a world-wide scale by regulating international trade in wild animals and plants among the 130 or so participating countries. Regulated species are listed in CITES Appendices, with trade in those species being strictly regulated by required licenses and documents. CITES Appendix I lists endangered species; all trade in them is prohibited. Appendix II lists threatened/vulnerable species, those that are not yet endangered but may soon be; trade in them is strictly regulated. Appendix III lists species that are protected by laws of individual countries that have signed the CITES agreements. The USA's Endangered Species Act works in a similar way – by listing endangered and threatened species, and, among other provisions, strictly regulating trade in those animals.

Information in the Color Plate Sections

Pictures. Among amphibians, reptiles, and mammals, males and females of a species usually look alike, although often there are size differences. For many species of birds, however, the sexes differ in color pattern and even anatomical features. If only one individual is pictured, you may assume that male and female of that species look exactly or almost alike; when there are major sex differences, both male and female are depicted.

Name. I provide the common English name for each profiled species, the scientific, or Latin, name, and local Spanish names and their English translations, if known. I did not attempt to list animal names used by local indigenous peoples.

ID. Here I provide brief descriptive information that, together with the pictures, will enable you to identify most of the animals you see. The lengths of reptiles and amphibians given in this book are *snout–vent lengths* (SVLs), the distance from the tip of the snout to the vent, unless I mention that the tail is included. The *vent* is the opening on their bellies that lies approximately where the rear limbs join the body, and through which sex occurs and wastes exit. Therefore, long tails of salamanders and lizards, for instance, are not included in the reported length measurements, and frogs' long legs are not included in theirs. I made an exception for snakes, for which in most cases I give total lengths. For mammals, measurements I give are generally the lengths of the head and body, but do not include tails. Birds are measured from tip of bill to end of tail. For birds commonly seen flying, such as seabirds and hawks, I provide wingspan (wingtip to wingtip) measurements, if known. For most of the passerine birds (see p. 122), I use to describe their sizes the terms *large* (more than 30 cm, 12 in, long); *mid-sized* (between 15 or 18 cm, 6 or 7 in, and 30 cm, 12 in); *small* (10 to 15 cm, 4 to 6 in); and *very small* (less than 10 cm, 4 in).

Habitat/Locations. In these sections I give the regions and habitat types in which each species occurs, symbols for the habitat types each species prefers, and the Mexican states where each species may be found. The states listed for a particular species are (1) those in which the species is known to occur and also (2) those in which it probably occurs, based on the known range of the animal, the elevation and type of habitat it prefers, and the elevations and types of habitats known to be present in the various states.

In general, when I say an animal is usually found at *low elevations*, I mean at elevations of less than 500 m (1600 ft); at *middle elevations*, between 500 and 1200 m (1600 and 4000 ft); and at *high elevations*, above 1200 m (4000 ft).

Explanation of location symbols (see Maps 2, 3 and 4, pp. 15, 16, 18):

OAX-HI refers to the large central portion of Oaxaca, containing middle and high elevation terrain; it includes the entire area around Oaxaca City.

OAX-LO refers to low elevation regions of Oaxaca, primarily the lowlands along the Pacific Ocean, the northern swath adjacent to the state of Veracruz, and the Isthmus of Tehuantepec. Not all species that occur in the northern lowlands occur in the southern lowlands, and vice versa.

CHIAP-HI refers to the large central portion of Chiapas, containing middle and high elevation terrain; it includes the areas around Tuxtla Gutiérrez and San Cristóbal.

CHIAP-LO refers to low elevation regions of Chiapas, primarily the lowlands along the Pacific Ocean and the swath adjacent to Tabasco and northern Guatemala (the Petén). Not all species that occur in the northern lowlands occur in the southern lowlands, and vice versa.

TAB is Tabasco, the small state north of Chiapas, bordering the Gulf of Mexico (this state encompasses only lowlands).

CAM is Campeche, the western part of the Yucatán Peninsula (this state encompasses only lowlands).

YUC is Yucatán, the northern and central parts of the Yucatán Peninsula (this state encompasses only lowlands).

QRO is Quintana Roo, the eastern section of the Yucatán Peninsula (this state encompasses only lowlands).

Explanation of habitat symbols:

= Lowland wet forest.

= Lowland dry forest.

= Highland forest/cloud forest. Includes middle and higher elevation forests, including the pine and pine–oak forests of Oaxaca and Chiapas, and cloud forests.

= Forest edge/streamside. Some species typically are found along forest edges or near or along streams; these species prefer semi-open areas rather than dense, closed, interior parts of forests. Also included here: open woodlands, tree plantations, and shady gardens.

= Pastureland/savannah (grassland with scattered trees, shrubs)/non-tree plantations/gardens without shade trees/roadside. Species found in these habitats prefer very open areas.

= Freshwater. For species typically found in or near lakes, streams, rivers, marshes, swamps.

= Saltwater/marine. For species usually found in or near the ocean or ocean beaches.

Example

Plate 48b

Keel-billed Toucan
Ramphastos sulfuratus
Tucán Pico-multicolor = multi-colored-bill toucan
Tucán Cuello Amarillo = yellow-necked toucan
Tucán Real = royal toucan

ID: A large, mostly black bird with yellow face and chest, yellowish green skin around eye, and that amazing, rainbow-colored (green-orange-blue) toucan's bill; to 56 cm (22 in).

HABITAT: Low and middle elevation forests; found in tree canopy in more open habitats such as forest edges, tree plantations, along rivers and streams.

LOCATIONS: OAX-HI, OAX-LO, CHIAP-HI, CHIAP-LO, TAB, CAM, YUC, QRO

Chapter 6

Amphibians

General Characteristics and Natural History

AMPHIBIANS first arose during the mid part of the Paleozoic Era, 400 million years ago, developing from fish ancestors that had lungs and thus could breathe air. The word *amphibian* refers to an organism that can live in two worlds, and that is as good a definition of the amphibians as any: most stay in or near the water, but many spend at least portions of their lives on land. In addition to lungs, amphibians generally have wet, thin skins that aid in gas exchange (breathing). Amphibians were also the first animals to develop legs for walking on land, the basic design of which has remained remarkably constant for all other land vertebrates. Many have webbed feet that aid locomotion in the water.

Approximately 4500 species of living amphibians have been described. (Owing to their all-terrestrial existence, almost all reptiles, birds, and mammals living on Earth have been identified but, most experts agree, many more amphibians, with their aquatic ways, remain to be discovered.) They are separated into three groups. The *salamanders* (Order Caudata, or "tailed" amphibians) comprise 450 species (about 90 in Mexico), the mysterious *caecilians* (sih-SIL-ians; Order Gymnophiona) number about 160 species (two in Mexico), and the *frogs* and *toads* (Order Anura, "without tails") make up the remainder, and the bulk, of the group (with about 200 Mexican species).

Most amphibians live in the water during part of their lives. Typically a juvenile stage is spent in the water and an adult stage on land. Because amphibians need to keep their skin wet, even when on land they are mostly found in moist habitats – in marshes and swamps, around the periphery of bodies of water, in wet forests. Adults of most species return to the water to lay eggs, which must stay wet to develop. Some amphibians – toads and some salamanders – are entirely terrestrial, laying eggs on land in moist places.

Relatively little is known of the caecilians. They are confined to tropical forests of Africa, Southeast Asia, and Central and South America, the latter two locations being their principal home (there are no Sicilian caecilians). They are legless with slender bodies up to 40 cm (15 in) long and up to 2.5 cm (1 in) in diameter, with ringed creases along their lengths – resembling nothing so much as a cross between a snake and an earthworm (Plate 1). Their skin is studded with small scales. Caecilian eyes are tiny and most are blind as adults. For sensing their environment they possess a pair of small tentacles at the head end that are continually extended out of and then withdrawn into the body. They feed on worms and other small invertebrate animals – termites, beetles, etc. – that they find underground. Because they spend much of their lives beneath the ground, caecilians are very rarely seen by people. However, they commonly leave their burrows at night during rains, moving along the ground, presumably foraging. Occasionally they are found in meadows and forests under logs and rocks, or buried deep in the leaf litter. I remember finding one, apparently bloodied and dropped by a predator, on a dirt road in South America. It was the first one another biologist and myself had seen outside of photographs, and it took quite a while for us to decide just what it was (a mutant worm from outer space, our first guess, turned out to be incorrect). Caecilians do not lay eggs, but give birth to live young.

Salamanders, intriguingly, are not much interested in the tropics. They are an almost exclusively temperate zone group. In the tropics they usually inhabit moist, cool areas, such as middle and higher-elevation cloud forests. About 200 species are completely terrestrial. Females of these species lay eggs in moist places and then stay with the eggs, protecting them until hatching, sometimes coiling themselves around the eggs. *Newts* are a group of aquatic salamanders, usually with rough-textured skin, that generally lay their eggs singly on pond bottoms or on vegetation; other aquatic salamanders lay their eggs in large, jelly-covered clusters, which they attach to sticks or plants in the water. Aquatic salamanders do not protect their eggs.

Most species of frogs and toads live in tropical or semi-tropical areas but some groups are abundant in temperate latitudes. Frogs can be either mostly aquatic or mostly terrestrial; some live primarily in vegetation on land. A few species of frogs live in deserts by staying underground, remaining moist enough to survive. At the arrival of periodic rains, they climb to the surface and breed in temporary ponds, before scampering back underground. Eggs hatch and the froglets burrow underground before the ponds dry. *Toads* constitute a group of frogs that have relatively heavy, dry skin that reduces water loss, permitting them to live on land. Most frogs and toads leave their eggs to develop on their own, but a few guard nests or egg masses, and some species actually carry their eggs on their backs or in skin pouches.

Frogs and toads are known for the vocal behavior of males, which during breeding periods call loudly from the edges of lakes and ponds or on land, attempting to attract females. Each species has a different type of call. Some species breed *explosively* in synchronous groups – on a single night thousands gather at forest ponds, where males call and compete for females and females choose from among available suitors. Many frogs, such as *bullfrogs*, fight fiercely for the best calling spots and over mates. Because frogs really have no weapons to fight with, such as teeth or sharp claws, size usually determines fight outcomes. Some species of frogs and toads have developed *satellite* strategies to obtain matings. Instead of staking out a calling spot and vocalizing themselves (which is energy-draining and risky because calling attracts predators), satellite males remain silent but stay furtively near calling males, and attempt to intercept and mate with approaching females. Smaller males are more likely to employ such "sneaky-mater" tactics.

All amphibians as adults are predatory carnivores – as far as it is known, there is not a vegetarian among them. Many animals eat amphibians, although it is not as easy as you might think. At first glance amphibians appear to be among the most defenseless of animals. Most are small, many are relatively slow, and their teeth and claws are not the types appropriate for aggressive defense. But a closer look reveals an array of ingenious defenses. Most, perhaps all, amphibians produce toxins in the skin, many of which are harmless to humans and so not very noticeable, but a few of which are quite poisonous and even lethal (the toxins produced by South America's *poison-dart frogs*, for instance). These toxins deter predators. Most amphibians are cryptically colored, often being amazingly difficult for people, and presumably predators, to detect in their natural settings. The jumping locomotion of frogs probably evolved as an anti-predator strategy – it is a much more efficient way of escaping quickly through leaves, dense grass, thickets, or shrubby areas than are walking or running. Some frogs hiss loudly and inflate their throat sacs when approached, which presumably makes predators think twice about attacking. Last, some frogs give loud screams when grabbed, which are startling to predators and so create opportunities for escape.

In general, amphibians are under less direct threat from people than are other vertebrate groups, there being little commercial exploitation of the group (aside from certain peoples' inexplicable taste for the limbs of frogs). However, amphibians are very sensitive to habitat destruction and, particularly, to pollution, because aquatic eggs and larvae are very susceptible to toxic substances. Many amphibian populations have been noticeably declining in recent years, although the reasons are not entirely clear. Much research attention is currently focused on determining whether these reported population changes are real and significant and, if so, their causes. Not all ecologists agree yet on what is going on (see Close-Up, p. 88).

Seeing Amphibians in Tropical Mexico

Oaxaca and Chiapas and, to a lesser extent, the Yucatán Peninsula support a rich variety of amphibian species, many of which are observable, provided some effort is taken. First, an admonition: snakes have our respect, due to the evil reputations of the venomous ones, but many people, with little knowledge of tropical amphibians, improperly do not show them the same respect. Skin secretions of some terrestrial frogs are toxic. The toxins can be dangerous if absorbed into the body through, for instance, a cut in your skin. The skin secretions of many frogs and salamanders, even if not harmful to humans, can sting badly. Take-home message: when amphibians are located, enjoy them visually and leave their handling to experts.

With that warning in mind, here are the best ways to see amphibians. Look for them, of course, in moist habitats – wet forests, near bodies of water, in small pools and puddles, along streams. If you dare, and using your shoes, turn over rocks and logs in these habitats, or look through and under leaf litter. Most adult salamanders and most frogs are nocturnal, so a night walk with flashlight or head-lamp is often a good way to see amphibians. The best times are during a wet night (although not during torrential downpours) or after rain, especially during the rainy season, when most species breed. The calls of male frogs can be of assistance then in locating the little beasts.

Family Profiles

1. Salamanders

Salamanders are long, generally slender amphibians with four limbs. They loosely resemble lizards in body plan, but their skin is smooth and wet – both from the aquatic or moist locales that they inhabit and the skin secretions that keep their skin wet for breathing. World-wide they number about 450 species, most of them being found within temperate zone regions. The group's largest family, however, containing about half the known species, is well represented in the American tropics. This is the Family Plethodontidae, the lungless salamanders. All or most of the salamanders that occur in the region covered by this book are members of this group. About 69 *plethodontid* species occur in Mexico, 50 of which are endemic. Being lungless, almost all gas exchange is through the moist skin (some occurs through the membranes of the mouth and throat). Most of the pletho-dontids are primarily terrestrial, inhabiting all but the very driest regions within their range.

Plethodontid salamanders vary quite a bit in size and shape, from ones typi-cally only 3 cm (1 in) long to those 25 cm (10 in) long. Some have webbed hands and feet. Color patterns are often striking, and can vary within the same species between locations.

Salamanders are secretive and usually nocturnal in their activities; hence, unless searched for, they are only infrequently encountered.

Natural History

Ecology and Behavior
In contrast to frogs, with their vegetarian tadpole phase, all salamanders, both as juveniles and adults, are carnivores. They feed opportunistically, essentially snap-ping at and trying to ingest whatever squirms and will fit into their mouths – gen-erally, on land, insects and spiders, but also small frogs and other salamanders. Because studying salamanders in the wild is difficult, relatively little is known about what eats them, but the list includes snakes, salamanders, frogs, fish, and a few birds and mammals. Although most of the plethodontids are considered ter-restrial, some are *semi-arboreal*, regularly climbing into trees and shrubs, and others are *semi-fossorial*, usually burrowing into the leaf litter or underneath rocks or logs.

Breeding

Salamanders are known for their complex mating behavior. Unlike frogs, salamander males do not call to advertise territories or attract mates, but males do fight each other for females. Losers employ various strategies to try to interfere with the winners' mating – either physically inserting themselves between a female and a courting male, or approaching the mating pair, which distracts the male, causing him to pause or stop.

How sperm gain access to eggs in terrestrial salamanders is different from the way one might expect. Following a courtship ritual, the male deposits a tiny, cone-shaped packet containing sperm, called a *spermatophore*, on the ground. The female maneuvers until she can pick it up with her *vent* (the opening on her belly that provides entry and exit for both reproductive and excretory systems), which permits the sperm to gain access to her reproductive tract. She can then store the sperm for many months, having it always available to fertilize her eggs.

In the terrestrial salamanders, females typically lay a clutch of eggs – for instance, a clutch of 63 eggs was found in a MEXICAN SALAMANDER (Plate 1) – in moist areas under logs, moss, or rocks. Then they coil themselves around the eggs and stay with them, protecting them and keeping them moist with their skin secretions. Males have also been found protecting eggs in this way, but it appears to be mainly a female role. Egg care is apparently necessary; eggs not protected by an adult are unlikely to hatch. Eggs hatch in four to five months. The young emerge looking like miniature adults and must fend for themselves from the moment of birth. Growth proceeds very slowly, with many salamanders attaining sexual maturity only at six or more years of age.

Ecological Interactions

Salamanders are usually inconspicuous and hidden to casual observers. Actually, however, they are integral parts of the forest animal community, often being very abundant in good habitat. Therefore, they are one of the predominant predators on forest insects, acting to keep many bug populations in check.

In addition to hiding, running and, in some species, biting, to prevent themselves from becoming another animal's dinner, salamanders employ chemical weapons. All of them apparently have skin secretions that are sufficiently toxic or distasteful to render them unpalatable to most potential predators, and some of the secretions contain potent poisons. Unfortunately for the salamanders, a few predators, such as some snakes, have evolved immunity from the secretions' noxious effects.

Lore and Notes

Many salamanders apparently lead long lives, as do several other groups of cold-blooded animals. They have low metabolisms and they grow slowly in cool habitats, at times eating infrequently and going through long periods of inactivity. In captivity, some salamanders – for instance, one species from Japan – live more than 50 years. Many New World species have lifespans that commonly range up to 20 to 25 years.

Status

None of the plethodontid salamanders of southern Mexico are thought to be rare or threatened; on the other hand, little is known of the biology of most of them. Several species of this group are threatened, however, in the USA, and five are endangered, including DESERT SLENDER, SHENANDOAH, and TEXAS BLIND SALAMANDERS.

Profiles

Mexican Salamander, *Bolitoglossa mexicana*, Plate 1b
Yucatán Salamander, *Bolitoglossa yucatana*, Plate 1c
Rufescent Salamander, *Bolitoglossa rufescens*, Plate 1d

2. Toads

Scientists sometimes have trouble formally differentiating toads from frogs, but not so non-scientists, who usually know their toads: they are the frog-like animals with a rough, lumpy, wart-strewn appearance, not built for speed, that one finds on land. Actually, toads and frogs are both included in the amphibian group Anura; toads are a kind of frog. They have some special skeletal and reproductive traits that are used to set them apart (for instance, frogs lay eggs in jellied clusters, but toads lay them in jellied strings), but for our purposes, a modification of the common definition will do: toads are squat, short, terrestrial frogs with thick, relatively dry skins that prevent rapid water loss, short limbs, and glands that resemble warts spread over their bodies. A few families of frogs are called toads, but the predominant group, hugely successful, is the *true* toads, Family Bufonidae (*bufo* is Latin for "toad"), which spread naturally to all continents except Australia (and has been introduced by people to that continent; all the toads occurring in the USA are in this family). *Bufonids* usually have two prominent "warts" on each side of the neck or shoulder area, called *parotoids*. Often shades of olive or brown, toads vary in size, with the largest in southern Mexico, the MARINE TOAD (Plate 2), being anywhere from 9 to 20 cm (3.5 to 8 in) long, and weighing in at up to 1200 g (2.5 lb). World-wide there are perhaps 350 bufonid toad species. About 24 species occur in Mexico, of which 10 are endemic. Ten species occur in Oaxaca and Chiapas but only two in the Yucatán Peninsula.

Natural History

Ecology and Behavior

Although their relatively heavy, dry skin (as compared with that of other frogs) permits adult toads a permanently terrestrial existence, they experience some water loss through their skin, so unless they stay near water or in moist habitat, they dry out and die in just a few days. Although they are freed from an aquatic life, water still governs their existence. Many toads are primarily nocturnal, avoiding the sun and its drying heat by sheltering during the day under leaf litter, logs, or rocks, coming out to forage only after sundown. Toad tadpoles are vegetarians, feeding on green algae and bacteria in their aquatic habitats, but adult toads are all carnivorous, foraging for arthropods, mostly insects, amid the leaf litter. As one researcher defines the toad diet, "if it's bite-sized and animate it is food, no matter how noxious, toxic, or biting/stinging" (G. Zug 1983). Beetles and ants are frequent prey, as are small vertebrates such as small frogs, salamanders, and lizards.

Slow-moving toads (with their short legs they are capable of covering only very short distances per hop) have two methods to escape being eaten. They can be extremely hard to detect in their habitats, concealing themselves with their cryptic coloration and habit of slipping into crevices, under leaves, or actually burying themselves in the earth. Also, apparently quite effectively, they exude noxious fluids from their skin glands – the warts are actually a defense mechanism. If grabbed, a viscous, white fluid oozes from the warts. The fluid is very irritating to mucous membranes, such as those found in a predator's mouth and nose. Toads also have muscle control over the poison glands and some can squeeze them to spray the poison more than 30 cm (1 ft). Most predators that pick up a toad probably do not do it twice; peoples' four-legged pets that put toads into their mouths have been killed by the poisons. A few predators, however, such as raccoons and opossums, having learned their way around toad anatomy, avoid the warts on the back and legs by eating only the inside of the toad, entering through the mostly poison-gland-free belly.

Breeding
Many toads in the New World tropics breed at any time of year (for example, MARINE TOADS), but often there are breeding surges when seasons change. Fertilization is external, which means that it happens outside of the animal, in water. To breed, males migrate to ponds and streams, calling to attract females. After appropriate mating maneuvers of both sexes, sperm are released by a male in a cloud into the water, followed by a female releasing her eggs into the cloud. The eggs are laid within jellied strings, the jelly protecting the eggs physically and also, because it contains toxins, discouraging consumption by potential predators. Depending on species and size, a female may lay from 100 to 25,000 eggs at a time. Eggs hatch in only a few days, releasing young in the larval feeding stage known commonly as *tadpoles*. They feed, grow, and develop, transforming themselves into *toadlets* after a few weeks, which then swarm up the banks and disappear into their terrestrial existences. Toads generally reach sexual maturity in a year or two, although the period before breeding is longer in some species.

Ecological Interactions
The MARINE TOAD, the region's largest toad, has somehow become semi-domesticated, and is now commonly found around human settlements at much higher densities than in wild areas. Adapting nicely to people's behavior, these toads will eat dog and cat food left outdoors for pets (which is quite a feat, given that most frogs will only eat live, moving prey). These toads have also been introduced by humans to areas outside their natural range, including Hawaii, the Philippines, and Australia, and they have become pests in some of their new homes. Because of their abundance around dwellings, they probably are the toad most likely to be seen by visitors to southern Mexico.

Lore and Notes
The claim that a person will contract warts by handling toads is not true. Human warts are caused by viruses, not amphibians. The glands on toads' skin that resemble warts release noxious fluids to discourage predators. Various poisons have been identified in these fluids that, among other effects, cause increased blood pressure, blood vessel constriction, increased power of heartbeat, heart muscle tissue destruction, and hallucinations. Because these fluids minimally are irritants, a smart precaution is to avoid handling toads or, after such handling, make sure to wash your hands. Caustic irritation will result if the fluids are transferred from hands to eyes, nose or mouth. Some reports have it that voodoo practitioners in Haiti use the skin secretions of toads in their zombie-making concoctions.

Several hallucinogenic chemicals, which cause LSD-type effects when swallowed, have been isolated from the skin glands of Central American toads, for instance, from Marine Toads. Cultural historians suspect that these substances were known to the ancient Mayans and were used during religious and ceremonial occasions.

Status

About five New World bufonid species are currently endangered and a few more are threatened. The only endangered species known to occur in Mexico is the WESTERN TOAD, which also occurs in the USA and Canada. YOSEMITE, HOUSTON, and AMARGOSA TOADS, are all endangered in the USA. Costa Rica's GOLDEN TOAD is considered critically endangered and may now be extinct. Preservation of Golden Toads was one of the original reasons for the creation of the Monteverde Cloud Forest Reserve, one of Costa Rica's major ecotourism draws.

Profiles

Marine Toad, *Bufo marinus*, Plate 2a
Gulf Coast Toad, *Bufo valliceps*, Plate 2b
Marbled Toad, *Bufo marmoreus*, Plate 2c
Mexican Burrowing Toad, *Rhinophrynus dorsalis*, Plate 2d

3. Rainfrogs

The *rainfrogs*, Family Leptodactylidae, also known as the *tropical frogs*, constitute a large New World group that is distributed from southern North America to throughout most of South America and the Caribbean. Primarily frogs of low and middle elevations, they occupy many different types of habitats, but especially forests. Often they are the most abundant kind of amphibian found in a given habitat. Species are variously land-dwelling, aquatic, or arboreal. About 60 of the 800 living species occur in Mexico, 37 of them being endemic. More than 20 species occur in Oaxaca and Chiapas, but only four live in the Yucatán Peninsula.

Rainfrogs vary extensively in size and color patterns, so much so that a general description of the group is difficult. One distinctive trait is the greatly enlarged finger disks – little suction cups – characteristic of many, the better to get around arboreal habitats. Sizes vary from small species that at their largest are only 2.5 to 3.5 cm (1 in) long, to others that range up to 9 cm (3.5 in). Rainfrogs tend to be various shades of tan, beige, brown, gray, or pale yellow, often with dark markings such as bars or spots, particularly on the legs. Telling species apart by their markings is rendered more difficult because of the large amount of variation present within some species. For instance, in Costa Rica, one rainfrog species varies so extensively among individuals – even among those found in the same small area – in color, spotting pattern, and skin texture, that for many years even scientists were confused, repeatedly naming new species of rainfrogs although they were actually all the same species (the way a newcomer to Earth might mistakenly assign species status to each breed of domestic canine, which, although they look very different, are all of the same species of marginal intelligence that we call dogs). The explanation for such extreme variability in color schemes within a single species is unknown, but is suspected to have to do with the advantages each kind has in blending in with various types of surroundings on the forest floor.

Natural History

Ecology and Behavior

Rainfrogs in southern Mexico are primarily terrestrial, encountered most often on the forest floor or in or adjacent to pools, ponds, or streams. Several small species typically coexist in the same area, making up a large portion of the "leaf litter amphibian community."

Many rainfrogs are known as ant specialists, but typically a variety of insects, other arthropods, and even small vertebrates, are included in the diet – mites, beetles, spiders, small lizards, and others. Rainfrogs are *sit-and-wait* predators. Some live in protective rock crevices or ground burrows, venturing out only to forage around the opening (or to breed), escaping back into the shelter at the first sign of trouble. Rainfrogs, especially the nocturnal ones, have reputations for being quite shy, which seems surprising for a night-active animal until one realizes that one of the chief predators on these frogs is also nocturnal – bats.

Breeding
These frogs breed in a variety of ways. One group (genus *Eleutherodactylus*) has become completely terrestrial in its breeding, females laying small clutches of eggs, often between 10 and 30, on land in moist sites. Embryos pass through a tadpole stage while still in the eggs, which hatch out fully formed but tiny froglets. Another group (genus *Leptodactylus*) deposits its eggs in foam nests which are placed in pockets in the ground near water. Males use their hindfeet to whip into foam a concoction of air, water, mucus from the female, and semen. Tadpoles that hatch out from eggs within the foam are washed into nearby temporary ponds by rainwater (perhaps providing the common name for these frogs), later to emerge as fully developed frogs. Many rainfrogs breed at any time of year, but the ones that rely on rains to wash tadpoles into ponds often breed before heavy rains. Clutches vary in size from about 10 to several hundred eggs. Most rainfrogs reproduce at one year of age or younger.

Lore and Notes
Ancient American cultures such as the Mayans doubtless associated rainfrogs and other frogs with rain because these amphibians often appeared in large numbers in forests and ponds during heavy rains – when breeding occurs. Central America's Maya Rainfrog, *Eleutherodactylus chac*, is named for the Mayan rain god, Chac.

Rainfrogs, like most amphibians, secrete poisons of varying degrees of toxicity from skin glands to discourage predation. The SMOKY JUNGLE FROG, found over much of Central America, has a particularly potent poison, named leptodactylin, after the genus name, *Leptodactylus*. Its effects include blocking some neuromuscular activity and stimulating parts of the nervous system, particularly those that control blood vessels and, hence, blood pressure. The secretions, very irritating to mucous membranes of the mouth, nose, and throat, presumably compel some predators to release the offending frog in their mouth.

Status
No rainfrogs of Mexico are currently thought to be threatened, but as is the case with most amphibian groups, adequate monitoring on most species has yet to be done. Many rainfrogs of South America, particularly in Chile and Argentina, where a great number of species occur, are threatened or endangered, as are several from Jamaica and Puerto Rico.

Profiles
Tungara Frog, *Physalaemus pustulosus*, Plate 2e
Alfredo's Rainfrog, *Eleutherodactylus alfredi*, Plate 3a
Polymorphic Robber Frog, *Eleutherodactylus rhodopis*, Plate 3b

Black-backed Frog, *Leptodactylus melanonotus*, Plate 3c
White-lipped Frog, *Leptodactylus labialis*, Plate 3d

4. Treefrogs

The *treefrogs*, Family Hylidae, are an intriguing array of animals that, somewhat anomalously for amphibians, but owing to big eyes and bright colors, have joined toucans, parrots and other glamorous celebrities to become poster animals for tropical forests. Rarely these days is a book, poster, or calendar printed with pictures of tropical animals that does not include at least one of a treefrog. They are distributed on all continents except Antarctica, but are most abundant and diverse in the American tropics. There are about 700 species of *hylids* world-wide, with more than 70 species occurring within Mexico; about 44 of them are endemic to the country. SPRING PEEPERS and CHORUS FROGS, their vocalizations familiar to so many North Americans, are hylids.

Most hylid treefrogs are small and elegantly slender-wasted. They have relatively large heads with long, rounded snouts, conspicuously large, bulging eyes, and long but skinny thighs and legs. Many have prominent, enlarged toe pads – sucking disks on the tips of their toes that permit them to cling to and climb tree leaves and branches. Most are between 3.5 and 8.5 cm long (1.5 to 3.5 in). Coloring is highly variable. Some are very cryptic, clad in greens and browns that produce virtual invisibility in their arboreal lodgings. Others, particularly ones that are poisonous, are brightly colored, presumably to warn away potential predators. For instance, the RED-EYED LEAF FROG (Plate 3), up to 7 cm (3 in) in length, is right out of a cartoon studio – bright leaf green above, with blue to purple sides, bright blue and green thighs, a white belly, orange hands and feet, and red eyes. Some species also have what biologists refer to as *flash* coloring – patches of bright blue, red, orange, or yellow on the hind legs or groin area that are concealed at rest but exposed as brilliant flashes of color when the frog moves. The purpose of flash coloring presumably is to startle predators as they attack, allowing a moment to escape.

Natural History

Ecology and Behavior

As their name suggests, most treefrogs are arboreal creatures, spending much of their time jumping and climbing among tree leaves and twigs. Although their skinny legs suggest weak muscles, they are good jumpers. Their adhesive toe pads allow them to climb even vertical surfaces. Some species are more ground-dwelling and some are fully aquatic. The majority are nocturnal, passing the daylight hours sheltered in tree crevices or among the plants that grow on trees (epiphytes), especially bromeliads. They emerge at night to forage for insects, their predominant food, and to pursue their romantic interests. Most species have the ability to change their skin color rapidly, like chameleons.

Breeding

Even the most jaded observer of animal behavior would be forced to admit that treefrog reproduction has some unusual features. As in most amphibians, one or both of the non-adult life stages, eggs and/or larvae (tadpoles), occur in water, but not just in any pond or stream. Females of one group of species lay eggs only in what ecologists term *arboreal* water – small, stagnant puddles present in tree cavities, on logs, or at the center of bromeliads and other epiphytic plants that have central *cisterns*, or bucket areas, that always hold rainwater. The eggs hatch in a few days, releasing tiny tadpoles that feed and develop within these isolated, protected ponds, eventually turning into *froglets*.

Another group of species – now, tell me this isn't fascinating – attach their eggs to tree leaves over water and, as the eggs hatch, the tadpoles drop into the water below. For example, in the RED-EYED LEAF FROG, a male in a reproductive mood moves to a tree branch overlooking a temporary pond or puddle and proceeds to call. A female approaches the vocalizing male and, after appropriate behaviors, the male clings to the female's back in the mating position. The female, with the male on her back, then climbs down to the pond to soak up some water into her bladder. She climbs back up into the tree and deposits a clutch of 30 to 50 eggs onto leaves and releases water over them; the male deposits sperm onto the wetted eggs as the female releases them, and fertilization occurs then. She climbs down again with the male to get some more water and repeats the process, sometimes three to five times in a night. In species that typically deposit eggs over streams, the eggs are particularly large, so that the tadpoles that hatch out and drop into the moving water are large and powerful enough to stand up to the currents. A number of other breeding methods exist within the hylid family as well – this is currently a hot area of animal behavior research in Central America. Some species lay 2000 or more eggs per clutch. Eggs generally hatch in two to six days. Sexual maturity is reached usually at about a year of age. Breeding can be at any time of year.

Ecological Interactions
Why do treefrogs live on land and often breed at least partially away from water? Evolutionary biologists believe that the most likely reason that some groups of frogs, such as the hylids, have developed terrestrial behavior is not that terrestrial existences are inherently better than aquatic ones, but that terrestrial life means a great reduction in the loss of eggs and tadpoles to predators. The chief culprit driving such predator-avoidance is thought to be fish, which eat copious amounts of frog eggs and tadpoles – when they can reach them. Eggs laid on leaves are obviously out of fish range, as are eggs and tadpoles that develop in fishless ponds and small, ephemeral ponds and puddles. A study of the comparative distributions of Costa Rican fish and frogs provided evidence for the claim: above about 1500 m (5000 ft) in elevation there are no native fish but there is a large variety of stream-breeding treefrogs, which are distinctly rare below that elevation. This pattern suggests that where fish are present, treefrogs have evolved more terrestriality and its accompanying breeding away from streams; but where fish are absent, stream-breeding is less costly in terms of lost eggs and young, and so it continues. Avoiding fish predation by depositing eggs on land may be a good idea, but it is not without its own hazards: some snakes feed frequently on these egg clusters.

Lore and Notes
Treefrogs employ a number of methods to deter predation, including cryptic coloring that allows them to blend into their environments, rapid leaping (what must be, aside from flying or gliding, the most efficient way to move about quickly in trees), loud, startling screams or squalls emitted when grabbed by predators, and poisons. Some species are cryptic but not poisonous, some are brightly colored and poisonous, and some are both cryptically colored and poisonous. One hylid species produces in its skin secretions toxins that cause

sneezing in people, even if they are not handling a frog, and that cause pain and paralysis when ingested or when penetrating the body via an open wound.

Status

No hylid treefrogs outside of Australia are positively known to be immediately threatened or endangered. However, for many New World species, there is currently insufficient information to determine population sizes and trends. Some Brazilian and Caribbean species are thought to be rare and, hence, vulnerable, as is the USA's PINE BARRENS TREEFROG. As far as it can be known, Mexican treefrogs are not currently threatened.

Profiles

Red-eyed Leaf Frog, *Agalychnis callidryas*, Plate 3e
Variegated Treefrog, *Hyla ebraccata*, Plate 4a
Loquacious Treefrog, *Hyla loquax*, Plate 4b
Yellow Treefrog, *Hyla microcephala*, Plate 4c
Cricket Treefrog, *Hyla picta*, Plate 4d
Veined Treefrog, *Phrynohyas venulosa*, Plate 4e
Stauffer's Treefrog, *Scinax staufferi*, Plate 5a
Mexican Treefrog, *Smilisca baudinii*, Plate 5b
Blue-spotted Treefrog, *Smilisca cyanosticta*, Plate 5c
Casque-headed Treefrog, *Triprion petasatus*, Plate 5d

5. Other Frogs

There are a number of groups of frogs that, although not represented by many species in Mexico, nonetheless are important members of the region's amphibian community and are often conspicuous to observant visitors. Frogs of the Family Ranidae are known as *true frogs* not because they are better frogs than any other but because they are the most common frogs of Europe, where the early classification of animals took place. Over 650 species strong, the true frogs now have an almost world-wide distribution (the ones in Australia were introduced there by people). Most are in Africa, 250 species occur in the New World, about 23 occur in Mexico (14 are endemic), and about 10 of them occur in southern Mexico (only two in the Yucatán Peninsula). True frogs include many of what most people regard as typical frogs – green ones that spend most of their lives in water. Among frogs familiar to many North Americans, BULLFROGS and LEOPARD FROGS are members of the Ranidae. Typically *ranids* are streamlined, slim-wasted frogs with long legs, webbed back feet, and thin, smooth skin. They are usually shades of green. Size varies extensively. VAILLANT'S FROG (Plate 6), a typical true frog, reaches lengths of 11 cm (4.5 in), excluding the long legs, of course.

Frogs of the Family Centrolenidae, known as *glass frogs* owing to their transparent abdominal skin, are limited in their distributions to humid regions of the Americas from southern Mexico to Bolivia and northern Argentina. There are about 100 species in the group, only one of which occurs in southern Mexico. These arboreal frogs are usually small, most in the range of 2.5 to 3 cm (an inch or so) long, and most come clad in a variety of greens, including lime, blue-green, and dark green. Various body organs and bones can be seen through the transparent skin of their bellies.

The *narrow-mouth toads*, Family Microhylidae, are a group of about 300 species that are distributed over broad swaths of the Southern Hemisphere, including the

Neotropics, sub-Saharan Africa, and Southeast Asia. They are highly variable in the way they look and also in their habits. Only a few species occur in southern Mexico, and only two over the Yucatán Peninsula. They are both small, squat frogs with small heads, pointed snouts, and fairly smooth skins.

Natural History

Ecology and Behavior

True frogs. These are aquatic frogs that are good swimmers and jumpers. With their webbed toes and long, muscular hind legs, they are built for speed on land or in water. This is fortunate for the frogs because most lack the poison glands in their skins that many other types of frogs use to deter predation. They have thin skin through which water evaporates rapidly and thus they tend to remain in or near the water, often spending much of their time around the margins of ponds or floating in shallow water. Except during breeding seasons, they are active only during the day. True frogs feed mainly on bugs, but also on fish and smaller frogs. In turn, they are common food items for an array of beasts, such as wading birds, fish, turtles, and small mammals. Because they are such tasty morsels to so many predators, these frogs are very alert to their surroundings, attempting escape at the slightest movement or noise; the splashing heard as you walk along a pond or lakeshore is usually made by these frogs on the shore leaping into the water.

Glass frogs. Glass frogs are treefrogs of tropical moist forests. They are found along streams in lowland areas and all the way up to 3800 m (12,000 ft). Typically active during the day, glass frogs eat mainly insects, opportunistically taking whatever bugs they encounter. Males are highly territorial, defending sites to which, with their calls, they will attract females for mating. When another male intrudes into a resident's territory and will not leave, frog fights ensue.

Narrow-mouth Toads. These are terrestrial frogs that often spend large portions of the day hiding under the leaf litter or in ground holes or burrows. Much of their activity apparently is at night, when they roam the ground in search of their favorite foods – ants, termites and assorted other bugs. SHEEP FROGS (Plate 6) commonly appear at night on and along roads after heavy rains.

Breeding

True frogs. These frogs reproduce under what most would consider the standard amphibian plan. Eggs are released by the female into the water in ponds or streams, and are fertilized then by sperm released by the male. Eggs and tadpoles develop in the water. In some species, the jellied egg masses float, but in others, they are attached to the undersides of rocks in streams. Clutch size varies tremendously, but large females can release thousands of eggs. Sexual maturity occurs in from one to four years.

Glass frogs. Glass frog males call, attracting females to their territories. Females deposit their eggs on leaves that hang over puddles and temporary pools. Some species of glass frogs stay near their eggs, guarding them. Males have been observed to sit on eggs during the night and near them during the day. Eggs hatch in 8 to 20 days, the released tadpoles dropping into the water below. Often hatching occurs during heavy rain, which is thought to help increase the chances that tadpoles make it to the pools. Tadpoles develop in the pools, finally emerging as frogs.

Narrow-mouth Toads. Breeding occurs usually after the start of the rainy season, primarily in temporary pools – roadside ditch pools, small ponds, flooded pastures. Eggs are deposited into a pool and then fertilized. They hatch quickly, sometimes in only a few days, so that the tadpoles have time to develop and leave the pools before they dry.

Ecological Interactions

The way that biologists view how animals distribute themselves within an area has been partially worked out by thinking about how true frogs might arrange themselves in a pond. Predation is a major worry for many animals, including aquatic frogs, and it doubtless influences many aspects of their lives. Often, there is safety in numbers, the rationale being that, although a group is larger than a single animal and therefore more easily located by a predator, the chances of any one animal in the group being taken is low, and so associating with the group, instead of striking out on one's own, is advantageous to an individual. But within the group, where should an individual frog, say one named Fred, position himself to minimize his chances of being eaten? Imagine a round pond around which frogs tread in the shallow water near shore or sun themselves on the bank. If the predator is a snake or a bird that for lunch will take one frog, the best place for Fred is between two other frogs, so that his left and right are "protected," the neighbor frogs there to be eaten first. To gain such advantageous positioning, clearly Fred's best move would be to find two frogs near each other and move into the gap between them. If such a strategy exists, then the neighboring frogs should move in turn, also trying to fit into small gaps. Eventually, all the frogs in the pond, if playing this game, should end up in a small heap, the best protected frogs at its center. And indeed, as any small child who hunts frogs will tell you, they are often found in tight groups at the edges of lakes and ponds. This *selfish herd* explanation for the formation of frog and other animal aggregations was first proposed about 30 years ago.

Lore and Notes

The true frogs have always provided a minor source of protein to people throughout the world – a bit revolting to this writer, but there's no accounting for national and ethnic taste. These frogs are often non-toxic, abundant, and large enough to make harvesting and preparation economically profitable. In much of Europe and parts of the USA, the long muscles of the rear legs, often dusted with flour and fried in butter, are eaten by otherwise civilized people. Tastes like chicken breast, so I gather. Some Mayan groups today include frog legs in their diet, particularly those of the RÍO GRANDE LEOPARD FROG (Plate 6).

Status

As far as it is known, no species of true, glass, or microhylid frog are currently endangered in Mexico or other parts of the Americas. At least three species of true frogs are threatened in the USA. As is the case for many of the Amphibia, relatively little is known about the sizes and health of the populations of many of these frogs; therefore, authoritative assurance that some are not endangered is impossible to provide.

Profiles

Fleischmann's Glass Frog, *Hyalinobatrachium fleischmanni*, Plate 5e
Vaillant's Frog, *Rana vaillanti*, Plate 6a
Río Grande Leopard Frog, *Rana berlandieri*, Plate 6b

Forrer's Grass Frog, *Rana forreri*, Plate 6c
Highland Frog, *Rana maculata*, Plate 6d
Sheep Frog, *Hypopachus variolosus*, Plate 6e

Environmental Close-Up 1
Frog Population Declines: Amphibian Armageddon or Alarmist Absurdity?

The Problem

Recently, there has been an outpouring of articles and stories about amphibians in the popular and scientific media, bearing titles such as "The Silence of the Frogs," "Where Have All the Frogs Gone?," "Why Are Frogs and Toads Croaking?," "Playing it Safe or Crying Frog?," and "Chicken Little or Nero's Fiddle?" What's stimulating all the silly titles? The issue traces back a few years, to 1989 to be precise, to a conference in Britain attended by many of the world's leading scientists who study reptiles and amphibians in the wild. Gabbing in the hallways of the convention center, as scientists will do, relating stories about their research and the respective species that they study, they noticed a common thread emerging: many of the populations of frogs, toads, and salamanders that had been monitored for at least several years seemed to be declining in numbers, often drastically. Where some species had been common 20 years before, they were now rare or near extinction. Suspecting that something important was afoot, the scientists met formally the next year in California to discuss the subject, compare notes, and try to reach some preliminary conclusions or, minimally, to phrase some preliminary questions. Such as:

1   Are amphibian populations really declining over a broad geographic area? Or are the stories just that – anecdotal accounts concerning a few isolated populations that, even if they are in decline, do not portend a general trend?
2   If amphibian population declines are, in fact, a general phenomenon, are they over and above those of other kinds of animals (many of which, after all, given the alarming rate of natural habitat destruction occurring over the globe, are also thought to be declining – albeit usually at a gentler pace than that ascribed to the amphibian drops); that is, is the amphibian decline a special case, happening for a reason unrelated to general loss of biodiversity?
3   If there really is a generalized worldwide amphibian problem, why is it happening?

Most of the conferring scientists agreed that a widespread pattern of amphibian declines was indicated by available data: there were reports from the USA, Central America, the Amazon Basin, the Andes, Europe, and Australia. They speculated about two possible main causes. First, that habitat loss – continued destruction of tropical forests, etc. – was almost certainly contributing to general declines in population sizes of amphibians and other types of animals. That is, that amphibian declines could at least partly be attributed to world-wide biodiversity loss. Second, that owing to their biology, amphibians were doubtless more sensitive than other groups to pollution, especially acid rain (rain that is acidified by various atmospheric pollutants, chiefly from engine emissions, leading to lake and river water being more acidic). Amphibians are more vulnerable to this kind of pollution than, say, reptiles, because of their thin, wet skins, through which they breathe and through which polluting chemicals may gain entry to their bodies, and also owing to their highly exposed and so vulnerable early life stages – eggs and tadpoles in lakes, ponds, and streams. Changes in the acidity of their aquatic environments, or even changes in water temperature, are known to have dire consequences for egg and larval (tadpole) development – and therefore such changes could be the main culprits behind amphibian declines. Plainly, if their reproduction were compromised, population numbers would plunge precipitously. Although it sounded a bit far-fetched, some scientists also suggested that the increased bombardment of the Earth by ultraviolet (UV) light, a direct result of the thinning, protective atmospheric ozone layer, might likewise be taking a disproportionate toll on biologically vulnerable amphibians. (Indeed, recent studies have proved to the satisfaction of some, but not all, scientists that increased UV light in natural situations destroys frog eggs and can interact chemically with diseases and acid rain to increase amphibian mortality rates.)

The Controversy

As might be expected, not all herpetologists agree that pollution, UV light, or some other environmental factor is currently exerting lethal effects world-wide on frogs and salamanders. Some biologists point out that the amphibians that they study in some of the world's most pristine environments, having little or no detectable pollution, also have experienced catastrophic population crashes. One of the prime examples is a Costa Rican resident, the Golden Toad. It had been monitored for many years in the same five breeding ponds at the Monteverde Cloud Forest Reserve, and until the late 1980s, more than 1500 adults had bred each year in the ponds; but from 1988 through 1990, fewer than 20 individuals could be found each year and the toad is now thought by many to be extinct. In total, populations of 20 of the approximately 50 frog species that inhabit the Cloud Forest Reserve crashed dramatically during the late 1980s and, as of 1996, still had not recovered. Pollution was probably not a factor, although UV light and weather could have had effects – average water temperature had risen a bit and there was a decrease in rainfall during the late 1980s.

Still other scientists, those who study natural size oscillations in animal populations, point out that science is aware of many animals whose populations cycle between scarcity and abundance. For instance, several small rodents of the arctic tundra, *voles* and *lemmings*, are famous for one year being at low population densities (a few per acre or hectare) but three or four years hence, being at very high densities (thousands per acre or hectare, so many that it's difficult to walk without stepping on them). These biologists point out that, unless those sounding the alarm of amphibian declines can show that the declines are not part of natural cycles, it is too early to panic. Further, that the only way to know about possible natural population cycles is to monitor closely amphibian populations during long-term studies, which at present are few and far between.

The side of the debate that takes the amphibian declines as established fact and suspects an *anthropogenic* cause (scientific jargon for *people-caused*), while agreeing that long-term studies are necessary, believes that it would be a grave mistake to wait for the conclusions of those studies, 10 or 20 years down the road, before decisions are reached about whether amphibians are declining on a broad scale and what to do about it – because at that point, they believe, it will be too late to do anything but record mass extinctions.

The Future

What will happen now is that the controversy will continue as long-term population studies are conducted. In one novel approach, current populations of frogs and toads in California's Yosemite National Park were compared with notes on their population sizes from a survey conducted in 1915; all seven native species were found to have declined, but the causes are unclear. A group of Australian researchers proposed in 1996 that the catastrophic declines in frogs in eastern Australian rainforests were caused by an unknown virus – a rapidly spreading, water-borne, epidemic disease. They further suspect that the same disease has spread over the past 15 or 20 years from continent to continent, perhaps in water transported with aquarium fish, and, in concert with pollution and other environmental stresses that make amphibians more vulnerable to diseases, may be responsible for the world-wide amphibian die-offs. This is a controversial idea and other scientists will have to test it before it is widely accepted as a correct explanation. However, recent work in Panama shows that many frogs there are dying of a protozoan disease and findings from Central America and Australia suggest a fungus is killing many frogs in those regions, supporting disease explanations for amphibian declines.

A major problem is that, even if the scientific consensus right now was that disease, fungi, pollution, and the increased incidence of UV light were harming amphibians around the world, the will and resources are currently lacking to do anything about it on the massive scale that conservation would require. Unless peoples' worldview changes, preservation of amphibians and reptiles, save special cases like *sea turtles* (p. 97), will always lag behind preservation efforts made on the behalf of the "cuddlies" – birds and mammals. The best hope is that the recent conservation emphasis on preserving entire ecosystems, rather than particular species, will eventually benefit the disappearing amphibians.

One positive element of the amphibian declines, if it can be called that, is that some scientists have suggested that if it is determined that amphibians are particularly sensitive to pollutants, then perhaps they can be used as indicators of environmental health: ecosystems with healthy amphibian populations could be deemed relatively healthy, those with declining amphibians could be targeted for hasty improvement. As one biologist phrased it, amphibians could be used as "yardsticks for ecosystem vitality" (R. B. Primack 1993), just as canaries in cages indicated to the coal miners who carried them something of vital importance about their environments.

Chapter 7

Reptiles

General Characteristics and Natural History

Most people journeying to southern Mexico to experience tropical habitats and view exotic wildlife harbor ambivalent feelings about reptiles. On one hand, almost all people react with surprise, fear, and rapid withdrawal when suddenly confronted with reptiles anywhere but the zoo (which is quite understandable, given the dangerous nature of some reptiles). But on the other hand, these creatures are fascinating to look at and contemplate – their threatening primitiveness, their very "dinosaur-ness" – and many have highly intriguing lifestyles. Most reptiles are harmless to people and, if discovered going about their daily business, are worth a look. Unfortunately, to avoid predation, most reptiles are inconspicuous both in their behavior and color patterns and often flee when alerted to people's presence; consequently most reptiles are never seen by people during a brief visit to a region. Exceptions are reptiles that achieve a great deal of safety by being very large – crocodiles and iguanas come immediately to mind. Also, some small lizards are very common on forest trails and among stone ruins. But overall, you should expect to see relatively few reptile species during any brief trip to the Neotropics. Still, it is a good idea to keep a careful watch for them, remembering not to get too close to any that you find, and count yourself lucky for each one you see.

Reptiles have been around since the late portion of the Paleozoic Era, some 300 million years ago. Descendants of those first reptiles include 7000+ species that today inhabit most regions of the Earth, with a healthy contingent in the American tropics. Chief reptile traits, aside from being scary-looking, are that (1) their skin is covered with tough scales, which cuts down significantly on water loss from their body surface. The development of this trait permitted animals for the first time to remain for extended periods on dry land, and most of today's reptiles are completely terrestrial (whereas amphibians, which lack a tough skin, need always remain in or near the water or moist places, lest they dry out). (2) Their heart is divided into more chambers that increase efficiency of circulation over that of the amphibians, allowing for a high blood pressure and thus the sustained muscular activity required for land-living. (3) Some employ *oviparous* reproduction, placing their fertilized eggs in layers of tough membrane or in hard shells and then expelling the eggs to the external environment, where development of the embryos occurs; whereas others are *ovoviviparous* – eggs are not shell-encased or laid, but remain within the mother until "hatching," the young being born "live;" still others are *viviparous*, in which developing embryos are connected to mom via a type of placenta and derive nourishment from her until being born live. Most reptiles do not feed or protect their young, but desert their eggs after they are laid.

Reptile biologists usually recognize three major groups.

The *turtles* and *tortoises* (land turtles) constitute one reptile group, with about 260 species world-wide. Some turtles live wholly on land, the sea turtles live out their lives in the oceans (coming ashore only to lay eggs), but most turtles live in lakes and ponds. Although most eat plants, some are carnivorous. Turtles are easily distinguished by their unique body armor – tough plates that cover their back and belly, creating wrap-around shells into which head and limbs are retracted when danger looms.

The *crocodiles* and their relatives, large predatory carnivores that live along the shores of swamps, rivers, and estuaries, constitute a small second group of about 20 species.

Last, and currently positioned as the world's dominant reptiles, the 3300 *lizard* species and 3500 *snakes* comprise a third group (lizards and snakes have very similar skeletal traits, indicating a very close relationship).

Lizards walk on all four limbs, except for a few that are legless. Most are ground-dwelling animals, but many also climb when the need arises; a fair number spend much of their lives in trees. Almost all are capable of moving quite rapidly. Most lizards are insectivores, but some, especially larger ones, eat plants, and several prey on amphibians, other lizards, mammals, birds, and even fish. Lizards are hugely successful and are often the most abundant vertebrate animals within an area. Ecologists suspect that they owe this ecological success primarily to their ruthlessly efficient predation on insects and other small animals and to low daily energy requirements.

Most lizards of southern Mexico are insectivores, also opportunistically taking other small animals such as spiders and mites. Lizards employ two main foraging strategies. Some, such as the small whiptail lizards, are *active searchers*. They move continually while looking for prey, for instance nosing about in the leaf litter of the forest floor. *Sit-and-wait* predators, highly camouflaged, remain motionless on the ground or on tree trunks or branches, waiting for prey to happen by. When they see a likely meal – a caterpillar, a beetle – they reach out to snatch it if it is close enough or dart out to chase it down.

Many lizards are territorial, defending territories from other members of their species with displays, such as bobbing up and down on their front legs and raising their head crests. Lizards are especially common in deserts and semi-deserts, but they are numerous in other habitats as well. They are active primarily during the day, except for many of the gecko species, which are nocturnal.

Mexico is unique in that it has the world's only two venomous lizards. The GILA MONSTER occurs in the deserts of northern Mexico and the southwestern USA. The BLACK-BEADED LIZARD (Plate 19) is distributed along Mexico's Pacific Coast and occurs in Oaxaca and Chiapas. Beaded Lizards are large, stout animals with large heads, thick, shortish tails, and bead-like scales. They are found usually on or beneath the ground in forest or scrub areas, and are active mainly during cooler parts of the day or at night. If these lizards are similar in habits to Gila Monsters, which have been more studied, they eat lizards, turtle and bird eggs, and young mammals such as baby rodents and rabbits. Beaded Lizards usually move sluggishly and are not at all aggressive. If bothered, however, they can turn and maneuver rapidly and bite. Their venom is *neurotoxic* (that is, it disrupts nerve function), causing intense pain near the bite, possibly paralysis, bleeding of the wound, nausea and vomiting, and other effects; few people, however, die of these bites. Venom is manufactured in glands in the lower jaw, and passes into a wound along grooves in some of the lizards' larger teeth. Biologists believe the venom of these lizards evolved for protection, not for subduing prey. Beaded lizards live 20+ years in captivity.

Snakes probably evolved from burrowing lizards, and all are limbless. Snakes are all carnivores, but their methods of capturing prey differ. Several groups of species have evolved glands that manufacture poisonous venom that is injected into prey through the teeth. The venom immobilizes and kills the prey, which is then swallowed whole. Other snakes pounce on and wrap themselves around their prey, constricting the prey until it suffocates. The majority of snakes are non-venomous, seizing prey with their mouths and relying on their size and strong jaws to subdue it. Snakes generally rely on vision and smell to locate prey, although members of two families have thermal sensor organs on their heads that detect the heat of prey animals.

More snake species probably exist now than species of all other reptiles combined. This success is thought to be attributable to their ability to devour prey that is larger than their heads (their jaw bones are highly mobile, separating partially and moving around prey as it is swallowed). This unique ability provides snakes with two great advantages over other animals: because they eat large items, they have been able to reduce the frequency with which they need to search for and capture prey; and owing to this, they can spend long periods hidden and secluded, safe from predators. Like lizards, snakes use either active searching or sit-and-wait foraging strategies.

Snakes are themselves prey for hawks and other predatory birds, as well as for some mammals. While many snakes are quite conspicuous against a solid color, being decorated with bold and colorful skin patterns, against their normal backdrops, such as a leaf-strewn forest floor, they are highly camouflaged. They rely on their cryptic colorations, and sometimes on speed, to evade predators.

Mating systems and behaviors of snakes in the wild are not well known. In some, males during the breeding period remain in mating areas, where interested females tend to gather. Males may fight each other to gain the right to mate with a particular female. Male size seems to matter most in determining the outcomes of these fights. Males and females are known to engage in multiple matings with different individuals.

Seeing Reptiles in Tropical Mexico

As with amphibians, many reptiles are difficult to observe. They spend most of their time concealed or still. Most do not vocalize like birds or frogs, so you cannot use sound to find them. The superb cryptic coloration of snakes, including venomous ones, makes a motionless snake a dangerous snake. Because of the difficulty people have of seeing snakes before getting very close to them, the rule for exploring any area known to harbor venomous snakes or any area for which you are unsure, is never, *never* to place your hand or foot anywhere that you cannot see first. Do not climb rocks or trees, do not clamber over rocks where your hands or feet sink into holes or crevices; do not reach into bushes or trees. Walk carefully along trails and, although your attention understandably wanders as new sights and sounds are taken in, try to watch your feet and where you are going.

With safety in mind, if you want to see reptiles, there are a few ways to increase the chances. Knowing about activity periods helps. Lizards and most snakes are active during the day, but some snakes are active at night. Thus, a night walk with flashlights that is organized to find amphibians might also yield some reptile sightings. Weather is also important – snakes and lizards are often more active in sunny, warm weather. If all else fails, one may look for small snakes and lizards by carefully moving aside rocks and logs with a robust stick or with one's boots, although such adventures are not for the faint-hearted.

Particularly good places to spot lizards are at the many archaeological sites that are commonly visited by travellers in southern Mexico. Many species of small and mid-sized lizards are easily spotted at the Monte Albán ruins in Oaxaca and at Palenque in Chiapas. Small lizards as well as large Spiny-tailed Iguanas (Plate 16) are frequently seen at many of the Mayan ruin sites on the Yucatán Peninsula, particularly Uxmal, Ek Balam, and Chichén Itzá. Happy hunting!

Family Profiles

I. Crocodilians

Remnants of the age when reptiles ruled the world, today's crocodilians (*alligators, caimans,* and *crocodiles*), when seen in the wild, generally inspire awe, respect, a bit of fear, and a great deal of curiosity. Recent classification schemes include a total of 22 species, distributed over most tropical and sub-tropical areas of the continents. Three are found in southern Mexico: The AMERICAN CROCODILE (Plate 7), which is the most widely distributed of the four New World crocodile species, MORELET'S CROCODILE (Plate 7), and the SPECTACLED CAIMAN. There is a small population of American Crocodiles in southern Florida (USA), but mainly they range from Mexico, Central America and the Caribbean islands south to Colombia, Venezuela, Peru, and Ecuador. American Crocodiles can reach lengths of 7 m (21.5 ft), as large as any crocodilian species, but in the wild individuals over 4 m (13 ft) are now rare. American Crocodiles occur mostly in both brackish and freshwater habitats such as river mouths and mangrove swamps, but sometimes spread up rivers farther inland. Morelet's Crocodile is found only in southern Mexico, Belize, and Guatemala. These crocs are smaller than American crocs, most now obtaining lengths of no more than 2.5 m (8 ft). Morelet's Crocodiles mainly inhabit inland freshwater sites such as lagoons, marshy parts of lakes, and sluggish rivers. A third Mexican species, the CAIMAN, or Spectacled Caiman, is probably the most abundant crocodilian in the New World. It is smaller than the other two Mexican species and in Oaxaca and Chiapas it occupies coastal regions and marshes; caiman do not occur on the Yucatán Peninsula.

Generally, differences in coloring are not a good way to distinguish crocodilians – most are shades of brown or olive-brown. Anatomy and location are more useful clues. Snouts of alligators and caiman tend to be broad and rounded, whereas those of crocodiles are longer and more pointed. Also, in crocodiles, the fourth tooth of the lower jaw projects upwards outside the mouth and can be seen above the upper jaw. Morelet's and American Crocodiles can be distinguished by location and, where they might overlap in distribution, by the Morelet's wider snout and generally smaller size. Male crocodilians, in general, are larger than females of the same age.

Natural History

Ecology and Behavior

Although not amphibians, crocodilians are amphibious animals. They usually move slowly over land but in short bursts can cover ground rapidly. Most of their time, however, is spent in the water. They adore basking in the sunshine along the banks of rivers, streams, and ponds. Crocodilians in the water are largely hidden, resembling from above floating logs. This unassuming appearance allows them to move close to shore and seize animals that come to the water to drink. Crocodilians are meat-eaters. The foods taken depend on their age and size. Juveniles eat primarily aquatic insects and other small invertebrates such as snails, although young AMERICAN CROCODILES also feed on small aquatic and terrestrial vertebrates. Adult American Crocodiles, which often forage at night, specialize on fish and turtles.

American Crocodiles sometimes excavate burrows along waterways, into which they retreat to escape predators and, when water levels fall too low, to *estivate* (sleep until water conditions improve). Crocodilians may use vocal signals extensively in their behavior, in communicating with one another, but their sounds have been little studied. It is known that juveniles give alarm calls when threatened, and that parents respond by quickly coming to their rescue.

One might guess that among such primitive reptiles, *parental care* would be absent – females would lay eggs, perhaps hide them, but at that point the eggs and hatchlings would be on their own. Surprisingly, however, crocodilians show varying degrees of parental care. Nests are guarded and one or both parents often help hatchlings free themselves from the nest. In some species, such as the AMERICAN ALLIGATOR, parents also carry hatchlings to the nearest water. Females may also remain with the young for up to two years, protecting them. This complex parental care in crocodilians is sometimes mentioned by scientists who study dinosaurs to support the idea that dinosaurs may have exhibited complex social and parental behaviors.

Crocodilians are long-lived animals, many surviving 60+ years in the wild and up to 90+ years in zoos.

Breeding

During courtship, male crocodilians often defend aquatic territories, giving displays with their tails – up-and-down and side-to-side movements – that probably serve both to defend the territory from other males and to court females. Typically the female makes the nest by scraping together grass, leaves, twigs, and sand or soil, into a pile near the water's edge. She then buries 20 to 30 eggs in the pile that she, and sometimes the male, guard for about 70 days until hatching. Nests of the AMERICAN CROCODILE are a bit different. The female digs a hole in sandy soil, deposits her eggs, then covers them with sand, which she packs down. She guards the nest and often helps the young emerge. As in the turtles and some lizards, the sex of developing crocodilians is determined largely by the temperature of the ground around the eggs: males develop at relatively high temperatures, females at lower temperatures (see p. 100). Crocodile young from a brood may remain together in the nest area for up to 18 months. Breeding seasons for crocodilians vary, with nesting observed during both wet and dry parts of the year; however, eggs are commonly laid in the Yucatán region from March through June.

Ecological Interactions

Somewhat surprisingly, crocodilians are prey for a number of animals. Young, very small crocodiles are eaten by a number of predators, including birds such as herons, storks, egrets and anhingas, and mammals such as raccoons and possibly foxes. Large adults apparently have only two enemies: people and large anaconda snakes. Slow-movers on land, crocodiles are sometimes killed by automobiles. Cases of cannibalism have been reported.

Lore and Notes

Larger MORELET'S and, especially, AMERICAN CROCODILES are potentially dangerous to people, but they are not considered particularly aggressive species. Morelet's are usually inoffensive, most being below a size where they try to eat land mammals. There are few documented cases of American Crocodiles killing people. Some species, such as the NILE CROCODILE, *are* known to be aggressive. In one famous, historical collecting trip, 444 large ones were shot randomly and opened; four had human remains in their stomachs. True, this number represents only 1% of the crocodile population, but still .…

Owing to their predatory nature and large size, crocodilians play large roles in the history and folklore of many cultures, going back at least to ancient Egypt, where a crocodile-headed god was known as Sebek. The Egyptians apparently welcomed crocodiles into their canals, possibly as a defense from invaders. It may have been believed by Egyptians and other African peoples that crocodiles caused blindness, probably because the disease called river blindness results from infestation with a river-borne parasitic roundworm. To appease the crocodiles during canal construction, a virgin was sacrificed to the reptiles. Indeed, providing crocodiles with virgins seems to have been a farily common practice among several cultures, showing a preoccupation with these animals. Even today, carvings of crocodiles are found among many relatively primitive peoples, from South America to Africa to Papua New Guinea. The ancient Olmecs of eastern Mexico also had a crocodile deity. Crocodile remains have been found at Mayan ruin sites in both southern Mexico (on Cozumel Island, for instance) and in Belize, suggesting that these ancient groups made use of the large reptiles for food and/or ceremonially.

Status

Most crocodilian species world-wide were severely reduced in numbers during this century. Several were hunted almost to extinction for their skins. In the USA, hunting almost caused AMERICAN ALLIGATORS to go extinct. In 1961 hunting alligators was made illegal, but poaching continued. Thanks to the 1973 Endangered Species Act, which gave protection to the alligators, they have returned to most of the areas from which they were eliminated. Crocodile and alligator farms (with captive-bred stock) and ranches (wild-caught stock) in many areas of the world now permit skins to be harvested while wild animals are relatively unmolested. Many of the Latin American crocodilians were hunted heavily during the first half of the 20th century. MORELET'S CROCODILE, in particular, is one of the more endangered species, listed by CITES Appendix I. AMERICAN CROCODILES, considered by some agencies as vulnerable to threat, are also listed as endangered by CITES Appendix I. Today, only the COMMON CAIMAN is hunted in large numbers, particularly in the Pantanal region of Brazil. All crocodilians are listed by the international CITES agreements, preventing or highly regulating trade in their skins or other parts, and their numbers have been steadily rising during the past 20 years. However, most of the 22 crocodilian species are still threatened or endangered.

Profiles

American Crocodile, *Crocodylus acutus*, Plate 7a
Morelet's Crocodile, *Crocodylus moreleti*, Plate 7b

2. Turtles

It is a shame that *turtles* in the wild are relatively rarely encountered reptiles (at least at close range) because they can be quite interesting to watch and they are generally innocuous and inoffensive. It is always a pleasant surprise stumbling across a turtle on land, perhaps laying eggs, or discovering a knot of them basking in the sunshine on rocks or logs in the middle of a pond. The 260 living turtle species are usually grouped into 12 families that can be divided into three types by their typical habitats. Two families comprise the *sea turtles*, ocean-going animals whose females come to shore only to lay eggs. The members of nine families, containing most of the species, live in freshwater habitats – lakes and ponds – except for the exclusively terrestrial *box turtles*. Finally, one family contains the *land tortoises*, which are completely terrestrial. About 38 turtle species occur in Mexico, 16 in the Yucatán region.

Turtles all basically look alike: bodies encased in tough shells (made up of two layers – an inner layer of bone and an outer layer of scale-like plates); four limbs, sometimes modified into flippers; highly mobile necks; toothless jaws; and small tails. This body plan must be among nature's best, because it has survived unchanged for a long time; according to fossils, turtles have looked more or less the same for at least 200 million years. Enclosing the body in heavy armor above and below apparently was an early solution to the problems vertebrates faced when they first moved onto land. It provides both rigid support when outside of buoyant water and a high level of protection from drying out and from predators.

Turtles come, for the most part, in a variety of browns, blacks and greens, with olive-greens predominating. They range in size from tiny terrapins 11.5 cm (4.5 in) long to 250-kg (550-lb) GALÁPAGOS TORTOISES and giant LEATHERBACK SEA TURTLES that are nearly 2 m (7 ft) long, 3.6 m (12 ft) across (flipper to flipper), and that weigh 550+ kg (1200+ lb). Leatherbacks, which occur off southern Mexico's Pacific coast, are the heaviest living reptiles. In many turtle species, females are larger than males. Turtle sexes are sometimes distinguishable by an curved indentation in the bottom shell of the male, which fits over the female's top shell during sex.

Natural History

Ecology and Behavior

The diet of freshwater turtles changes as they develop. Early in life they are carnivorous, eating almost anything they can get their jaws on – snails, insects, fish, frogs, salamanders, reptiles. As they grow the diet of most changes to herbivory. Turtles are slow-moving on land, but they can retract their heads, tails, and limbs into their shells, rendering them almost impregnable to predators – unless they are swallowed whole, such as by crocodiles. Long-lived animals, individuals of many turtle species typically live 25 to 60 years in the wild. (GALÁPAGOS TORTOISES routinely live 100 years, the record being 152 years.) Many turtle species grow throughout their lives.

Snapping turtles (Family Chelydridae; Plate 7) are large freshwater species with long tails, hooked, mean-looking jaws, and usually bad dispositions – they will bite people. They are found mainly in marshes, ponds, lakes, rivers, and streams. They are omnivorous, and often quite inconspicuous: sometimes algae grows on their backs, camouflaging them as they lunge at and snatch small animals that venture near. They remain predatory throughout their lives, even taking birds and small mammals, but they also eat aquatic vegetation.

Aquatic, or *pond*, turtles (Family Emydidae) occupy a variety of habitats. Some spend most of their time in lakes and ponds, but others leave the water frequently to bask in the sun during the day and to forage on land at night. Because they feed on land, they can occupy rivers that lack vegetation. FURROWED WOOD TURTLE and COMMON SLIDER are common pond turtles that are illustrated here (Plates 7, 8).

Semi-aquatic, *mud turtles* (Family Kinosternidae) are generally found in swamps, ponds, and slow streams, and only occasionally on land. Many appear to favor a diet of aquatic snails, but they have been also observed feeding on land. TABASCO MUD, WHITE-LIPPED MUD, SCORPION MUD, and MEXICAN GIANT MUSK TURTLES from this family are illustrated here (Plate 8).

Sea turtles are large reptiles that live in the open oceans, with the result that, aside from their beach nesting habits, relatively little is known of their behavior. Their front legs have been modified into oar-like flippers, which propel them through the water. Although they need air to breathe, they can remain submerged for long periods. At first, all sea turtles were assumed to have similar diets, probably sea plants. But some observations of natural feeding, as well as examinations of stomach contents, reveal a variety of specializations. GREEN TURTLES (Plate 9) eat bottom-dwelling sea grasses and algae, HAWKSBILL TURTLES eat bottom sponges, LOGGERHEAD TURTLES feed predominantly on mollusks (snails, etc.) and crustaceans (crabs, etc.), and KEMP'S RIDLEY TURTLES eat mainly crabs. Some other species eat jellyfish.

Breeding

Courtship in turtles can be quite complex. In some, the male swims backwards in front of the female, stroking her face with his clawed feet. In the tortoises, courtship seems to take the form of some between-the-sexes butting and nipping. All turtles lay their leathery eggs on land. The female digs a hole in the earth or sand, deposits the eggs into the hole, then covers them over and departs. It is up to the hatchlings to dig their way out of the nest and navigate to the nearest water. Many tropical turtles breed at any time of year.

Although the numbers of eggs laid per nest varies extensively among southern Mexico freshwater turtles (from 1 to about 100), in general, these turtles, specialized for life in the tropics, lay small clutches, often only one to four eggs. The reason seems to be that, because of the continuous warm weather, they need not breed in haste like their northern cousins, putting all their eggs in one nest. The danger with a single nest is that if a predator finds it, a year's breeding is lost. Tropical turtles, by placing only one or a few eggs in each of several nests spread through the year, are less likely to have predators destroy their total annual breeding production. Also, it may pay to lay a few big eggs rather than many small ones because bigger hatchlings can run faster to the water and its comparative safety from predators.

All sea turtle species breed in much the same way. Mature males and females appear offshore during breeding periods (for example, GREENS, July to October; LOGGERHEADS, June through September). After mating, females alone come ashore on beaches, apparently the same ones on which they were born, to lay their eggs. Each female breeds probably every 2 to 4 years, laying from two to eight clutches of eggs in a season (each clutch being laid on a different day). All within about an hour, and usually at night, a female drags herself up the beach to a suitable spot above the high-tide line, digs a hole with her rear flippers (a half meter or more, 2 ft, deep), deposits about 100 golfball-sized eggs, covers them with sand, tamps the sand down, and heads back to the ocean. Sometimes females emerge from the sea alone, but often there are mass emergences, with hundreds of females nesting on a beach in a single night. Eggs incubate for about 2 months, then hatch simultaneously. The hatchlings dig themselves out of the sand and make a dash for the water (if tiny turtles can be said to be able to "dash"). Many terrestrial and ocean predators devour the hatchlings and it is thought that only between 2% and 5% survive their first few days of life. The young float on rafts of sea vegetation during their first year, feeding and growing, until they reach a size when they can, with some safety, migrate long distances through the world's oceans. When sexually mature, in various species from 7 to 20+ years later, they undertake reverse migrations, returning to their birth sites to breed.

Ecological Interactions

If turtles can make it through the dangerous juvenile stage, when they are small and soft enough for a variety of predators to take them, they enjoy very high year-to-year survival – up to 80% or more of an adult population usually survives from one year to the next. However, there is very high mortality in the egg and juvenile stages. Nests are not guarded, and many kinds of predators, such as crocodiles, lizards, and, especially, armadillos, dig up turtle eggs or eat the hatchlings. Although adult turtles have few predators because they are difficult to kill and eat, some turtles have additional defenses: within the mud turtle family is a group known as *musk turtles*. When grabbed or handled they give off a musky smell from scent glands located on the sides of their bodies; in North America they are known as *stinkpots*.

There is an intriguing relationship between turtle reproduction and temperature that nicely illustrates the intimate and sometimes puzzling connections between animals and the physical environment. For many vertebrate animals, the sex of an individual is determined by the kinds of sex chromosomes it has. In people, if each cell has an X and a Y chromosome, the person is male, and if two Xs, female. In birds, it is the opposite. But in most turtles, it is not the chromosomes that matter, but the temperature at which an egg develops. The facts are these. In most turtles, eggs incubated at constant temperatures above 30 °C (86 °F) all develop as females, whereas those incubated at 24 to 28 °C (75 to 82 °F) become males. At 28 to 30 °C (82 to 86 °F), both males and females are produced. In some species, a second temperature threshold exists – eggs that develop below 24 °C (75 °F) again become females. (In the crocodiles and lizards, the situation reverses, with males developing at relatively high temperatures and females at low temperatures.) The exact way that temperature determines sex is not clear although it is suspected that temperature directly influences a turtle's developing brain. This method of sex determination is also mysterious for the basic reason that no one quite knows why it should exist; that is, is there some advantage of this system to the animals that we as yet fail to appreciate? Or is it simply a consequence of reptile structure and function, some fundamental constraint of their biology?

Lore and Notes

Turtles apparently were used by the ancient Mayans for both food and for religious/ceremonial purposes. Turtle remains from an array of species, often with markings that suggest cooking, were found at various archaeological sites, including many of those on the Yucatán Peninsula. Identified remains include fragments of CENTRAL AMERICAN RIVER TURTLES (Plate 7), COMMON SLIDERS (Plate 8), FURROWED WOOD TURTLES (Plate 7), mud turtles (Plate 8), and sea turtles such as GREEN TURTLES (Plate 9). Several of these turtles are still eaten today by indigenous peoples.

Some turtles are quite aggressive toward people. When disturbed, SNAPPING TURTLES, which are often described by experts as "vicious," can use their large, powerful jaws to wreak considerable damage to human flesh. MEXICAN GIANT MUSK TURTLES, likewise, have reputations for vile tempers. Local stories tell of these turtles being hauled into small boats only to have the human occupants jump overboard to avoid the biting reptiles (J. C. Lee 1996).

SNAPPING TURTLES occur in southern Canada, the eastern half of the USA, in southern (but not northern) Mexico, and in Central and northern South America, and so enjoy one of the longest north–south distributions of any New World reptile. Until recently, all were considered the same species. But new DNA comparisons of individuals from the various populations suggest that there are actually three separate species, one in North America and two in Central and South America.

Status

The ecology and status of populations of most freshwater and land turtle species are still poorly known, making it difficult to determine whether population numbers are stable or declining. However, it is mainly sea turtles, rather than freshwater or terrestrial turtles, that are exploited by people, and therefore that are most threatened. Sea turtle eggs are harvested for food in many parts of the world, including southern Mexico, and adults are taken for meat (only some species) and for their skins. Many adults also die accidentally in fishing nets and collisions with boats. One of the sea turtles, the HAWKSBILL, is the chief provider of tortoiseshell, which is carved for decorative purposes. The Hawksbill is under international protection, but some are still hunted. Other species of sea turtle – eggs and adults – are still eaten in the Yucatán region. KEMP'S RIDLEY is considered critically endangered. It's primary nesting beach is in Mexico, at Rancho Nuevo, on the Gulf. All sea turtles are listed as endangered by CITES Appendix I.

Other than sea turtles, southern Mexico's only endangered turtle is the CENTRAL AMERICAN RIVER TURTLE (CITES Appendix II and USA ESA listed). The species is the only surviving representative of a formerly widely distributed and diverse turtle family that traces back to the Jurassic Period, 195 million years ago. Unfortunately, these turtles are still hunted for their meat throughout their limited range – in southern Mexico, Guatemala, and Belize. They make for particularly easy prey during dry seasons, when river levels fall. In northern Belize, a Hickatee, as the turtle is known locally, is commonly the basis for Easter dinner. The only other turtle endangered in Mexico is the AQUATIC BOX TURTLE, which is endemic to the northern state of Coahuila; three other species are threatened.

Profiles

Central American River Turtle, *Dermatemys mawii*, Plate 7c
Snapping Turtle, *Chelydra serpentina*, Plate 7d
Furrowed Wood Turtle, *Rhinoclemmys areolata*, Plate 7e
Common Slider, *Trachemys scripta*, Plate 8a
Mexican Giant Musk Turtle, *Staurotypus triporcatus*, Plate 8b
Tabasco Mud Turtle, *Kinosternon acutum*, Plate 8c
White-lipped Mud Turtle, *Kinosternon leucostomum*, Plate 8d
Scorpion Mud Turtle, *Kinosternon scorpioides*, Plate 8e
Green Sea Turtle, *Chelonia mydas*, Plate 9a
Hawksbill Sea Turtle, *Eretmochelys imbricata*, Plate 9b
Loggerhead Sea Turtle, *Caretta caretta*, Plate 9c
Kemp's Ridley Sea Turtle, *Lepidochelys kempii*, Plate 9d

3. Colubrids: Your Regular, Everyday Snakes

All snakes, particularly tropical varieties, are thought by many people to be poisonous, and hence to be avoided at all cost. This "reptile anxiety" is serious when it prevents people who are paying initial visits to tropical forests from enjoying the many splendors around them. The phobia is understandable because of the lethal nature of a small number of snake species, snakes' abilities to blend in with surroundings and to move and strike so rapidly, and the long history we have, dating back to the origins of the Bible and beyond, of evil snake stories and legends. It is much more than Western societies' legends that cause our fear of snakes. All people it seems, irrespective of geography or culture, recoil from images of snakes and regard the real thing as dangerous, indicating a long conflict over evolutionary time between snakes and people. The fact is, however, that everywhere, except unfortunate Australia, the majority of snakes are *not* poisonous. Also, poisonous snakes in the American tropics tend to be nocturnal and secretive. Therefore, with a modicum of caution, visitors should be able to enjoy their days in southern Mexico without worrying unduly about poisonous snakes, and, again with caution, be able to watch snakes that cross their paths. Many of them are beautiful organisms, fairly common, and worth a look.

The largest group of snakes are those of the Family Colubridae – the *colubrid* snakes. Most of these are non-poisonous or, if venomous, dangerous only to small prey, such as lizards and rodents; in other words, they are only mildly venomous. This is a world-wide group comprising over 1500 species, including about three-quarters of the New World snakes. About 250 colubrid species occur in Mexico, about half of them being endemic; 60 species occur in the Yucatán region. Most of the snakes with which people have some familiarity, such as *water, brown, garter, whip, green, rat,* and *king snakes,* among a host of others, are colubrids, which have a wide variety of habits and lifestyles. It is not even possible to provide a general physical description of colubrid snakes because of the great variety of shapes and colors that specialize each for their respective lifestyles. Most people will not get close enough to notice, but an expert could identify colubrids by their anatomy; they have rows of teeth on the upper and lower jaws but they do not have hollow, venom-injecting fangs in front on the upper jaw.

Natural History

Ecology and Behavior

Because colubrids vary so much in their natural history, I shall concentrate on the habits of the species illustrated (Plates 10–13), which are representative of several general types. You should keep in mind that many snakes have not been much studied in nature. Such research is difficult for several reasons. Most snake species are not plentiful in the wild – even members of healthy populations often are found only few and far between – making it difficult to locate and study simultaneously more than a few individuals. Snakes also spend long periods being inactive and they feed infrequently, which means that to collect enough observations, studies of behavior need to be of long duration.

Typical lifestyles of various colubrids are terrestrial, burrowing, arboreal, and aquatic. Arboreal snakes spend most of their time in trees and shrubs. *Vine snakes* (Plate 11), for instance, are slender, elegant, grayish, green, or brown snakes that inhabit dryer areas, feeding usually on lizards taken in trees. These snakes have a mild venom that helps subdue their victims. Their thin, long bodies look much like vines and if not moving, these snakes are very difficult to see. They rely on their camouflage for both hunting and protection: they freeze in place when alerted to danger. *Blunt-headed snakes* (Plate 11) are also arboreal. They possess exceptional heads – broad and squarish, relative to a long, thin body – and large, bulging eyes. They forage at night for small frogs and lizards that they locate in their special hunting preserve, at the very outer layer of leaves and branches of trees and shrubs. These light-weight snakes, which are slightly sideways flattened, can move from branch to branch over open gaps that are half the length of their bodies. They hide during the day in trees.

The INDIGO SNAKE (Plate 10) is common throughout much of the American lowland tropics, inhabiting riverbeds, swamps, and marshes. It moves slowly over the ground and sometimes through bushes or shallow water, searching for its varied prey, which includes small turtles, frogs, mammals, birds, fish, eggs, and even other snakes. These beige, brown, or greenish-brown snakes range up to 4 m (13 ft) in length. Having no venom, they simply grab prey with their strong jaws, hold it, and swallow it. ROADGUARD, BROWN RACER, SPECKLED RACER, and RIBBON SNAKES (PLATES 10, 12) are all mid-sized snakes, between 40 and 100 cm

(16 to 40 in) long, found usually on the ground. They are all active, fast-moving snakes, and hunt during the day. Brown Racers and Roadguards eat lizards or lizards and snakes, while Speckled Racers and Ribbon Snakes, often found near water, eat frogs, toads, and, in the case of Ribbon Snakes, fish also. The Roadguard, named for its habit of sunning itself on the smooth black surfaces upon which we pilot automobiles (it enjoys more naturalistic open habitats as well, such as beaches and savannas), is mildly venomous and, according to the experts, one of the hardest snakes to hold in the standard behind-the-head grip without being bitten.

Breeding
Relatively little is known of the breeding particulars of most colubrids. For the group as a whole, the best information is available for the North American garter snakes. In these snakes, group mating assemblages, or *mating balls*, occur on warm days in spring. Many males – dozens and sometimes hundreds – and a few females swarm together and mating occurs. Some males are successful in mating with one or more females, while others are not. Each female probably mates with several males, and DNA tests on snake babies indicate that a single brood may have more than one father. Some of the males within a mating ball actually mimic females by releasing female courtship pheromones, which distract other males from the real females; the mimics then obtain more matings themselves. In other colubrid species, monogamous mating seems to be the case. Males make prolonged searches for mates, using their chemical senses to detect female pheromones. Once a female is located, a male may spend several days courting her before mating occurs. Because larger female snakes lay more eggs per clutch, males, when given a choice, prefer to mate with larger females. The typical number of eggs per clutch varies from species to species, but some, such as the BLUNT-HEADED TREE SNAKE (Plate 11), lay small clutches of one to three eggs. Most snakes that lay eggs deposit them in a suitable location and depart; the parents provide no care of the eggs or young. A few snakes guard their eggs.

Female snakes do not necessarily breed every year – some breed every 2 years or even 5 years. After mating, a female can store sperm and sometimes use it years later. An Indigo Snake in captivity, for instance, had fertile eggs 4 years after her last conjugal visit.

Ecological Interactions
Body shape. Body shape of snakes nicely demonstrates how a single body scheme, in this case cylindrical and legless, can be modified superbly through evolution to cope with a variety of habitats and lifestyles. The colubrid *worm snakes* live primarily underground. Their bodies are for the most part uniformly round, like earthworms. The majority of colubrids are long and very thin, at least partly as an adaptation for speedy movement – they escape from predators by moving rapidly. This very light, slender body plan also permits many arboreal colubrids to cross broad gaps between tree branches. In contrast to the mainly slender colubrids, most of the *vipers* (see p. 105) are heavy-bodied and rely on their bites, rather than speed, for protection and hunting. Being mostly sit-and-wait predators, they move around a lot less than do the colubrids. Finally, *sea snakes* (Plate 13), which spend their time in the water, have flattened tails that help with aquatic propulsion (N. G. Hairston 1994).

Body temperature. Biologists who study snakes know that temperature regulates a snake's life, and is the key to understanding their ecology. Snakes are cold-blooded animals – they inhabit a world where the outside temperature governs their activity. Unlike birds and mammals, their body temperature is determined primarily by how much heat they obtain from the physical environment. Simply put, they can only be active when they gather sufficient warmth from the sun. They have some control over their body temperature, but it is behavioral rather than physiological – they can lie in the sun or retreat to shade to raise or lower their internal temperatures to within a good operating range, but only up to a point: snakes must "sit out" hours or days in which the air temperature is either too high or too low. This dependence on air temperature affects most aspects of snakes' lives, from date of birth, to food requirements, to the rapidity with which they can strike at prey. For instance, in cold weather, snakes are less successful at capturing prey (they move and strike more slowly) and have less time each day when the ambient temperature is within their operating range, and so within the range in which they can forage. On the other hand, their metabolisms are slower when they are cold, which means that they need less food to survive these periods. At lower temperatures, snakes also probably grow slower, reproduce less often, and live longer (C. R. Peterson *et al.* 1993).

Lore and Notes

Snakes' limbless condition, their manner of movement, and the venomous nature of some of them, have engendered for these intriguing reptiles almost universal hatred from people, stretching back thousands of years. Myths about the evil power and intentions of snakes are, as they say, legion. But one need go no farther than the Old Testament, where the snake, of course, plays the pivotal role of Eve's corrupt enticer, responsible for people's expulsion from the Garden of Eden. The ensuing enhanced evil reputation of snakes came down through the ages essentially intact – so much so that even people who should have known better, such as Linnaeus, the 18th century botanist who began the scientific system we currently use to name and categorize plants and animals, considered them an abomination. Linnaeus, lumping snakes with other reptiles and the amphibians, referred to them in his writings as "these foul and loathsome animals" that are "abhorrent because of their cold body, pale color, cartilaginous filthy skin, fierce aspect, calculating eye, offensive smell, harsh voice, squalid habitation and terrible venom." He concluded that, owing to their malevolence, "their Creator has not exerted His powers to make many of them."

Scientists who study the ancient Mayan civilization suspect that some larger colubrid snakes were considered a good source of animal protein. Remains of species such as Indigo and Tropical Rat Snakes (Plates 10, 12) have been found at several Mayan ruin sites in the Yucatán region.

Status

At this time, no colubrid snakes of Mexico are known to be threatened. Most experts will concede, however, that little is known about the biology and population numbers of most snakes. Long-term studies are necessary to determine if population sizes are stable or changing. Because individual species of snakes normally are not found in great numbers, the truth is that it will always be difficult to tell when they are threatened. Worldwide, about 20 colubrids are listed as vulnerable, threatened, or endangered. The leading threats are habitat destruction and the introduction by people of exotic animals that prey on snakes at some point in their life cycles, such as fire ants, cane toads, cattle egrets, and armadillos. Considered a top priority now for research and conservation are the snakes of the West Indies. The islands of the Caribbean have hundreds of unique snake species, many of which appear to be declining rapidly in numbers – the result of habitat destruction and predation by mongooses, which were imported to the islands to control venomous snakes.

Profiles

Snail-eating Thirst Snake, *Dipsas brevifacies*, Plate 10a
Black-striped Snake, *Coniophanes imperialis*, Plate 10b
Roadguard, *Conophis lineatus*, Plate 10c
Brown Racer, *Dryadophis melanolomus*, Plate 10d
Indigo Snake, *Drymarchon corais*, Plate 10e
Speckled Racer, *Drymobius margaritiferus*, Plate 10f
Blunt-headed Tree Snake, *Imantodes cenchoa*, Plate 11a
Green Tree Snake, *Leptophis ahaetulla*, Plate 11b
Green-headed Tree Snake, *Leptophis mexicanus*, Plate 11c
Neotropical Vine Snake, *Oxybelis aeneus*, Plate 11d
Green Vine Snake, *Oxybelis fulgidus*, Plate 11e
Cat-eyed Snake, *Leptodeira frenata*, Plate 12a
Terrestrial Snail Sucker, *Sibon sartorii*, Plate 12b
Tropical Rat Snake, *Spilotes pullatus*, Plate 12c
Scorpion-eating Snake, *Stenorrhina freminvillei*, Plate 12d
Ribbon Snake, *Thamnophis proximus*, Plate 12e
Red Coffee Snake, *Ninia sebae*, Plate 13a
Tropical Kingsnake, *Lampropeltis triangulum*, Plate 13b

4. Dangerous Snakes and Boas

In this section I group together what are usually considered the more dangerous snakes, those that are highly poisonous and large ones that kill by squeezing their prey. Mexico has the most species of venomous snakes (about 60) of any country in the Western Hemisphere. However, few short-term visitors to tropical Mexico encounter a poisonous snake because most are well camouflaged, secretive in their habits, or nocturnal and, therefore, they are really outside the scope of this book. But most people are extremely leery of snakes and want to be well informed about them just in case … .

Vipers. Vipers, of the Family Viperidae, comprise most of the New World's poisonous snakes. Among all snakes they have the most highly developed venom-injection mechanisms: long, hollow fangs that inject poison into prey when they bite. The venom is often *neurotoxic*; that is it interferes with nerve function, causing paralysis of the limbs and then respiratory failure. Other venoms cause hemorrhaging both at the site of the bite and then internally, leading to cardiovascular shock and death. (The answer to the question of why venomous snakes are not harmed by their own venom is that they are immune.) Typically vipers coil prior to striking. They vary considerably in size, shape, color pattern, and lifestyle. Many of the *viperids* are referred to as *pit-vipers* because they have heat-sensitive "pits," or depressions, between their nostrils and eyes that are sensory organs. Pit-vipers occur from southern Canada to Argentina, as well as in the Old World. The familiar venomous snakes of North America are pit-vipers – rattlesnakes, copperheads, water moccasins – as are most of the poisonous snakes of southern Mexico.

About 45 viperid species occur in Mexico, with about 20 of them being endemic. The deadly FER-DE-LANCE (Plate 14) is abundant over most of the Yucatán Peninsula and parts of Oaxaca and Chiapas. Most are shorter than the maximum length of 2.5 m (8 ft). They are slender snakes with lance- or spear-shaped heads (hence the name, which means "iron spear"). EYELASH VIPERS (Plate 14) are a kind of *palm viper*, common arboreal snakes of lowland areas. There are several species, all small (as long as 1 m, or 3 ft, but most are only half that long) with prehensile tails and large wide heads. Their color schemes vary extensively. The TROPICAL RATTLESNAKE (Plate 14) occurs over much of southern Mexico. Reaching a length of 1.5 m (5 ft), it is a heavy-bodied animal with a slender neck and broad, triangular head. The rattles consist of loosely interlocking segments of a horn-like material at the base of the tail. A new segment is added each time the snake sheds its skin.

Coral Snakes. The family Elapidae contains what are regarded as the world's deadliest snakes, the Old World *cobras* and *mambas*. In the Western Hemisphere, the group is represented by the coral snakes – small, often quite gaily attired in bands of red, yellow and black, and, unfortunately, possessed of a very powerful neurotoxic venom. Fourteen species occur in Mexico, about five of them in southern Mexico; only one, a fairly common but secretive terrestrial snake known as the VARIABLE CORAL SNAKE (Plate 13), is distributed over the Yucatán Peninsula. Coral snakes rarely grow longer than a meter (3 ft).

Boas. The Family Boidae, members of which kill by constriction, encompasses about 65 species that are distributed throughout the world's tropical and subtropical regions. They include the Old World *pythons* and the New World *boas* and *anacondas*, the pythons and anacondas being the world's largest snakes. About four boas occur in Mexico, but only one, the BOA CONSTRICTOR (Plate 13), is common over the southern part of the country. It occurs over a wide range of habitat types from sea level up to almost 1000 m (3300 ft), and is the only boa of the Yucatán Peninsula. The snake reaches lengths of about 6 m (19 ft), but typical specimens are only 1.5 to 2.5 m (5 to 8 ft) long. They have shiny, smooth scales and a back pattern of dark, squarish shapes that provides good camouflage against an array of backgrounds.

Within a species male and female snakes usually look alike, although in many there are minor differences between the sexes in traits such as color patterns or the sizes of their scales.

Natural History

Ecology and Behavior

Vipers. FER-DE-LANCE as adults are terrestrial, but partially arboreal as juveniles. They inhabit moist forests but also some dryer areas. They eat mammals such as opossums, and birds. EYELASH VIPERS move through trees, along vines and twigs, searching when hungry for treefrogs, lizards, and mice. These snakes are often seen sunning themselves on leaves and branches. TROPICAL RATTLERS are denizens of the forest floor and of more open areas. They eat primarily lizards, birds, and rodents. Like other pit-vipers (and some other snakes), they can sense the heat radiated by prey animals, which aids their foraging. Searching by heat detection probably works for both warm-blooded prey (birds, mammals) as well as cold-blooded (lizards, etc.), as long as the prey is at a higher temperature than its surroundings.

Coral Snakes. Coral snakes are usually secretive and difficult to study; consequently relatively little is known about their ecology and behavior in the wild. They apparently forage by crawling along slowly, intermittently poking their heads into the leaf litter. They eat lizards, amphibians including caecilians (p. 75), and small snakes, which they kill with their powerful venom. They are often found under rocks and logs.

Boas. Boas are mainly terrestrial but they are also good climbers, and young ones spend a good deal of time in trees. When foraging, boas apparently search for good places to wait for prey, such as in a mammal's burrow or in a tree, near fruit. The diet includes lizards, birds, and mammals, including domesticated varieties. Prey, recognized by visual, smell (chemical), or heat senses, is seized with the teeth after a rapid, open-mouth lunge. As it strikes, the boa also coils around the prey, lifting it from the ground, and then constricts, squeezing the prey. The prey cannot breathe and suffocates. When the prey stops moving, the boa swallows it whole, starting always with the head.

Breeding

Details of the breeding in the wild of most tropical snakes are not well known. Many of the vipers may follow the general system of North American rattlesnakes, which have been much studied. Females attract males when they are ready to mate by releasing pheromones (odor chemicals) into the air and also, through the skin of their sides and back, onto the ground. Males search for females, following their odor trails. When one is located, the male accompanies and courts her for several days before mating occurs. Fighting between males for the same female is probably uncommon, because it is rare for two males to locate the same female at the same time. North American rattlers have distinct breeding periods, but many tropical vipers may breed at almost any time of year. Most snakes that lay eggs deposit the eggs in a suitable location and depart; the parents provide no care of the eggs or young. A few snakes guard their eggs.

Vipers. Most of the vipers give birth to live young. The FER-DE-LANCE has a reputation as a prolific breeder, females giving birth to between 20 and 70+ young at a time. Each is about a third of a meter (1 ft) long at birth, fully fanged with active poison glands, and dangerous. EYELASH VIPERS have clutches of up to 20 live young, and TROPICAL RATTLERS, 20 to 40.

Coral Snakes. Coral snakes lay eggs, up to 10 per clutch.

Boas. In some species of the boa family, males are known to fight over access to a single female. BOA CONSTRICTORS give birth to live young. Litters vary in size between 12 and 60. Each snakelet at birth is about half a meter (1.5 ft) long.

Ecological Interactions

Coral snakes hold a special place in snake biology studies because a number of non-venomous or mildly venomous colubrid snakes (p. 101), as well as at least one caterpillar species, mimic the bright, striking coral snake color scheme – alternating bands of red, yellow (or white), and black. About 10 species of colubrid snakes in the Yucatán region (which includes Belize and northern Guatemala) imitate – to varying degrees – the color patterns of coral snakes (see Plate 13 for examples). The function of the mimicry apparently is to take advantage of the quite proper respect many predatory animals show toward the lethal coral snakes. Ever since this idea was first proposed more than 100 years ago, the main argument against it has been that it implied either that the predators had to be first bitten by a coral snake to learn of their toxicity and then survive to generalize the experience to all snakes that look like coral snakes, or that the predators were born with an innate fear of the coral snake color pattern. It has now been demonstrated experimentally that several bird predators on snakes (motmots, kiskadees, herons, and egrets) need *not* learn that a coral snake is dangerous by being bitten – they avoid these snakes instinctively from birth. Thus, many snakes have evolved as defensive mechanisms color schemes that mimic that of coral snakes. (However, some biologists argue that this explanation falls apart because, they say, the alternating color bands of the coral snake's body function as camouflage and *not* to warn predators away, and also because the snakes' mammalian predators lack color-vision and therefore could not make use of the patterns to avoid the mimics.)

Non-venomous snakes also mimic some of the behavior of poisonous snakes, the most obvious example being that a good many snakes, when threatened, coil up and wiggle the tips of their tails, as do rattlesnakes and some other vipers.

Lore and Notes

These snakes are dangerous! Remember: snakes, like traffic cops, lack a sense of humor. All of the venomous snakes discussed in this section, if encountered, should be given a wide berth. Watch them only from a distance. Very few visitors to Latin America are bitten by poisonous snakes, even those that spend their days tramping through forests. The biology of snake bites is an active area of study. Venomous snakes can bite without injecting any venom, and they can also vary the amount of venom injected – even if bitten, one does not necessarily receive a fatal dose. Within the same species, the toxicity of a snake's venom varies geographically, seasonally, and from individual to individual.

Vipers. Because of its aggressive nature, the FER-DE-LANCE is often said to be the most feared and dangerous of Central American snakes – when approached, it is more likely to bite than retreat. Biologists have noted Fer-de-lances, when attempting to escape capture, reverse direction when they are out of sight under a large leaf on the forest floor, as if to ambush any pursuers (H. W. Greene 1997). Stories abound attesting to the potency of its venom. One, out of Honduras, tells the sorry tale of a railway worker and his wife. The man, bitten by a Fer-de-lance at work, was brought home and ministered to by his wife. The venom killed the man two hours later and the wife the next day – she had scraped her finger the previous day while cooking, and the poison had entered through the open cut as she dressed his wound. These snakes killed so many sugar cane workers in Jamaica in the early 1900s that mongooses from the Old World were shipped to the island to kill them. They were somewhat successful, but unfortunately, the mongooses also killed many of the isle's non-venomous snakes, not to mention domestic fowl. Also, unhappily for the mongooses, it turns out that the New World pit-vipers are a good deal faster striking than are cobras, their Old World nemeses. A documentary motion-picture maker trying to film a mongoose killing a Fer-de-lance had several dead mongooses on his hands before obtaining the pictures he wanted. Somewhat amazingly, there is a Central American colubrid snake, the MUSURANA, that kills and eats the Fer-de-lance by first injecting it with its own considerably less potent venom and then by constricting it.

EYELASH VIPERS, although small, pack a potent venom and are a serious threat because of their arboreal habits and highly effective cryptic coloration.

They move and coil themselves among a tree's leaves and branches. Some species of palm vipers have the habit of coiling their tails around a branch and hanging down, then turning their body in the air so that it is parallel to the ground, their head in a position to strike at a passing animal. Therefore, pushing through vegetation or clearing a path with a machete can be dangerous pursuits. The JUMP-ING VIPER (Plate 14), although possessing a weaker venom than others of its ilk, is particularly scary and dreaded by local people; when threatened, these snakes sometimes jump nearly a meter (2 or 3 ft) into the air, and as they jump, they bite. TROPICAL RATTLERS are usually more aggressive than the North American rattlers, and their venom is stronger and faster-acting. For some reason, they do not always rattle when approached. If threatened, a Tropical Rattler will coil, raise its head high and, if necessary, attack.

Based on the frequency with which they are depicted on art and artifacts recovered at Mayan ruin sites, several viper species (especially the Fer-de-lance and Tropical Rattler) plainly were very important to the ancient Mayans.

Coral Snakes. These small, pretty snakes are rarely seen by people because of their secretive habits. Reports are that they are usually quite docile and seldom go out of their way to bite people. However, if threatened they give a scary defensive display: "the body is flattened and erratically snapped back and forth ... the head ... is swung from side to side with the mouth open, and any object that is contacted is bitten ... the tail is coiled, elevated, and waved about" (H. Greene & R. Seib 1983). Their venom is very powerful and coral snakes have killed a good many incautious people.

Boas. Boa personalities appear to vary, but some individuals are notoriously bad-tempered and aggressive. A BOA CONSTRICTOR may hiss loudly at people, draw its head back with its mouth open in a threat posture, and bite. They have large sharp teeth that can cause deep puncture wounds. Therefore, even though boas present no real threat to most people, keeping a respectful distance is advised.

Status

None of the Mexican vipers or coral snakes are considered threatened by people (for once, we can say that it's the other way around!). BOA CONSTRICTORS may be threatened in some regions of the Yucatán by habitat destruction and by capture for the pet trade; otherwise the boas seem to do well living near people and are still common in many parts of southern Mexico, as well as in the rest of the American tropics. They are listed by CITES Appendix II, as are all boas and pythons. Several other boa species, particularly some island endemics, are endangered. The only New World vipers known to be threatened or endangered are two island species: Brazil's GOLDEN LANCEHEAD and the ARUBA ISLAND RATTLER.

Profiles

Boa Constrictor, *Boa constrictor*, Plate 13c
Coral Snake, *Micrurus diastema*, Plate 13d
Pelagic Sea Snake, *Pelamis platurus*, Plate 13e
Yucatán Cantil, *Agkistrodon bilineatus*, Plate 14a
Fer-de-lance, *Bothrops asper*, Plate 14b
Eyelash Viper, *Bothriechis schlegelii*, Plate 14c
Jumping Viper, *Atropoides nummifer*, Plate 14d
Yucatán Hognosed Pit-viper, *Porthidium yucatanicum*, Plate 14e
Tropical Rattlesnake, *Crotalus durissus*, Plate 14f

5. Geckos

Geckos are most interesting organisms because, of their own volition, they have become "house lizards" – probably the only self-domesticated reptile. The family, Gekkonidae, is spread throughout tropical and subtropical areas the world over, 800 species strong. In many regions, geckos have invaded houses and buildings, becoming ubiquitous adornments of walls and ceilings. Ignored by residents, they move around dwellings chiefly at night, munching insects. To first-time visitors from northern climes, however, the way these harmless lizards always seem to position themselves on ceilings directly above one's sleeping area can be a bit disconcerting. About 30 species occur in Mexico, 14 or so being endemic (one, the OAXACAN LEAF-TOED GECKO, is endemic to a single state, Oaxaca). More than 20 species occur in southern Mexico, about eight of them in the Yucatán region.

Geckos are fairly small lizards, usually gray or brown, with large eyes. They have thin, soft skin covered often with small, granular scales, often producing a slightly lumpy appearance, and big toes with well developed claws that allow them to cling to vertical surfaces and even upside-down on ceilings. The way geckos manage these feats has engendered over the years a fair amount of scientific detective work. Various forces have been implicated in explaining the gecko's anti-gravity performance, from the ability of their claws to dig into tiny irregularities on man-made surfaces, to their large toes acting as suction cups, to an adhesive quality of friction. The real explanation appears to lie in the series of miniscule hair-like structures on the bottom of the toes, which provide attachment to walls and ceilings by something akin to surface tension – the same property that allows some insects to walk on water.

Adult geckos mostly report in at only 5 to 10 cm (2 to 4 in) in length, tail excluded; tails can double the length. Because lizard tails frequently break off and regenerate (see p. 116), their length varies tremendously; gecko tails are particularly fragile. Lizards, therefore, are properly measured from the tip of their snouts to their *vent*, the urogenital opening on their bellies, usually located somewhere near to where their rear legs join their bodies. The geckos' 5 to 10 cm length, therefore, is their range of "SVLs," or snout–vent lengths.

Natural History

Ecology and Behavior

Although most lizards are active during the day and inactive at night, nearly all gecko species are nocturnal. In natural settings, they are primarily ground dwellers, but, as their behavior in buildings suggests, they are also excellent climbers. Geckos feed on arthropods, chiefly insects. In fact, it is their ravenous appetite for cockroaches and other insect undesirables that renders them welcome house guests in many parts of the world. Perhaps the only "negative" associated with house geckos is that, unlike the great majority of lizards, which keep quiet, geckos at night are avid little chirpers and squeakers. They communicate with each other with loud calls – surprisingly loud for such small animals. Various species sound different; the word *gecko* approximates the sound of calls from some African and Asian species.

Geckos are *sit-and-wait* predators; instead of wasting energy actively searching for prey that is usually highly alert and able to flee, they sit still for long periods, waiting for unsuspecting insects to venture a bit too near, then lunge, grab, and swallow.

Geckos rely chiefly on their *cryptic coloration* and their ability to flee rapidly for escape from predators, which include snakes during the day and snakes, owls, and bats at night. When cornered, geckos give threat displays; when seized, they give loud calls to distract predators, and bite. Should the gecko be seized by its tail, it breaks off easily, allowing the gecko time to escape, albeit tail-less; tails regenerate rapidly. Some geckos when seized also secrete thick, noxious fluids from their tails, which presumably discourages some predators. Almost all geckos can lighten or darken their skin coloring (see p. 114). The primary function of such color changing is often thought to be increased camouflage – to make themselves more difficult to see against various backgrounds.

Breeding
Geckos are egg-layers. Mating occurs after a round of courtship, which involves a male displaying to a female by waving his tail around, followed by some mutual nosing and nibbling. Clutches usually contain only a few eggs, but a female may lay several clutches per year. There is no parental care – after eggs are deposited, they and the tiny geckos that hatch from them are on their own.

Lore and Notes
Indigenous peoples in parts of Mexico and Central America regard some geckos, erroneously, as poisonous, as suggested by the common name given them: "escorpión."

The world's smallest reptile, at 4 cm (1.5 in) long, is a gecko, the CARIBBEAN DWARF GECKO. As reptile biologists like to say, it is shorter than its name.

Status
More than 25 gecko species are listed by conservation organizations as rare, vulnerable to threat, or endangered, but they are almost all restricted to the Old World. The MONITO GECKO, found only on Monito Island, off Puerto Rico, and Venezuela's PARAGUANAN GROUND GECKO, are endangered. No Mexican species are currently threatened.

Profiles
Yucatán Banded Gecko, *Coleonyx elegans*, Plate 15a
House Gecko, *Hemidactylus frenatus*, Plate 15b
Mediterranean Gecko, *Hemidactylus turcicus*, Plate 15c
Dwarf Gecko, *Sphaerodactylus glaucus*, Plate 15d
Central American Smooth Gecko, *Thecadactylus rapicauda*, Plate 15e

6. Iguanids

The Iguanidae, a large family of lizards, has an almost exclusively New World distribution. There are more than 700 species. About 175 species occur in Mexico, 105 of them being endemic; 20+ species occur in the Yucatán region. (A new classification splits the iguanids into several smaller families, but it's easier for our purposes here to use the old grouping.) Most of the lizards commonly encountered by ecotourists or that are on their viewing wish-lists are members of this group. It includes the very abundant *anolis* lizards, the colorfully named Jesus Christ lizards, and the spectacular, dinosaur-like GREEN IGUANAS (Plate 16).

The *iguanids* are a rich and varied group of diverse habits and habitats. Many in the family are brightly colored and have adornments such as crests, spines, or throat fans. They range in size from tiny anolis lizards, or *anoles*, only a few centimeters in total length and a few grams in weight, to Green Iguanas, which are up to 2 m (6.5 ft) long. SPINY-TAILED IGUANAS (Plate 16) range up to a meter (3 ft) in total length and can weigh up to a kilogram (2 lb). The *basilisks*, or Jesus Christ lizards, likewise range up to a meter (3 ft) long and can weigh more than half a kilogram (1 lb). Most of the length in iguanid lizards resides in the long, thin tail; hence the paradoxically low weight for such long animals. The basilisk of southern Mexico, the STRIPED BASILISK (Plate 16), is usually fairly small, only 12 to 15 cm (5 to 6 in) long, minus the tail. *Spiny lizards* are a large group (80 species) of small to moderate-sized iguanids that range from southern Canada to Panama. The ones in southern Mexico are usually 5 to 9 cm (2 to 3.5 in) long, excluding the tail. They are quite common in natural areas and also around human habitations. Scales on their backs are often overlapping and pointed, which yields a bristly appearance.

Natural History

Ecology and Behavior

Green Iguana. You won't mistake this animal; it's the large one resembling a dragon sitting in the tree near the river. They are common inhabitants of many Neotropical rainforests, in moist areas at low to middle elevations. Considered semi-arboreal, they spend most of their time in trees, usually along waterways. They don't move much, and when they do it's often in slow-motion. They are herbivores as adults, eating mainly leaves and twigs and, more occasionally, fruit; insects are favorites of youngsters. They are fun to discover, but boring to watch. When threatened, an iguana above a river will drop from its perch into the water, making its escape underwater; they are good swimmers. During their breeding season, males establish and defend mating territories on which live one to four females.

Spiny-tailed Iguana. These lizards are sometimes confused with GREEN IGUANA, but they are darker-colored, lack the Green's conspicuous head spines and crest, and are often found in dryer habitats, such as fields, farms, scrubland, dry woodlands, savannas, and roadsides. Like Greens, they are semi-arboreal, spending considerable time in trees, feeding and basking; when chased on the ground, they often run to a tree and climb to escape. They also burrow into the ground and under rocks. Spiny-tails are territorial, each one defending its shelter and perch sites from all others. They are predominantly vegetarians, eating flowers, fruits, agricultural crops, but also the odd lizard or small mammal. Juvenile Spiny-tails eat insects and, in turn, are preyed upon by a variety of animals, including snakes, hawks, jays, skunks, and raccoons. These large lizards are frequently spotted at archaeological ruin sites.

Basilisk. Basilisks are medium to large, active lizards commonly found along watercourses in lowland areas. Often they are very abundant – one study in Central America estimated their numbers to be more than 500 per hectare (200 per acre)! They are classified as terrestrial but are also semi-arboreal – possessed of a tendency to climb. They are omnivorous, eating a variety of invertebrate and vertebrate animals (especially hatchling lizards), some flowers, and a good deal of fruit. The name "Jesus Christ lizard" refers to their ability to run over the surface of ponds or streams, really skipping along, for distances up to 20 m (60 ft) or more. They do it in an upright posture, on their rear legs, further inviting the divine comparison. The trick is one of fluid dynamics: some of the force to support the lizard comes from resistance of the water when the large rear feet are slapped onto the surface; the rest of the upward force stems from the compression of the water that occurs as the lizard's feet move slightly downwards into the water (their feet move so rapidly up and down that they are actually pulled from below the surface before the water can close over them). Juveniles are better at water-running than are adults, which, when too heavy to be supported on the surface, escape predation by diving in and swimming underwater.

Anole. Anoles are small, often arboreal lizards. Nearly 50 species occur in Mexico, 11 of them in the Yucatán region. Several species are frequently encountered, but others, such as ones that live in the high canopy, are rarely seen. Some are ground dwellers, and others spend most of their time on tree trunks perched head toward the ground, visually searching for insect prey. The SILKY ANOLE (Plate 17) is commonly found in grass along roadsides; when startled, it aligns itself on a blade of grass to try to appear inconspicuous. Anoles are known especially for their territorial behavior. Males defend territories on which one to three females may live. In some species males with territories spend up to half of each day defending their territories from males seeking to establish new territories. The defender will roam his territory, perhaps 30 sq m (325 sq ft), occasionally giving territorial advertisements – repeatedly displaying his extended throat sac, or *dewlap*, and performing *push-ups*, bobbing his head and body up and down. Trespassers that do not exit the territory are chased and even bitten. Anoles are chiefly sit-and-wait predators on insects and other small invertebrates. Anoles themselves, small and presumably tasty, are frequent prey for many birds (motmots, trogons, and others) and snakes.

Spiny Lizard. Most spiny lizard species in the wild are ground-dwelling or semi-arboreal, but around buildings they are climbers of fences, walls, and rooftops. Sit-and-wait predators, they eat mainly insects, and many also consume some plant materials.

Breeding

Green Iguana. Breeding occurs during the early part of the dry season. These large lizards lay clutches that probably average about 40 eggs. They are laid in burrows that are 1 to 2 m (3 to 6.5 ft) long, dug by the females. After laying her clutch, the female fills the burrow with dirt, giving the site a final packing down with her nose. It has been said that a female digging her nest burrow probably engages in the most vigorous activity performed by these sluggish reptiles.

Spiny-tailed Iguana. Female Spiny-tails annually lay a single clutch of between 12 and 88 eggs, older females producing more. Eggs are laid in burrows in open, sunny areas. They hatch in the Yucatán region from May through August.

Basilisk. Female Basilisks produce clutches of 2 to 18 eggs several times each year. Smaller, younger females, have fewer eggs per clutch. Eggs hatch in about 3 months. Hatchling lizards, weighing only 2 g, are wholly on their own; in one study, only about 15% of them survived the first few months of life.

Anole. Female Anoles lay small clutches of eggs throughout the year; an individual female may produce eggs every few weeks.

Spiny Lizard. In contrast to most iguanids, some spiny lizard species are *viviparous*, giving birth to live young once per year. Average brood size in one study was six.

Ecological Interactions

Via interactions between the external environment and their nervous and hormonal systems, many iguanids have the novel ability to change their body color. Such color changes presumably are adaptations that allow them to be more cryptic, to blend into their surroundings, and hence to be less detectable to and safer from predators. Also, alterations in color through the day may aid in temperature regulation; lizards must obtain their body heat from the sun, and darker colors absorb more heat. Color changing is accomplished by moving pigment granules within individual skin cells either to a central clump (causing that color to diminish) or spreading them evenly about the cell (enhancing the color). It is now thought that the stimulus to change color arises with the physiology of the animal rather than with the color of its surroundings. Spiny lizards change color with temperature or light intensity, some being very dark, even black, in the cool early morning and green at midday. Even the large SPINY-TAILED IGUANA has skin that lightens or darkens with changing temperature or activity levels. Anolis lizards also change color. North American species, particularly CAROLINA ANOLES, owing to their color-changing ways, are hawked in pet stores as "chameleons" even though the real chameleons are strictly Old World lizards.

Lore and Notes

The large iguanid lizards (GREEN and SPINY-TAILED IGUANAS) are not dangerous. They are not poisonous and they will not bite unless given no other choice. They are hunted by local people for invitation to the dinner table. Give them a try if the meat is offered; but it's an acquired taste. Their eggs are also eaten by present-day Mayan groups. Ancient Mayans also consumed these large lizards, as evidenced by the discovery of their remains around Mayan ruins. One theory is that, as Mayan populations increased in size about 1000 years ago, populations of local game animals (birds and mammals) became severely depleted, and so the Mayans turned to larger reptiles for meat.

Over parts of Central America, Spiny-tailed Iguana is thought to have great medicinal power, especially as an anti-impotence agent.

Status

None of the Mexican iguanids are currently considered threatened. Because SPINY-TAILED and GREEN IGUANAS are hunted for meat, they are scarce in some localities. Several iguanids of the Caribbean are endangered, such as the JAMAICAN GROUND IGUANA and VIRGIN ISLANDS (ANEGADA) ROCK IGUANA (both are USA ESA listed; all iguanas – that is, genus *Iguana* – are CITES Appendix II listed). At least three iguana species in the Galápagos Islands are threatened, and two iguanids in the USA are known to be endangered.

Profiles

Green Iguana, *Iguana iguana*, Plate 16a
Striped Basilisk, *Basiliscus vittatus*, Plate 16b
Spiny-tailed Iguana, *Ctenosaura similis*, Plate 16c
Helmeted Basilisk, *Corytophanes hernandezii*, Plate 16d
Ghost Anole, *Anolis lemurinus*, Plate 17a
Yucatán Smooth Anole, *Anolis rodriguezii*, Plate 17b
Brown Anole, *Anolis sagrei*, Plate 17c
Silky Anole, *Anolis sericeus*, Plate 17d
Yellow-spotted Spiny Lizard, *Sceloporus chrysostictus*, Plate 17e

Cozumel Spiny Lizard, *Sceloporus cozumelae*, Plate 18a
Blue Spiny Lizard, *Sceloporus serrifer*, Plate 18b
Teapen Rosebelly Lizard, *Sceloporus teapensis*, Plate 18c

7. Skinks and Whiptails

The *skinks* are a large family (Scincidae, with about 1000 species) of small and medium-sized lizards with a world-wide distribution. Over the warmer parts of the globe, they occur just about everywhere. Skinks are easily recognized because they look different from other lizards, being slim-bodied with relatively short limbs, and smooth, shiny, roundish scales that combine to produce a satiny look. Many skinks are in the 5 to 9 cm (2 to 4 in) long range, not including the tail, which can easily double an adult's total length. About 25 species occur in Mexico, half of which are endemic. GROUND SKINKS and SHINY SKINKS (Plate 18), profiled here, have fairly broad distributions over southern Mexico and are two of only four species that occur in the Yucatán area. They are common to rainforests at low to moderate elevations, and are especially prevalent in forest edge areas.

Whiptails, Family Tiidae, are a New World group of about 200 species, distributed throughout the Americas. Most are tropical residents, inhabiting most areas below 1500 m (5000 ft) in elevation. About 30 species occur in Mexico, about five in the Yucatán area. Whiptails are often quite abundant along trails, clearings, beaches, and roads, and, hence, are frequently conspicuous. They are small to medium-sized, slender lizards, known for their highly alert, active behavior. They have long, slender, whip-like tails, often twice the length of their bodies, which range from 7 to 12 cm (3 to 5 in) in length. Some whiptails are striped, others are striped and spotted.

Natural History

Ecology and Behavior
Skinks. Many skinks are terrestrial lizards, particularly appreciative of moist ground habitats such as sites near streams and springs, or of spending time under wet leaf litter. A few species are arboreal, and some are burrowers. Skink locomotion is surprising; they use their limbs to walk but when the need arises for speed, they locomote mainly by making rapid wriggling movements with their bodies, snake-fashion, with little leg assistance. Through evolutionary change, in fact, some species have lost limbs entirely, all movement now being snake-fashion.

Skinks are day-active lizards, most activity in the tropics being confined to the morning hours; they spend the heat of midday in sheltered, insulated hiding places, such as deep beneath the leaf litter. Some skinks are sit-and-wait foragers, whereas others seek their food actively. They consume many kinds of insects, which they grab, crush with their jaws or beat against the ground, then swallow whole. Predators on skinks are snakes, larger lizards, birds, and mammals such as coati, armadillo, and opossum.

Skinks generally are not seen unless searched for. Most species are quite secretive, spending most of their time hidden under rocks, vegetation, leaf litter or, in the case of the SHINY SKINK, between stone slabs of Mayan ruins.

Whiptails. Whiptails actively search for their food, usually insects, but also small amphibians. Typically they forage by moving slowly along the ground, poking their nose into the leaf litter and under sticks and rocks. Although most are terrestrial, many also climb into lower vegetation to hunt. Whiptails have a characteristic gait, moving jerkily forward while rapidly turning their head from side to side.

Breeding

Skinks. Skinks are either egg-layers or live-bearers. GROUND SKINKS lay eggs, usually small clutches of one to three. They probably breed year round, with individual females laying several clutches annually.

Whiptails. Southern Mexican whiptails may breed throughout the year. They are egg-layers, females producing small clutches that average three to five eggs; individuals produce two or more clutches per year.

Ecological Interactions

Many lizards, including the skinks, whiptails, and geckos, have what many might regard as a self-defeating predator escape mechanism: they detach a large chunk of their bodies, leaving it behind for the predator to attack and eat while they make their escape. The process is known as *tail autotomy* – "self removal." Owing to some special anatomical features of the tail vertebrae, the tail is only tenuously attached to the rest of the body; when the animal is grasped forcefully by its tail, the tail breaks off easily. The shed tail then wriggles vigorously for a while, diverting a predator's attention for the instant it takes the skink or whiptail to find shelter. A new tail grows quickly to replace the lost one.

Is autotomy successful as a lifesaving tactic? Most evolutionary biologists would argue that, of course it works, otherwise it could not have evolved to be part of lizards' present-day defensive strategy. But we have hard evidence, too. For instance, some snakes that have been caught and dissected have been found to have in their stomachs nothing but skinks – not whole bodies, just tails! Also, a very common finding when a field biologist surveys any population of small lizards (catching as many as possible in a given area to count and examine them) is that a hefty percentage, often 50% or more, have regenerating tails; this indicates that tail autotomy is common and successful in preventing predation.

Lore and Notes

Among the whiptails, a number of species exhibit what for vertebrate animals is an odd method of reproduction, one that is difficult for us to imagine. All individuals in these species are female; not a male amongst them. Yet they breed merrily away, by *parthenogenesis*. Females lay unfertilized eggs, which all develop as females that, barring mutations, are all genetically identical to mom. (Some fish also reproduce this way.) This hardly seems a happy state of affairs; many people would argue that something important is missing from such societies. It is likely that parthenogenetic species arise when individuals of two different but closely related, sexually reproducing, "parent" species mate and, instead of having hybrid young that are sterile (a usual result, as when horses and donkeys mate to produce sterile mules), have young whose eggs can produce viable females.

Status

Skinks and whiptails of southern Mexico, as far as is known, are secure; none are presently considered threatened. As is the case for many reptiles and amphibians, however, many species have not been sufficiently monitored to ascertain the true health of populations. Many skinks of Australia and New Zealand regularly make lists of vulnerable and threatened animals, and several Caribbean skinks and whiptails are endangered. In the USA, Florida's SAND and BLUETAIL MOLE SKINKS (USA ESA listed) and the ORANGE-THROATED WHIPTAIL are considered vulnerable or threatened species.

Profiles

Ground Skink, *Sphenomorphus cherriei*, Plate 18d
Shiny Skink, *Mabuya brachypoda*, Plate 18e
Barred Whiptail, *Ameiva undulata*, Plate 19a
Cozumel Whiptail, *Cnemidophorus cozumelae*, Plate 19b
Yucatán Whiptail, *Cnemidophorus angusticeps*, Plate 19c
Deppe's Whiptail, *Cnemidophorus deppii,* Plate 19d

Environmental Close-Up 2
Coffee, Birds, and Conservation

Modern agriculture, although somewhat necessary for the continuation of human life on Earth, has its drawbacks. Aside from afflicting ecosystems with pesticides and chemical fertilizers (*agrochemicals*), pollution, and massive soil erosion, the basic disruption is that large tracts of natural habitats are cleared of native vegetation and replaced with one or a few crop plant species. Such wholesale habitat alteration is bound to cause significant problems for wildlife adapted to the natural vegetation. Harmful effects of modern farming may be reduced using techniques of *sustainable agriculture*, a phrase used with increasing frequency by environmentalists and others concerned with the future condition of the Earth's habitats. The phrase means farming in ways that are economically profitable for the local economy yet not ecologically harmful; using techniques, in other words, that will not lead to significant ecosystem damage or to decline in biodiversity. Researchers are finding more and more ways that environmentally harmful farming practices can be altered to make them more "eco-friendly." Perhaps the most internationally publicized case concerns the modest brown coffee bean.

Coffee crops can be grown on plantations in environmentally damaging or environmentally enhancing (*green*) ways. Traditional coffee crops over much of southern Mexico and Central America were grown on small family farms where low coffee plants thrived beneath a canopy of taller trees, which provided shade as well as protective ground mulch. The trees selected to shade the coffee were often themselves productive: species that increased soil nutrients like nitrogen, food-producing fruit trees, or timber/fuel-wood trees. (Some coffee was also grown using *rustic* methods: by placing coffee plants under natural forest canopy after first clearing away the understory vegetation.) When large corporations went into the coffee business, they increased yields by switching to *sun coffee*, growing coffee plants alone on large plantations, using fertilizers and pesticides extensively to do for their crop what shade trees formerly did. With smaller coffee plants packed closely together, and with chemical fertilizers, sun coffee can produce up to three times as much coffee as shade plants in the same space. Now many growers, large and small, produce sun coffee. Unfortunately, monotonous rows of low coffee plants make a poor habitat for animal life, especially birds. The drive to get large companies to produce only *shade-grown coffee* is concerned with having them add canopy trees to their plantations, thus improving the coffee's taste (as some insist), decreasing the amounts of chemicals needed, and, not incidentally, greatly enhancing the habitat for wildlife use. You might not think that simply planting shade trees above coffee plants would make much of a difference, but it does! It turns out that plantations with trees and understory crop plants provide habitats sufficiently complex to attract abundant wildlife, particularly migratory birds.

Ecological researchers are beginning to look closely at the relationship between declining populations of migratory songbirds in North America and changes in agricultural practices in Latin America and the Caribbean over the past few decades. *Neotropical migrants* are songbirds that migrate each year between breeding areas in the north temperate zone and wintering areas in the Central and South America. (In fact, about 335 of the 650 or so bird species that regularly occur in the USA and Canada migrate in this way.) During the last 25 years, many species of these small birds have shown significant declines in the sizes of their populations; in some, the rate of decline is still accelerating. For instance, surveys of breeding birds each spring and summer in the USA indicate that some species of warblers (p, 191) have suffered declines of about 40%, as have some species of orioles (p. 193) and sparrows (p. 199). (But some very recent studies conclude that when total songbird numbers are considered over large sections of North America, many species have actually been stable or even increasing in population during the past few decades.) Ecologists who study these birds have pretty much concluded that the chief culprit behind most of the population crashes is loss of forest habitat to development, including loss of breeding sites in North America, loss of migratory stop-over habitat, and loss of wintering habitat in the Neotropics. Recently, a fourth kind of habitat loss was added to the list: the switch to sun coffee from shade coffee, and the accompanying loss of tree plantation habitat for the little migrants.

Many Neotropical migrants originally wintered in forest habitats, but as forests were cut for development and agriculture, the birds increasingly switched their wintering sites to plantations with trees, such as the traditional coffee farms. In fact, in some regions of Latin America, tree plantations comprise essentially the only surviving forest-like habitat. Ecologists term these kinds of places *refugia* – the last remaining patches of naturalistic habitat in which wildlife under siege can seek refuge. In southern Mexico – for instance, around the Chiapas highland town of Ocosingo – researchers found that many migrants even seemed to prefer coffee plantations over native forest – often finding many more species among plantation trees than in nearby forests. Likewise, a study in Guatemala found about twice as many bird species using shade coffee farms as were using sun coffee farms. In fact, the usual relationship is that the more recent the farming methods used on a plantation, the fewer bird species found there. The tree plantations also apparently provide good habitat for many other kinds of animals – especially abundant are snakes, bats, monkeys, and opossums, and insect biodiversity is high. (In fact, one theory of why so many birds seem to prefer shade coffee farms to forests is the great abundance of insect food, particularly cicadas, available on the farms.) Now these last refuges of the migrant birds and other animals are increasingly transformed to tree-less sun coffee plantations. As of the mid-1990s, it was estimated that at least half of Latin America's coffee production was sun coffee.

It seems that, from a conservation standpoint, shade-grown coffee is the way to go. Fortunately, there is an incentive for companies to produce shade-grown coffee: they can advertise it as such and, hence, charge more for it. Many surveys have demonstrated that consumers in export countries are willing to pay a premium price for green products such as shade-grown coffee. As this book is written, popular calls for shade-grown coffee, reported in the media, are becoming more frequent. Also, some environmental groups have begun their own certification processes. They give an "Eco-OK" seal of approval to particular coffee brands that they feel meet green criteria, such as the coffee being grown beneath a splendid green canopy of native trees.

Preservation of habitat for Neotropical migrants is especially important in southern Mexico. Because of the region's strategic subtropical position between the northern temperate zone (USA and Canada) and the Central and South American tropics, and its being just south of the Gulf of Mexico, over which many migrants pass, a huge number of migrants either winter there or pass through on their way to sites farther south. In fact, recent estimates suggest that of the between two and five billion birds that leave northern breeding areas each year to winter in the Neotropics, fully one-third of them either pass through or winter in Chiapas, the Yucatán Peninsula, or adjoining northern Guatemala. About 200 species of migrant birds live part of each year in the region, including two hummingbirds (p. 162), nine raptors (p. 143), six swallows (p. 160), 12 flycatchers (p. 183), six thrushes (p. 187), and more than 30 warblers (p. 191). These birds are not freeloaders on Mexican soil – they pay for their keep. They consume enormous numbers of insects that are pests to people and agriculture, and the fruit-eaters among them contribute to the growth of forests by being seed dispersers: they eat fruit and then drop the seeds far away from the parent tree, where seeds are more likely to germinate and grow successfully. Many of these migrants are forest birds and therefore conservation of the region's last remaining large stands of forest (particularly Montes Azules Biosphere Reserve in Chiapas, Calakmul and Sian Ka'an Biosphere Reserves on the Yucatán Peninsula, and northern Guatemala's Maya Biosphere Reserve) is essential for these birds to survive and be able to continue their migratory lifestyle.

Chapter 8

Birds

- Introduction
- Features of Tropical Birds
- Seeing Birds in Tropical Mexico
- Family Profiles

Introduction

Most of the vertebrate animals one sees on a visit to just about anywhere at or above the water's surface are birds, and southern Mexico is no exception. Regardless of how the rest of a trip's wildlife-viewing progresses – how fortunate one is to observe mammals or reptiles or amphibians – birds will be seen frequently and in large numbers. The reasons for this pattern are that birds are, as opposed to those other terrestrial vertebrates, most often active during the day, visually conspicuous and, to put it nicely, usually far from quiet as they pursue their daily activities. But why are birds so much more conspicuous than other vertebrates? The reason goes to the essential nature of birds: they fly. The ability to fly is, so far, nature's premier anti-predator escape mechanism. Animals that can fly well are relatively less predation-prone than those which cannot, and so they can be both reasonably conspicuous in their behavior and also reasonably certain of daily survival. Birds can fly quickly from dangerous situations, and, if you will, remain above the fray. Most flightless land vertebrates, tied to moving in or over the ground or on plants, are easy prey unless they are quiet, concealed, and careful or, alternatively, very large or fierce; many smaller ones, in fact, have evolved special defense mechanisms, such as poisons or nocturnal behavior.

A fringe benefit of birds being the most frequently encountered kind of vertebrate wildlife is that, for an ecotraveller's intents and purposes, birds are innocuous. Typically, the worst that can happen from any encounter is a soiled shirt. Contrast that with too-close, potentially dangerous meetings with certain reptiles (venomous snakes!), amphibians (frogs and salamanders with toxic skin secretions), and mammals (bears or big cats). Moreover, birds do not always depart with all due haste after being spotted, as is the wont of most other types of vertebrates. Again, their ability to fly and thus easily evade our grasp permits many birds, when confronted with people, to behave leisurely and go about their business (albeit keeping one eye at all times on the strange-looking bipeds), allowing us extensive time to watch them. Not only are birds among the safest animals to observe and the most easily discovered and watched, but they are among the most beautiful. Experiences with Mexico's birds will almost certainly provide some of any trip's finest, most memorable naturalistic moments.

General Characteristics of Birds

Birds are vertebrates that can fly. They began evolving from reptiles during the Jurassic Period of the Mesozoic Era, perhaps 150 million years ago, and explosive development of new species occurred during the last 50 million years or so. The development of flight is the key factor behind birds' evolution, their historical spread throughout the globe, and their current ecological success and arguably dominant position among the world's land animals. Flight, as mentioned above, is a fantastic predator evasion technique; it permits birds to move over long distances in search of particular foods or habitats; and its development opened up for vertebrate exploration and exploitation an entirely new and vast theater of operations – the atmosphere.

At first glance, birds appear to be highly variable beasts, ranging in size and form from 135 kg (300 lb) ostriches to 4 kg (10 lb) eagles to 3 g (a tenth of an ounce) hummingbirds. Actually, however, when compared with other types of vertebrates, birds are remarkably standardized physically. The reason is that, whereas mammals or reptiles can be quite diverse in form and still function as mammals or reptiles (think how different in form are lizards, snakes, and turtles), if birds are going to fly, they more or less must look like birds, and have the forms and physiologies that birds have. The most important traits for flying are: (1) feathers, which are unique to birds; (2) powerful wings, which are modified upper limbs; (3) hollow bones; (4) warm-bloodedness; and (5) efficient respiratory and circulatory systems. These characteristics combine to produce animals with two overarching traits – high power and low weight, which are the twin dictates that make for successful feathered flying machines. (Bats, the flying mammals, also follow these dictates.)

Classification of Birds

Bird classification is one of those areas of science that continually undergoes revision. Currently about 9000 separate species are recognized. They are divided into 28 to 30 orders, depending on whose classification scheme one follows, perhaps 170 families, and about 2040 genera. For purposes here, we can divide birds into *passerines* and *non-passerines*. Passerine birds (Order Passeriformes) are the perching birds, with feet specialized to grasp and to perch on tree branches. They are mostly the small land birds (or *songbirds*) with which we are most familiar – blackbirds, robins, wrens, finches, sparrows, etc. – and the group includes more than 50% of all bird species. The remainder of the birds – seabirds and shorebirds, ducks and geese, hawks and owls, parrots and woodpeckers, and a host of others – are divided among the other 20+ orders.

Features of Tropical Birds

The first thing to know about tropical birds is that they are exceedingly varied and diverse. There are many more species of birds in the tropics than in temperate or arctic regions. For instance, fewer than 700 bird species occur in North America north of Mexico, but about 3300 species occur in the Neotropics (southern Mexico, Central and South America), most of those in the tropical regions. More than 1000 species occur in Mexico (769 breed there, and about 257 spend winters there or migrate through), most of them, again, in the nation's tropical southern reaches: 475 species alone occur over the Yucatán Peninsula. Many families of birds, such as the toucans, motmots, manakins, and cotingas, are endemic to the Neotropics – they occur nowhere else on Earth.

Many tropical birds rely for food on insects or seeds, but it is fruit-eating, or *frugivory*, that really distinguishes birds in the tropics. Frugivory has reached its zenith among tropical species, and the relationships between the birds that eat fruit and the plants that produce it exert powerful effects on the biology of both (see p. 182).

The mating systems of tropical birds range from typical, familiar *monogamy*, a male and a female pairing and cooperating to raise a brood of young, to *polygamy*, in which a single member of one sex mates with multiple members of the other, to the so-called *promiscuity* of manakins, some hummingbirds and others, in which males of the species gather in groups called *leks* to display and advertise for mates. Females attracted to the leks choose males to mate with and then depart to nest and raise their young themselves. The social systems of birds in the tropics are also quite variable. Many are *territorial*, aggressively defending parcels of real estate from other members of their species (*conspecifics*), typically during the breeding season, although some seem to exhibit year-round territoriality. Often it is a mated pair that keeps a territory, but in some species (some of the cotingas, jays, wrens, tanagers, and woodpeckers), small family groups stay together throughout the year, even engaging in *cooperative breeding* in which all members of the group assist with a single nest. A number of species, such as the oropendolas, stay in small colonial associations, and build their nests together in the same tree. Tropical birds often participate in *mixed-species foraging flocks*, spending non-breeding periods travelling around a large territory or semi-nomadically in the company of many other species – searching, for example, for trees bearing ripe fruit. Some of these flocks contain primarily *insectivorous* birds, others primarily *granivores* (seedeaters), and still others typically follow swarms of army ants, feasting on the insects and other small animals that bolt from cover at the approach of the predatory ants.

Breeding seasons in the tropics tend to be longer than in other regions. The weather is more conducive to breeding for longer periods. Also, unlike many temperate zone birds, those in the tropics need not hurry through breeding efforts because, being resident all year, they do not face migration deadlines. Breeding in the tropics is closely tied to wet and dry seasons. Many southern Mexico birds breed from March through July or August, timed to coincide with the greatest abundance of food for their offspring. Spring is usually when the months-long dry season ends, and showers bring heavy concentrations of insect life and ripening of fruit.

One notable aspect of bird breeding in the tropics that has long puzzled biologists is that clutches are usually small, most species typically laying two eggs per nest. Birds that breed in temperate zone areas usually have clutches of three to five eggs. Possible explanations are that: (1) small broods attract fewer predators; (2) because such a high percentage of nests in the tropics are destroyed by predators it is not worth putting too much energy and effort into any one nest; and (3) with the increased numbers of hours of daylight in northern areas, temperate zone birds have more time each day to gather food for larger numbers of growing nestlings.

Last, tropical birds include the most gorgeously attired birds, those with bright, flashy colors and vivid plumage patterns, with some of the Neotropic's parrots, toucans, trogons, and tanagers, among others, claiming top honors. Why so many tropical birds possess highly colored bodies remains an area of ornithological debate.

Seeing Birds in Tropical Mexico

Selected for illustration in the color plates are 250 species that are among the region's most frequently seen birds. The best way to spot these birds is to follow three easy steps:

1 Look for them at the correct time. Birds can be seen at any time of day, but they are often most active, and vocalize most frequently, during early morning and late afternoon, and so can be best detected and seen during these times.
2 Be quiet as you walk along trails or roads, and stop periodically to look around carefully. Not all birds are noisy, and some, even brightly colored ones, can be quite inconspicuous when they are directly above you, in a forest canopy. Trogons, for instance, beautiful medium-sized birds with green backs and bright red or yellow bellies, are notoriously difficult to see among branches and leaves.
3 *Bring binoculars* on your trip. You would be surprised at the number of people who visit tropical areas with the purpose of viewing wildlife and don't bother to bring binoculars. They need not be an expensive pair, but binoculars are essential to bird viewing.

A surprise to many people during their first trip to a tropical rainforest is that hordes of birds are not immediately seen or heard. During large portions of the day, in fact, the forest is mainly quiet, with few birds noticeably active. Birds are often present, but many are inconspicuous – small brownish birds near the ground, and greenish, brownish, or grayish birds in the canopy. A frequent, at first discombobulating experience is that you will be walking along a trail, seeing few birds, and then, suddenly, a mixed foraging flock with up to 20 or more species swooshes into view, filling the trees around you at all levels – some hopping along the ground, some moving through the brush, some clinging to tree trunks, others in the canopy – more birds than you can easily count or identify – and then, just as suddenly, the flock is gone, moved on in its meandering path through the forest (see p. 178).

Most visitors to southern Mexico spend at least a bit of time in coastal areas and these are good places to spot seabirds, birds-of-prey, and other kinds of birds as well. It would be a shame to leave the region without seeing at least some of its spectacular forest birds, such as parrots, toucans, trogons, and tanagers. If you have trouble locating such birds, ask people – tourguides, resort employees, reserve personnel – about good places to see them. One of the best times for birding in the region is March through June, when many species are breeding and so sometimes a bit more conspicuous; other good times are during migrations – early spring and fall, when, in addition to resident birds, migratory species that are moving through the area can be spotted.

Family Profiles

1. Seabirds

Along Mexico's coasts, as along coasts almost everywhere, *seabirds*, many of them conspicuously large and abundant, reign as the dominant vertebrate animals of the land, air, and water's surface. Many seabirds in southern Mexico commonly seen by visitors from northern temperate areas are very similar to species found back home, but some are members of groups restricted to the tropics and subtropics and, hence, should be of interest to the ecotraveller. A few of these birds will be seen by almost everyone. As a group seabirds are incredibly successful animals, present often at breeding and roosting colonies in enormous numbers. Their success surely is owing to their incredibly rich food resources – the fish and invertebrate animals (crabs, mollusks, insects, jellyfish) produced in the sea and on beaches and mudflats. Furthermore, people's exploitation of marine and coastal areas has, in many cases, enhanced rather than hurt seabird populations. Many gull species, for instance, which make good use of human-altered landscapes, such as garbage dumps and agricultural fields, and human activities such as fishing, are almost certainly more numerous today than at any time in the past.

Three birds treated here (Plate 20) are members of the Order Pelecaniformes: the BROWN BOOBY of the Family Sulidae (nine species worldwide, with three occurring in southern Mexico), MAGNIFICENT FRIGATEBIRD of the Family Fregatidae (five species with mainly tropical distributions; only one occurs off southern Mexico), and BROWN PELICAN, Family Pelecanidae (eight species worldwide, with two occurring in the covered region). Two more species, the LAUGHING GULL and ROYAL TERN (Plate 20), are in Family Laridae (more than 80 species worldwide, with relatively few occurring in tropical climes), which is allied with the shorebirds, and is part of another order. *Boobies*, of which the Brown Booby is one of the most common, are large seabirds known for their sprawling, densely packed breeding colonies, spots of bright body coloring, and for plunging into the ocean from heights to pursue fish. They have tapered bodies, long pointed wings, long tails, long pointed bills, and often, brightly colored feet. The Magnificent Frigatebird is a very large soaring bird, mostly black, with huge, pointed wings that span up to 2 m (6 ft) or more, and a long, forked tail. Males have red throat pouches that they inflate, balloonlike, during courtship displays. Brown Pelicans are large heavy-bodied seabirds, and, owing to their big, saggy throat pouches, are perhaps among the most recognizable of birds. They have long wings, long necks, large heads, and long bills from which hang the flexible, fish-catching pouches. Laughing Gulls are mid-sized seabirds with long, narrow wings and heavy bills; they are largely white and grayish, but adults' heads are black, or *hooded*, during the breeding season. Similarly, ROYAL TERNS are mid-sized white and gray seabirds. They are also hooded during breeding, and they have long orange bills.

Natural History

Ecology and Behavior

Most seabirds feed mainly on fish, and have developed a variety of ways to catch them. Boobies, which also eat squid, plunge-dive from the air or surface dive to catch fish underwater. Sometimes they dive quite deeply, and they often take fish unawares from below, as they rise toward the surface. Frigatebirds feed on the wing, sometimes soaring effortlessly for hours at a time. They swoop low to catch flying fish that leap from the water (the fish leap when they are pursued by larger, predatory fish or dolphins), and also to pluck squid and jellyfish from the wave-tops. Although their lives are tied to the sea, frigatebirds cannot swim and rarely, if ever, enter the water voluntarily; with their very long, narrow wings, they have difficulty lifting off from the water. To rest, they land on remote islands, itself a problematic act in high winds. *Pelicans* eat fish almost exclusively. BROWN PELI-CANS, in addition to feeding as they swim along the water's surface, are the only pelicans that also plunge from the air, sometimes from considerable altitude, to dive for meals. While underwater, the throat sac is used as a net to scoop up fish (to 30 cm, 1 ft, long). Captured fish are quickly swallowed because the water in the sac with the fish usually weighs enough to prohibit the bird's lifting off again from the water. Pelicans, ungainly looking, nonetheless are excellent flyers, and can use air updrafts to soar high above in circles for hours. These are large, hand-some birds; a flight of them, passing low and slow overhead on a beach, in per-fect V-formation or in a single line, is a tremendous sight. Of special note is that adult pelicans, so far as it is known, are largely silent. *Gulls* and *terns* are highly gregarious seabirds – they feed, roost, and breed in groups. They feed on fish and other sealife that they snatch from shallow water, and on crabs and other inver-tebrates they find on mudflats and beaches. Also, they are not above visiting garbage dumps or following fishing boats to grab whatever goodies that fall or are thrown overboard.

Breeding

Seabirds usually breed in large colonies on small islands (where there are no mam-mal predators) or in isolated mainland areas that are relatively free of predators. Some breed on slopes, cliffs, or ledges (boobies), some in trees or on tops of shrubs (pelicans, frigatebirds), and some on the bare ground (gulls and terns, and also, if trees are unavailable, pelicans and frigatebirds). Most species are monogamous, mated males and females sharing in nest-building, incubation, and feeding young. In some groups, such as the pelicans, the male gathers sticks and stones for a nest, but the female carries out the actual construction (perhaps the male supervises). High year-to-year fidelity to mates, to breeding islands, and to partic-ular nest sites is common. BROWN BOOBY females lay one or two eggs, which are incubated for about 45 days. Usually only a single chick survives to fledging age (one chick often pecks the other to death). MAGNIFICENT FRIGATEBIRDS lay a single egg that is incubated for about 55 days; male and female spell each other during incubation, taking shifts of up to 12 days. Young remain in and around the nest, dependent on the parents, for up to 6 months. BROWN PELICANS lay two or three eggs, which are incubated for 30 to 37 days; usually only one young is raised successfully. Gulls and terns typically lay one to three eggs, and both sexes incubate for 21 to 35 days; young fledge after 28 to 35 days at the nest. In most seabirds, young are fed when they push their bills into their parents' throats, in effect forcing the parent to regurgitate food stored in its *crop* – an enlargement of the top portion of the esophagus. Seabirds reach sexual maturity slowly (in 2 to 5 years; 7+ years in frigatebirds) and live long lives (pelicans in the wild probably average 15 to 25 years, and some live 50+ years in zoos; frigatebirds and boobies live 20+ years in the wild).

Ecological Interactions
Frigatebirds, large and beautiful, are a treat to watch as they glide silently along coastal areas, but they have some highly questionable habits – in fact, patterns of behavior that among humans would be indictable offenses. Frigatebirds practice *kleptoparasitism*: they "parasitize" other seabirds, such as boobies, frequently chasing them in the air until they drop recently caught fish. The frigatebird then steals the fish, catching it in mid-air as it falls. Frigatebirds are also common predators on baby sea turtles (p. 99), scooping them from beaches as the reptiles make their post-hatching dashes to the ocean.

Lore and Notes
Boobies are sometimes called *gannets*, particularly by Europeans. The term *booby* apparently arose because the nesting and roosting birds seemed so bold and fearless toward people, which was considered stupid. Actually, the fact that these birds bred on isolated islands and cliffs meant that they had few natural predators, so had never developed, or had lost, fear responses to large mammals, such as people. Frigatebirds are also known as *man-of-war* birds, both names referring to warships, and to the birds' kleptoparasitism; they also steal nesting materials from other birds, furthering the image of avian pirates.

Status
Most of the seabirds that occur along southern Mexico's shores, including boobies, frigatebirds, gulls, and terns, are quite abundant and not considered threatened. The BROWN PELICAN, although common in some areas, is considered an endangered species (USA ESA listed) over parts of its range. Likewise, the LEAST TERN, closely related to the ROYAL TERN but about half its size, is common in some coastal regions of southern Mexico, but is considered endangered (USA ESA listed) over parts of its range, particularly in California (USA). Some seabirds in other parts of the world are highly endangered. For example, within the pelican family, both the DALMATIAN PELICAN of Eurasia (CITES Appendix I listed) and the SPOT-BILLED PELICAN of India and Sri Lanka have only a few thousand breeding pairs remaining. Also, ABBOT'S BOOBY (endangered, CITES Appendix I and USA ESA) is now limited to a single, small breeding population on the Indian Ocean's Christmas Island.

Profiles
Brown Booby, *Sula leucogaster*, Plate 20a
Magnificent Frigatebird, *Fregata magnificens*, Plate 20b
Laughing Gull, *Larus atricilla*, Plate 20c
Royal Tern, *Sterna maxima*, Plate 20d
Brown Pelican, *Pelecanus occidentalis*, Plate 20e

2. Waterbirds

Included under the vague heading of *waterbirds* is a group of medium to large birds, all with distinctive, identifying bills, that make their livings in freshwater and coastal saltwater habitats. Visitors to marsh or river areas in southern Mexico are highly likely to encounter some of these species. Some, such as *cormorants* and *anhingas*, swim to catch fish, whereas others, *storks* and *ibises*, wade about in marshes, shallow water areas, and fields in search of a variety of foods. Cormorants (Family Phalacrocoracidae) inhabit coasts and inland waterways over much of the world; there are 28 species, but only two occur in southern Mexico. Anhingas (Family Anhingidae), closely related to cormorants, are fresh- and brackish-water birds mostly of tropical and subtropical regions; there are four species, one of which occurs in the covered region. Storks (Family Ciconiidae) are wading birds that occur worldwide in tropical and temperate regions; there are 17 species, but only three occur in the New World, two of them in southern Mexico. The 33 species of ibises and *spoonbills* (Family Threskiornthidae), also wading birds, are globally distributed; two ibises and a single spoonbill species occur in southern Mexico. The five species of *flamingos* (Family Phoenicopteridae) are large pink wading birds with a broad but highly specialized distribution: they occur in parts of Mexico, South America, and the Old World, but only in shallow lagoons and lakes with very high salt concentrations. One species occurs in the Yucatán region.

Cormorants are medium-sized birds, usually black, with short legs, long tails, and longish bills with hooked tips. ANHINGAS (Plate 21) are similar to cormorants, blackish with long tails, but they have very long, thin necks and their bills are longer and end with sharp points. Storks are huge, ungainly-looking – and so unmistakable – wading birds. Those in Mexico are white with black or black-and-red heads; the head and neck area are featherless. Storks have very large, heavy bills. The JABIRU (Plate 21) is one of the largest of the world's storks, up to 1.5 m (5 ft) tall. Ibises and spoonbills resemble storks, but are smaller and have shorter necks. Most ibises are white, brown, or blackish, and in many, the head is bare of feathers. Ibis bills are long, thin, and curved downwards. Spoonbills are named for their straight, flat, spoon-shaped bills. The ROSEATE SPOONBILL (Plate 22), large, pink, and spoon-billed, must be one of the easiest birds on Earth to identify. The AMERICAN FLAMINGO (Plate 22), also a shocking pink, has a long neck, long legs, and a heavy, conspicuously down-curved bill. In most of these waterbirds, the sexes are similar in plumage, but males are often a bit larger.

Natural History

Ecology and Behaviour

Diving from the surface of lakes, rivers, lagoons, and coastal saltwater areas, OLIVACEOUS CORMORANTS (Plate 21) and ANHINGAS pursue fish underwater. Cormorants, which take crustaceans also, catch food in their bills; Anhingas, which also take young turtles, baby crocodiles, and snakes, use their sharply pointed bills to spear fish. Cormorants are social birds, foraging, roosting, and nesting in groups, but Anhingas are somewhat territorial, defending resting and feeding areas from other birds. Both Anhingas and cormorants are known for standing on logs, trees, or other surfaces after diving and spreading their wings, presumably to dry them (they may also be warming their bodies in the sun following dives into cold water.) Storks feed by walking slowly through fields and marshy areas, looking for suitable prey, essentially anything that moves: small rodents, young birds, frogs, reptiles, fish, earthworms, mollusks, crustaceans, and insects. Food is grabbed with the tip of the bill and swallowed quickly. Storks can be found either alone or, if food is plentiful in an area, in groups. These birds are excellent flyers, often soaring high overhead for hours during hot afternoons. They are known to fly 80 km (50 miles) or more daily between roosting or nesting sites and feeding areas.

Ibises are gregarious birds that insert their long bills into soft mud of marshes and shore areas and poke about for food – insects, snails, crabs, frogs, tadpoles. Apparently they feed by touch, not vision; whatever the bill contacts that feels like food is grabbed and swallowed. Spoonbills, likewise, are gregarious birds that feed in marsh or shallow-water habitats. They lower their bills into the water and sweep them around, stirring up the mud, then grab fish, frogs, snails, or crustaceans that touch their bills. Spoonbills are soaring birds, but ibises, although good flyers, do not soar. A flamingo eats by lowering its bill into the water, resting it upside down on the bottom, and sucking in water, mud and bottom debris. The materials are pushed through comb-like bill filters, which catch the flamingo's meal: tiny invertebrate animals such as brine shrimp. Flamingos are highly social, occurring sometimes, where they have healthy populations, in groups of thousands – which, given their coloring, makes for quite a sight.

Breeding

Most of these waterbirds breed in colonies of various sizes, although ANHINGAS sometimes breed alone, and JABIRUS usually do so. Mating is monogamous, with both male and female contributing to nest-building, incubation, and feeding offspring. Cormorants, which begin breeding when they are 3 or 4 years old, construct stick nests in trees or on ledges. Two to four eggs are incubated for about 4 weeks, and young fledge 5 to 8 weeks after hatching. Anhingas breed at 2 or 3 years of age. They have stick and leaf nests in trees or bushes. Three to five eggs are incubated for 4 to 5 weeks, and chicks fledge 5 weeks after hatching. Storks first breed when they are 3 to 5 years old. They build platforms of sticks in trees or on ledges. Sometimes they add new material each year, which results eventually in enormous nests. Two to four eggs are incubated for 4 to 5 weeks and chicks remain in the nest for 50 to 90 days after hatching. Ibises and spoonbills also make stick nests, mixed with green vegetation, in trees. Two to four eggs are incubated for about 3 weeks, and young fledge 6 to 7 weeks after hatching. Flamingos nest in colonies. Each pair builds a mud-and-stone mound nest on top of which the single egg is placed. Both male and female incubate the egg, for about 4 weeks, and both feed the chick, also for about 4 weeks, with a milky fluid produced by glands in their digestive tracts.

Lore and Notes

AMERICAN FLAMINGOS occur along the Yucatán coast, breeding in late spring and summer primarily in and near the Ría Lagartos Special Biosphere Reserve, and wintering primarily to the west, particularly in and near the Ría Celestún Special Biosphere Reserve. These birds have long attracted notice – flamingos, in all probability from this population, were recorded as being prominent members of the Aztec Emperor Montezuma's menagerie. Although there are still many thousands of these birds, conservation organizations take special interest in them because their highly specialized habits make them vulnerable to drastic population declines and their numbers have fluctuated widely in the recent past. Their feeding method, dredging the bottom sediment from shallow lagoons and estuaries, makes them vulnerable to toxic chemicals in the mud, such as lead from the lead-shot used in shotguns. They feed only in areas with very special water conditions, which can change abruptly, as when, in 1988, Hurricane Gilbert apparently destroyed some of the flamingos' prime feeding areas, causing the birds to seek other sites and leading to many deaths. Also, their nests, on mudflats, are easy targets for egg and chick predators such as foxes and raccoons. Still, the Mexican flamingo population is considered fairly healthy, having risen from a low of between 8000 and 12,000 individuals during the 1970s (when the Ría Lagartos and Ría Celestún reserves were created, partially to protect the flamingos) to about 26,000 during the mid-1980s; Hurricane Gilbert was a setback to this population growth, but the flamingos have since recovered.

ANHINGAS are also known as *darters*, the name owing to the way the birds swiftly thrust their necks forward to spear fish. Because of their long necks, they are also called *snakebirds*. Cormorants were so common and conspicuous to early European mariners that they were given the name *Corvus marinus*, or "sea raven." (The word cormorant derives from the term corvus marinus.) Cormorants have been used for centuries in Japan, China, and Central Europe as fishing birds. A ring is placed around a cormorant's neck so that it cannot swallow its catch, and then, leashed or free, it is released into the water. When the bird returns or is reeled in, a fish is usually clenched in its bill. An ancient proverb is that after death, thieves return as ringed cormorants, condemned to fish forever for their masters.

Anhingas and cormorants are such successful fisherbirds that, believe it or not, sport fisherman and fish farmers in the USA have been complaining to the US Congress that the birds get more fish than they do, and they demand the legislators do something about it. They want hunting seasons established for Anhingas and the Double-crested Cormorant, probably the most numerous cormorant of North America.

Status

Most of the waterbirds considered here are widely or locally common in southern Mexico. JABIRUS, however, are fairly rare in the region, seen regularly probably only in coastal Quintana Roo and marshes of the Umacinta River in Tabasco and Chiapas. The species is considered threatened in Central America (CITES Appendix I listed even though not considered highly endangered); its likeness serves in several countries (notably in Belize and Costa Rica) as a conservation symbol. WOOD STORKS, although still somewhat common in the region, are threatened in other parts of their range, and are listed as endangered by USA ESA (there are small populations in southeast USA). Storks are declining because they are hunted for food over parts of their ranges (for instance, Jabiru are considered game birds in the Amazon Basin) and because nesting sites are disturbed by people. ROSEATE SPOONBILLS were almost eliminated in the southeastern USA in the late 19th and early 20th centuries when they were hunted to turn their pink wings into feather fans. Two species of New World flamingos, ANDEAN and JAMES FLAMINGOS, are considered very vulnerable to threat (CITES Appendix II listed, as are all flamingos): they occur only in high altitude lakes in Peru, Bolivia, Chile, and Argentina, and their numbers are declining.

Profiles

Olivaceous Cormorant, *Phalacrocorax brasilianus*, Plate 21a
Anhinga, *Anhinga anhinga*, Plate 21b
White Ibis, *Eudocimus albus*, Plate 21c
Jabiru, *Jabiru mycteria*, Plate 21d
Wood Stork, *Mycteria americana*, Plate 21e
Roseate Spoonbill, *Ajaia ajaja*, Plate 22a
American Flamingo, *Phoenicopterus ruber*, Plate 22b

3. Herons and Egrets

Herons and *egrets* are beautiful medium to large-sized wading birds that enjoy broad distributions throughout temperate and tropical regions of both hemispheres. Herons and egrets, together with the similar but quite elusive wading birds called *bitterns*, constitute the heron family, Ardeidae, which includes about 58 species. Sixteen species occur in southern Mexico, most of which also breed there. Herons frequent all sorts of aquatic habitats: along rivers and streams, in marshes and swamps, and along lake and ocean shorelines. They are, in general, highly successful birds, and some of them are among the region's most conspicuous and commonly seen waterbirds. Why some in the family are called herons and some egrets, well, it's a mystery; but egrets are usually all white and tend to have longer *nuptial plumes* – special, long feathers – than the darker-colored herons.

Most herons and egrets are easy to identify. They are the tallish birds standing upright and still in shallow water or along the shore, staring intently into the water. They have slender bodies, long necks (often coiled when perched or still, producing a short-necked, hunched appearance), long, pointed bills, and long legs with long toes. In Mexico herons range in height from 0.3 to 1.3 m (1 to 4 ft). Most are attired in soft shades of gray, brown, blue, or green, and black and white. From afar most are not striking, but close-up, many are exquisitely marked with small colored patches of facial skin or broad areas of spots or streaks; the *tiger herons*, in particular, have strongly barred or streaked plumages. Some species during breeding seasons have a few very long feathers (nuptial plumes) trailing down their bodies from the head, neck, back, or chest. The sexes are generally alike in size and plumage, or nearly so.

Natural History

Ecology and Behaviour

Herons and egrets walk about slowly and stealthily in shallow water and sometimes on land, searching for their prey, mostly small vertebrates, including fish, frogs, salamanders, and the occasional turtle, and small invertebrates like crabs. On land, they take mostly insects, but also other invertebrates and vertebrates such as small rodents. CATTLE EGRETS (Plate 23) have made a specialty of following grazing cattle and other large mammals, walking along and grabbing insects and small vertebrates that are flushed from their hiding places by the moving cattle. A typical pasture scene is a flock of these egrets intermixed with a cattle herd, with several of the white birds perched atop the unconcerned mammals. Many herons spend most of their foraging time as *sit-and-wait* predators, standing motionless in or adjacent to the water, waiting in ambush for unsuspecting prey to wander within striking distance. Then, in a flash, they shoot their long, pointed bills into the water to grab or spear the prey. They take anything edible that will fit into their mouths and down their throats, and then some. One particular heron that I recall grabbed a huge frog in its bill and spent the better part of half an hour trying to swallow it. Typically, the larger herons are easier to spot because they tend to stay out in the open while foraging and resting; smaller herons, easier prey for predators, tend to stay more hidden in dense vegetation in marshy areas. Most herons are day-active, but many of the subgroup known as *night herons* forage at least partly nocturnally. Most herons are social birds, roosting and breeding in colonies, but some, such as the tiger herons, are predominantly solitary.

Breeding

Many herons breed in monogamous pairs within breeding colonies of various sizes. A few species are solitary nesters and some are less monogamous than others. Herons are known for their elaborate courtship displays and ceremonies, which continue through pair formation and nest-building. Generally nests are constructed by the female of a pair out of sticks procured and presented to her by the male. Nests are placed in trees or reeds, or on the ground. Both sexes incubate the three to seven eggs for 16 to 30 days, and both feed the kids for 35 to 50 days before the young can leave the nest and feed themselves. The young are *altricial* – born helpless; they are raised on regurgitated food from the parents. Many herons breed during wet seasons, but some do so all year.

Ecological Interactions

Herons and egrets often lay more eggs than the number of chicks they can feed. For instance, many lay three eggs when there is sufficient food around to feed only two chicks. This is contrary to our usual view of nature, which we regard as having adjusted animal behavior through evolution so that behaviors are finely tuned to avoid waste. Here's what biologists suspect goes on: females lay eggs 1 or 2 days apart, and start incubating before they finish laying all their eggs. The result is that chicks hatch at intervals of one or more days and so the chicks in a single nest are different ages, and so different sizes. In years of food shortage, the smallest chick dies because it cannot compete for food from the parents against its larger siblings, and also because, it has been discovered, the larger siblings attack it (behavior called *siblicide*). The habit of laying more eggs than can be reared as chicks may be an insurance game evolved by the birds to maximize their number of young; in many years, true, they waste the energy they invested to produce third eggs that have little future, but if food is plentiful, all three chicks survive and prosper. Apparently, the chance to produce three surviving offspring is worth the risk of investing in three eggs even though the future of one is very uncertain.

The CATTLE EGRET is a common, successful, medium-sized white heron that, until recently, was confined to the Old World, where it made its living following herds of large mammals. What is so interesting about this species is that, whereas many of the animals that have recently crossed oceans and spread rapidly into new continents have done so as a result of people's intentional or unintentional machinations, these egrets did it themselves. Apparently the first ones to reach the New World were from Africa. Perhaps blown off-course by a storm, they first landed in northern South America in about 1877. Finding the New World to its liking, during the next decades the species spread far and wide, finding abundant food where tropical forests were cleared for cattle grazing. Cattle Egrets have now colonized much of northern South America, Central America, all the major Caribbean islands, and eastern and central North America, as far as the southern USA. We must assume that they have Chicago and New York City in their sights.

Lore and Notes

The story of the *phoenix*, a bird that dies or is burned but then rises again from the ashes, is one of the best-known bird myths of the Western world. One version, from about 2800 years ago, has it that one phoenix arrives from Arabia every 500 years. When it is old, it builds a nest of spices in which to die. From the remains a young phoenix emerges, which carries its parent's bones to the sun. Some authorities believe that the phoenix was a heron; in fact, the Egyptian hieroglyph for the phoenix appears to be a heron or egret.

Status

Some of Mexico's herons and egrets are fairly rare, but they are not considered threatened species because they are more abundant in other parts of their ranges, outside the country. The secretive AGAMI (or CHESTNUT-BELLIED) HERON, mid-sized and colorful, with a distribution from southern Mexico to northern South America, is considered near-threatened by some authorities. Elsewhere in the Neotropics, the FASCIATED TIGER-HERON, which ranges from Costa Rica to Argentina and Brazil, is also considered near-threatened.

Profiles

Reddish Egret, *Egretta rufescens*, Plate 22c
Little Blue Heron, *Egretta caerulea*, Plate 22d
Tricolored Heron, *Egretta tricolor*, Plate 22e
Snowy Egret, *Egretta thula*, Plate 23a
Cattle Egret, *Bubulcus ibis*, Plate 23b
Great Egret, *Ardea alba*, Plate 23c
Green Heron, *Butorides striatus*, Plate 23d
Bare-throated Tiger Heron, *Tigrisoma mexicanum*, Plate 23e
Yellow-crowned Night-heron, *Nycticorax violaceus*, Plate 24a

4. Marsh and Stream Birds

Marsh and *stream birds* are small and medium-sized birds adapted to walk, feed, and breed in swamps, marshes, wet fields, and along streams. The chief characteristics permitting this lifestyle usually are long legs and very long toes that distribute the birds' weight, allowing them to walk among marsh plants and across floating vegetation without sinking. *Jacanas* (zha-SAH-nahs or hah-SAH-nahs; Family Jacanidae) are small and medium-sized birds with amazingly long toes and toenails that stalk about tropical marshes throughout the world. There are eight species; only one occurs in southern Mexico, but it is found over most of the region wherever there is floating aquatic vegetation. The *rails* (Family Rallidae) are a large group of often secretive small and medium-sized swamp and dense vegetation birds, about 130 species strong. They inhabit most parts of the world save for polar regions. Twelve species occur in southern Mexico, including rails, wood-rails, crakes, coots, and gallinules.

The NORTHERN JACANA (Plate 24), with incredibly long toes, has a drab brown and black body but a bright yellow bill and forehead. When their wings are spread during flight or displays, jacanas expose large patches of bright yellow, rendering them instantly conspicuous. Female jacanas are larger than males. GRAY-NECKED WOOD-RAILS (Plate 24) are brown, olive, and gray, with long necks, legs, and toes. Although most rails, like the wood-rail, are colored to blend into their surroundings, the PURPLE GALLINULE (Plate 24) is a strikingly colored bird, purple and green, with a reddish bill. Rails generally have short wings and tails. Males are often larger than females. RUDDY CRAKES (Plate 24) are small reddish-brown birds with gray heads.

Natural History

Ecology and Behavior

Jacanas are abundant birds of marshes, ponds, lakeshores, and wet fields. They walk along, often on top of lily pads and other floating plants, picking up insects, snails, small frogs and fish, and some vegetable matter such as seeds. Likewise, rails stalk through marshes, swamps, grassy shores, and wet grasslands, foraging for insects, small fish and frogs, bird eggs and chicks, and berries. Typically they move with a head-bobbing walk. Many rails are highly secretive, being heard but rarely seen moving about in marshes. PURPLE GALLINULES, brightly colored and less shy than many other rails, are usually easy to see as they forage singly or in small family groups. RUDDY CRAKES are very common birds found in marshes and other freshwater habitats, including roadside ditches and rain puddles. They eat mainly seeds, aquatic vegetation, and insects.

Breeding

NORTHERN JACANAS are one of three or four jacana species that employ *polyandrous* breeding, the rarest type of mating system among birds. In a breeding season, a female mates with several males, and the males then carry out most of the breeding chores. Males each defend small territories from other males, but each female has a larger territory that encompasses two to four male territories. Males build nests of floating, compacted aquatic vegetation. Following mating, the female lays three or four eggs in the nest, after which the male incubates them for 21 to 24 days and then cares for (leads and protects) the chicks. Young are dependent on the father for up to 3 to 4 months. Meanwhile, the female has mated with other males on her territory and provided each with a clutch of eggs to attend. (Predation on jacana nests is very common, PURPLE GALLINULES often being the culprits.) In rails such as wood-rails and gallinules, the sexes contribute more equally to the breeding effort. Male and female build well-hidden nests in dense vegetation, sometimes a meter or two (several feet) off the ground, or floating nests of aquatic plants. Both sexes incubate the three to five or more eggs for 2.5 to 4 weeks and care for the young for 7 or 8 weeks until they are independent. RUDDY CRAKES build round grass nests; females lay three or four eggs per clutch.

Lore and Notes

Jacanas of Africa and Australia are also known as *lily-trotters* and *lotus-birds*. The word *jacana* is from a native Brazilian name for the bird.

Status

Although some of the region's rails are locally rare, most are not considered threatened because they are more numerous in other parts of their ranges. One species, the CLAPPER RAIL, which occurs in coastal salt marshes of the southern USA, northern Mexico, and the Yucatán, is threatened in some areas (USA ESA listed). All four species profiled here are locally common or abundant. Throughout the world, a number of rails, like the GUAM RAIL, are threatened or already endangered; several New World threatened species are Venezuela's PLAIN-FLANKED RAIL, Colombia's BOGOTA RAIL, Peru's JUNIN RAIL, and Cuba's ZAPATA RAIL.

Profiles

Ruddy Crake, *Laterallus ruber*, Plate 24b
Gray-necked Wood-rail, *Aramides cajanea*, Plate 24c
Purple Gallinule, *Porphyrula martinica*, Plate 24d
Northern Jacana, *Jacana spinosa*, Plate 24e

5. Ducks

Members of the Family Anatidae are universally recognized as *ducks*. They are water-associated birds that are distributed throughout the world in habitats ranging from open seas to high mountain lakes. The family includes about 150 species of ducks, geese, and swans. Although an abundant, diverse group throughout temperate regions of the globe, ducks, or *waterfowl* (*wildfowl* to the British), have only limited representation in most tropical areas. Forty species occur in Mexico, about half of them in the region covered here, and only a fraction of those are local breeders; the remainder are migratory, only passing through or wintering in the area. Only three species (Plate 25) are profiled here, the MUSCOVY DUCK, a bird of streams and ponds in lowland forests; the BLACK-BELLIED WHISTLING DUCK, which frequents freshwater marshes, ponds, and lagoons; and the BLUE-WINGED TEAL, a common northern migrant that winters in the area.

Ducks vary quite a bit in size and coloring, but all share the same major traits: duck bills, webbed toes, short tails, and long, slim necks. Plumage color and patterning vary, but there is a preponderance within the group of grays and browns, and black and white, although many species have at least small patches of bright color. In some species male and female look alike, but in others there are many differences between the sexes. The MUSCOVY DUCK is large, chunky in appearance, mostly black with white patches on its wings, and with bare patches of skin on its face; males are much larger than females. BLACK-BELLIED WHISTLING DUCKS are slender, medium-sized ducks, rust-colored with black bellies, and red bills and feet. BLUE-WINGED TEAL are undistinguished-looking small brown ducks with blue and green wing patches.

Natural History

Ecology and Behavior

Ducks are birds of wetlands, spending most of their time in or near the water. Many of the typical ducks are divided into *divers* and *dabblers*. Diving ducks plunge underwater for their food; dabblers, such as mallards, pintail, widgeon, and teal take food from the surface of the water or, maximally, put their heads down into the water to reach food at shallow depths. Ducks mostly eat aquatic plants or small fish, but some forage on land for seeds and other plant materials. The MUSCOVY DUCK, for instance, in addition to eating small fish and seeds of aquatic plants, feeds on terrestrial foods including crops such as corn, and insects. BLACK-BELLIED WHISTLING DUCKS eat seeds, leaves, and small invertebrates such as insects. This duck spends relatively little time in the water; it grazes on land, usually at night. During the day it roosts on riverbanks, or in trees adjacent to the water. BLUE-WINGED TEAL are typical dabbling ducks, rarely diving for meals. They occur in coastal areas, lakes, and wetlands, sometimes in very large numbers, and are particularly abundant in winter over the northern section of the Yucatán Peninsula.

Breeding

Ducks place their nests on the ground in thick vegetation or in holes. Typically nests are lined with downy feathers that the female plucks from her own breast. In many of the ducks, females perform most of the breeding duties, including incubation of the 2 to 16 eggs and shepherding and protecting the ducklings. Some of these birds, however, particularly among the geese and swans, have life-long marriages during which male and female share equally in breeding duties.

The young are *precocial*, able to run, swim and feed themselves soon after they hatch. BLUE-WINGED TEAL nest in thick vegetation, but both MUSCOVY and BLACK-BELLIED WHISTLING DUCKS most often nest in cavities. Females, after mating, incubate eggs and raise ducklings themselves.

Lore and Notes

Ducks, geese and swans have been objects of people's attention since ancient times, sometimes as cultural symbol (for instance, as a Chinese symbol of happiness), but chiefly as a food source. These birds typically have tasty flesh, are fairly large and so economical to hunt, and usually easier and less dangerous to catch than many other animals, particularly large mammals. Owing to their frequent use as food, several wild ducks and geese have been domesticated for thousands of years; the MUSCOVY DUCK, in fact, native to Mexico, in its domesticated form is now a common farmyard inhabitant in several parts of the world. Wild ducks also adjust well to the proximity of people, to the point of taking food from them – a practice that surviving artworks show has been occurring for at least 2000 years. Hunting ducks and geese for sport is also a long-practiced tradition. As a consequence of these long interactions between ducks and people, and the research on these animals stimulated by their use in agriculture and sport, a large amount of scientific information has been collected on the group; many of the ducks and geese are among the most well known of birds. The close association between ducks and people has even led to a long contractual agreement between certain individual ducks and the Walt Disney Company.

Status

None of the Mexican ducks, neither the local breeders nor winter migrants, are currently threatened or endangered. A few species are now fairly rare, but they are more abundant in other parts of their ranges. The MUSCOVY DUCK, because of hunting pressures and habitat destruction, has had its populations much reduced throughout its broad range in the Neotropics, so that it is now common only in restricted areas. The only critically endangered duck in the New World may be the BRAZILIAN MERGANSER, which occurs in Argentina, Brazil, and Paraguay.

Profiles

Black-bellied Whistling Duck, *Dendrocygna autumnalis*, Plate 25a
Muscovy Duck, *Cairina moschata*, Plate 25b
Blue-winged Teal, *Anas discors*, Plate 25c

6. Shorebirds

Spotting *shorebirds* is usually a priority only for visitors to southern Mexico who are rabid birdwatchers. The reason for the usual lack of interest is that shorebirds are often very common, plain-looking brown birds that most people are familiar with from their beaches back home. Still, it is always a treat watching shorebirds in their tropical wintering areas as they forage in meadows, along streams, on mudflats, and especially on the coasts, as they run along beaches, parallel to the surf, picking up food. Some of the small ones, such as SANDERLINGS (Plate 26), as one biologist wrote, resemble amusing wind-up toys as they spend hours running up and down the beach, chasing, and then being chased by, the outgoing and incoming surf (J. Strauch 1983). Shorebirds are often conspicuous and let themselves be watched, as long as the watchers maintain some distance. When in large flying groups, shorebirds such as *sandpipers* provide some of the most compelling sights in bird-dom, as their flocks rise from sandbar or mudflat to fly fast and low over the surf, wheeling quickly and tightly in the air as if they were a single organism, or as if each individual's nervous system were joined to the others'.

Shorebirds are traditionally placed along with the gulls in the avian order Charadriiformes. They are global in distribution and considered to be hugely successful birds – the primary reason being that the sandy beaches and mudflats on which they forage usually teem with their food. There are several families, four of which require mention. The sandpipers, Family Scolopacidae, are a world-wide group of approximately 85 species. About 35 species occur in Mexico, some being quite abundant during much of the year, yet they are nearly all migrants – one breeds in northern Mexico. The Sanderling, RUDDY TURNSTONE, and LESSER YELLOWLEGS (Plate 26) are all sandpipers. *Plovers* (Family Charadriidae), with about 60 species, likewise have a world-wide distribution. Eight species occur in the covered region, three of which breed there; the others are migrants from breeding sites to the north. The broadly distributed Family Recurvirostridae consists of about seven species of *stilts* and *avocets*, two of which occur in Mexico. Last, the Family Aramidae contains a single species, the LIMPKIN (Plate 25), which occurs from the southeastern USA to Argentina.

All shorebirds, regardless of size, have a characteristic "look." They are usually drably colored birds (especially during the non-breeding months), darker above, lighter below, with long, thin legs for wading through wet meadows, mud, sand, or surf. Depending on feeding habits, bill length varies from short to very long. Most of the Mexican sandpipers range from 15 to 48 cm (6 to 19 in) long. They are generally slender birds with straight or curved bills of various lengths. Plovers, 15 to 30 cm (6 to 12 in) long, are small to medium-sized, thick-necked shorebirds with short tails and straight, relatively thick bills. They are mostly shades of gray and brown but some have bold color patterns such as a broad white or dark band on the head or chest. The BLACK-NECKED STILT (Plate 25) is a striking, mid-sized black and white bird with very long red legs and long, fine bill. Limpkins are largish brown birds with white spots and long, thin bills. The sexes look alike, or nearly so, in most of the shorebirds.

Natural History

Ecology and Behavior

Shorebirds typically are open-country birds, associated with coastlines and inland wetlands, grasslands, and pastures. Sandpipers and plovers are excellent flyers but they spend a lot of time on the ground, foraging and resting; when chased, they often seem to prefer running to flying away. Sandpipers pick their food up off the ground or use their bills to probe for it in mud or sand – they take insects and other small invertebrates, particularly crustaceans. They will also snatch bugs from the air as they walk and from the water's surface as they wade or swim. Larger, more land-dwelling shorebirds may also eat small reptiles and amphibians, and even small rodents; some of the plovers also eat seeds. Usually in small groups, BLACK-NECKED STILTS wade about in shallow fresh and salt water, using their bills to probe the mud for small insects, snails, and crustaceans. LIMPKINS inhabit marshes, ponds, swamps, and riversides; they eat mainly snails.

Many shorebirds, especially among the sandpipers, establish winter *feeding territories* along stretches of beach; they use the area for feeding for a few hours or for the day, defending it aggressively from other members of their species. Many of the sandpipers and plovers are gregarious birds, often seen in large groups, especially when they are travelling. Several species make long migrations over large expanses of open ocean, a good example being the AMERICAN GOLDEN-PLOVER, a migrant seen occasionally in Mexico, which flies in autumn over the Western Atlantic, sometimes apparently non-stop from breeding grounds in northern Canada to Central and South America.

Breeding

Shorebirds breed in a variety of ways. Many species breed in monogamous pairs that defend small breeding territories. Others, however, practice *polyandry*, the least common type of mating system among vertebrate animals, in which some females have more than one mate in a single breeding season. This type of breeding is exemplified by the SPOTTED SANDPIPER (Plate 26). In this species, the normal sex roles of breeding birds are reversed: the female establishes a territory on a lakeshore that she defends against other females. More than one male settles within the territory, either at the same time or sequentially during a breeding season. After mating, the female lays a clutch of eggs for each male. The males incubate their clutches and care for the young. Females may help care for some of the broods of young provided that there are no more unmated males to try to attract to the territory.

Most shorebird nests are simply small depressions in the ground in which eggs are placed; some of these *scrapes* are lined with shells, pebbles, grass, or leaves. Sandpipers lay two to four eggs per clutch, which are incubated, depending on species, by the male alone, the female alone, or by both parents, for 18 to 21 days. Plovers lay two to four eggs, which are incubated by both sexes for 24 to 28 days. BLACK-NECKED STILTS breed in small colonies. The scrape, into which four eggs are placed, is lined with vegetation. Both sexes incubate for 22 to 26 days. LIMPKINS build a nest of leaves and twigs, into which the female lays four to six eggs; both sexes incubate. Shorebird young are *precocial*, that is, soon after they hatch they are mobile, able to run from predators, and can feed themselves. Parents usually stay with the young to guard them at least until they can fly, perhaps 3 to 5 weeks after hatching.

Lore and Notes

The manner in which flocks of thousands of birds, particularly shorebirds, fly in such closely regimented order, executing abrupt maneuvers with precise coordination, such as when all individuals turn together in a split second in the same direction, has puzzled biologists and engendered some research. The questions include: what is the stimulus for the flock to turn – is it one individual within the flock, a "leader," from which all the others take their "orders" and follow into turns? Or is it some stimulus from outside the flock that all members respond to in the same way? And how are the turns coordinated? Everything from "thought transference" to electromagnetic communication among the flock members has been advanced as an explanation. After studying films of DUNLIN, a North American sandpiper, flying and turning in large flocks, one biologist has suggested that the method birds within these flocks use to coordinate their turns is similar to how the people in a chorusline know the precise moment to raise their legs in sequence or how "the wave" in a sports stadium is coordinated. That is, one bird, perhaps one that has detected some danger, like a predatory falcon, starts a turn, and the other birds, seeing the start of the flock's turning, can then anticipate when it is their turn to make the turn – the result being a quick wave of turning coursing through the flock.

Status

None of the Mexican plovers, sandpipers, or stilts are threatened or endangered but two plovers that winter in the country are considered vulnerable owing to small or declining populations – the PIPING and MOUNTAIN PLOVERS. A major goal for conservation of shorebirds is the need to preserve critical migratory stopover points – pieces of habitat, sometimes fairly small, that hundreds of thousands of shorebirds settle into mid-way during their long migrations to stock up on food. For instance, one famous small patch of coastal mudflats near Grays Harbor, Washington State, USA, is a popular, traditional stopover point for millions of shorebirds. Its destruction or use for any other activity could cause huge losses to the birds' populations. Fortunately, it has been deemed essential, and protected as part of a national wildlife refuge.

Profiles

Black-necked Stilt, *Himantopus mexicanus*, Plate 25d
Limpkin, *Aramus guarauna*, Plate 25e
Spotted Sandpiper, *Actitis macularia*, Plate 26a
Sanderling, *Calidris alba*, Plate 26b
Black-bellied Plover, *Pluvialis squatarola*, Plate 26c
Ruddy Turnstone, *Arenaria interpres*, Plate 26d
Lesser Yellowlegs, *Tringa flavipes*, Plate 26e

7. Curassows, Quail, and Turkey

Large chickenlike birds strutting about the tropical forest floor or, somewhat to the surprise of visitors from temperate quarters, fluttering about in trees and running along high branches, are bound to be members of the *curassow* family. The family, Cracidae, contains not only the *curassows*, but their close relatives the *guans* and entertainingly named *chachalacas*. There are more than 40 species of *cracids*, which are limited in their distributions to warmer regions of the New World, most inhabiting moist forests at low and middle elevations. The curassow family is placed within the avian order Galliformes, which also includes the *pheasants, quail, bobwhite, partridges,* and *turkeys* (and which, for want of a better term, even professional ornithologists refer to technically as the "chickenlike birds").

Members of the curassow family in southern Mexico range in length from about 40 to 91 cm (16 to 36 in) – as large as small turkeys – and weigh up to 4 kg (9 lb). They have long legs and long, heavy toes. Many have conspicuous crests. The colors of their bodies are generally drab – gray, brown, olive, or black and white; some appear glossy in the right light. They typically have small patches of bright coloring such as yellow, red, or orange on parts of their bills, cheeks, or on a hanging throat sac, or *dewlap*. Male GREAT CURASSOWS (Plate 27), for instance, although all black above and white below, have a bright yellow "knob" on the top of their bill. Within the group, males are larger than females; the sexes are generally similar in coloring, except for the curassows, in which females are drabber than males. The OCELLATED TURKEY (Plate 27) is confined in its distribution to the Yucatán region. It is a very large, brightly colored, ground-dwelling bird with a bare-skin blue head. The name arises with the eye-like images ("ocelli") adorning the birds' plumage.

Natural History

Ecology and Behavior

The guans and curassows are birds of the forests, but the chachalacas prefer forest edge areas and even clearings – more open areas, in other words. The guans and chachalacas are mostly arboreal birds, staying high in the treetops as they pursue their diet of fruit, young leaves, and treebuds, and the occasional frog or large insect. For such large birds in trees, they locomote with surprising grace, running quickly and carefully along branches. One's attention is sometimes drawn to them when they jump and flutter upwards from branch to branch until they are sufficiently in the clear to take flight. While guans and chachalacas will occasionally come down to the forest floor to feed on fallen fruit, GREAT CURASSOWS are terrestrial birds, more in the tradition of turkeys and pheasants. They stalk about on the forest floor, seeking fruit, seeds, and bugs. Paired off during the breeding season, birds of this family typically are found at other times of the year in small flocks of 10 to 20 individuals. Chachalaca males provide some of the most characteristic background sounds of tropical American forests. In the evening and especially during the early morning, males in groups rhythmically give their very loud calls, described variously as "cha-cha-LAW-ka" or "cha-cha-lac." OCELLATED TURKEYS, usually observed in small groups, are primarily birds of low elevation wet forests and clearings, but they are also found in open, brushy areas. They feed on seeds, berries, nuts, and insects.

Breeding

The curassows, guans, and chachalacas are monogamous breeders, the sexes sharing reproductive duties. Several of the species are known for producing loud whirring sounds with their wings during the breeding season, presumably during courtship displays. Male and female construct a simple, open nest of twigs and leaves, placed in vegetation or in a tree within several meters of the ground. Two to four eggs are incubated for 22 to 34 days. The young leave the nest soon after hatching to hide in the surrounding vegetation, where they are fed by the parents (in contrast to most of the chickenlike birds of the world, which feed themselves after hatching). Several days later, the fledglings can fly short distances. The family group remains together for a time, the male leading the family around the forest. OCELLATED TURKEYS place their 7 to 18 eggs in a shallow scrape in a hidden place on the forest floor. The female only incubates for about 26 days. Young are born ready to leave the nest and feed themselves. Turkey nesting generally begins in April.

Ecological Interactions

In Central and South America, the curassows essentially take the ecological position of pheasants, which are only lightly represented in the Neotropics.

Lore and Notes

The indigenous Mayan people of tropical Mexico have a conservation ethic embedded in their culture, passed down through generations. Many of their myths and traditions focused on stewardship of the land and local animals, for instance, with regard to hunting. There were Lords of the Forest, spirits that guarded forest practices. One was the Giant Turkey Spirit, which was said to take revenge on villagers who killed more turkeys than needed.

Status

A variety of factors converge to assure that the curassows will remain a problem group into the forseeable future. They are chiefly birds of the forests at a time when Neotropical forests are increasingly being cleared. They are desirable game birds, hunted by local people for food. In fact, as soon as new roads penetrate virgin forests in Central and South America, one of the first chores of settlers is to shoot curassows for their dinners. Unfortunately, curassows reproduce slowly, raising only small broods each year. Exacerbating the problem, their nests are often placed low enough in trees and vegetation to make them vulnerable to a variety of predators, including people. In the face of these unrelenting pressures on their populations, curassows are among the birds thought most likely to survive in the future only in protected areas, such as biosphere reserves. The HORNED GUAN, restricted to southern Chiapas and Guatemala, is considered threatened or already endangered (CITES Appendix I and USA ESA listed), with small and declining populations; and several curassow and guan species of South America are currently on lists of endangered species. Owing to over-hunting and destruction of their forest habitats, OCELLATED TURKEYS are threatened throughout their range (CITES Appendix III listed by Mexico, Belize, and Guatemala). Large protected areas of the Yucatán Peninsula, such as Calakmul and Sian Ka'an Biosphere Reserves, are thought to provide the best chances for continued survival of the species.

Profiles

Plain Chachalaca, *Ortalis vetula*, Plate 27a
Crested Guan, *Penelope purpurascens*, Plate 27b
Highland Guan, *Penelopina nigra*, Plate 27c
Great Curassow, *Crax rubra*, Plate 27d
Ocellated Turkey, *Agriocharis ocellata*, Plate 27e
Black-throated Bobwhite, *Colinus nigrogularis*, Plate 28a

8. Tinamous

The *tinamous* are an interesting group of secretive, chickenlike birds that are occasionally seen walking along forest trails. They apparently represent an ancient group of birds, most closely related not to chickens or pheasants but to the rheas of South America – large, flightless birds in the ostrich mold. The family, Tinamidae, with about 45 species, is confined in its distribution to the Neotropics, from Mexico to southern Chile and Argentina. Four species occur in Mexico, mostly within the southern reaches of the country. They inhabit a variety of environments, including grasslands and thickets, but most commonly they are forest birds.

Tinamous are medium-sized birds, 23 to 45 cm (9 to 18 in) long, chunky-bodied, with fairly long necks, small heads, and slender bills. They have short legs and very short tails. The back part of a tinamou's body sometimes appears higher than it should be, a consequence of a dense concentration of rump feathers. Tinamous are attired in understated, protective colors – browns, grays, and olives; often the plumage is marked with dark spots or bars. Male and female look alike, with females being a little larger than males.

Natural History

Ecology and Behavior

Except for the locally uncommon GREAT TINAMOU, which sleeps in trees, tinamous are among the most terrestrial of birds, foraging, sleeping, and breeding on the ground. They are very poor flyers, doing so only when highly alarmed by a predator or surprised, and then only for short distances. They are better at running along the ground, the mode of locomotion ecologists call *cursorial*. The tinamou diet consists chiefly of fruit and seeds, but they also take insects such as caterpillars, beetles, and ants, and occasionally, small vertebrates such as mice. Some South American species dig to feed on roots and termites. Tinamous avoid being eaten themselves primarily by often staying still, easily blending in with surrounding vegetation, and by walking slowly and cautiously through the forest. If approached closely, tinamous will fly upwards in a burst of loud wing-beating and fly usually less than 50 m (160 ft) to a new hiding spot in the undergrowth, often colliding with trees and branches as they go. Tinamous are known for their songs, which are loud, pure-tone, melodious whistles, and which are some of the most characteristic sounds of Neotropical forests.

Except during breeding, tinamous lead a solitary existence. They are considered secretive and, with their cryptic colors, elusive. It is common to hear them but not see them, or to see the rear of one, up ahead on the trail, disappearing into the underbrush.

Breeding

The tinamous employ some unusual mating systems, the most intriguing of which is a kind of *group polyandry* (polyandry being a system in which one female mates with several males during a breeding season). One or more females will mate with a male and lay clutches of eggs in the same nest, for the male to incubate and care for. The females then move on to repeat the process with other males of their choosing. Apparently in all tinamous the male incubates the eggs and leads and defends the young. Nests can hardly be called that; they are simply slight indentations in the ground, hidden in a thicket or at the base of a tree. Up to 12 eggs, deposited by one or more females, are placed in the nest. The male incubates for 19 or 20 days; when the eggs hatch, he leads the troop of tiny tinamous about, defending them from predators. From hatching, the young feed themselves.

Lore and Notes

Outside of protected areas, all tinamous are hunted extensively for food. Tinamou meat is considered tender and tasty, albeit a bit strange-looking; it has been described variously as greenish and transparent.

Status

The tinamous' camouflage coloring and secretive behavior must serve the birds well because, although hunted for food, tinamous presently are able to maintain healthy populations. None of the Mexican species are considered vulnerable or threatened. Some of the tinamous are known to be able to move easily from old, uncut forest to *secondary*, recently cut, forest, demonstrating an impressive adaptability that should allow these birds to thrive even amid major habitat alterations such as deforestation. To be sure, some of the tinamous are in trouble, including especially the SOLITARY TINAMOU of Argentina and Brazil (CITES Appendix I and USA ESA listed) and the critically endangered MAGDALENA (RED-LEGGED)

TINAMOU of Colombia and KALINOWSKI'S TINAMOU of Peru. Unfortunately, dependable information on many other tinamous is scarce, so it is very difficult to to know their statuses accurately.

Profiles

Slaty-breasted Tinamou, *Crypturellus boucardi*, Plate 28b
Thicket Tinamou, *Crypturellus cinnamomeus*, Plate 28c

9. Raptors

Raptor is another name for *bird-of-prey*, birds that make their living hunting, killing, and eating other animals, usually other vertebrates. When one hears the term raptor, one usually thinks of soaring *hawks* that swoop to catch rodents, and of speedy, streamlined *falcons* that snatch small birds out of the air. Although these *are* common forms of raptors, the families of these birds are large, the members' behavior diverse. The two main raptor families are the Accipitridae, containing the *hawks*, *kites* and *eagles*, and the Falconidae, including the *true falcons, forest-falcons*, and *caracaras*. The reasons for classifying the two raptor groups separately mainly have to do with differences in skeletal anatomy and, hence, suspected differences in evolutionary history. (Owls, nocturnal birds-of-prey, can also be considered raptors.) Raptors are common and conspicuous animals in southern Mexico and, generally, in the Neotropics. Many are birds of open areas, above which they soar during the day, using the currents of heated air that rise from the sun-warmed ground to support and propel them as they search for meals. But raptors are found in all types of habitats, including within woodlands and closed forests.

The *accipitrids* are a world-wide group of about 200 species; they occur everywhere but Antarctica. Southern Mexico is home to about 30 species, some of them migratory: 19 hawks, eight kites, four eagles and hawk-eagles. *Falconids* likewise are world-wide in their distribution. There are about 60 species, 10 occurring in southern Mexico. Some falcons have very broad distributions, with the PEREGRINE FALCON found almost everywhere, that is, its distribution is *cosmopolitan*. Peregrines may have the most extensive natural distribution of any bird. Caracaras are falcons with distinctively long legs and unfeathered facial skin; several species are distributed throughout most of Mexico and Central America.

Raptors vary considerably in size and in patterns of their generally subdued color schemes, but all are similar in overall form – we know them when we see them. They are fierce-looking birds with strong feet, hooked, sharp claws, or *talons*, and strong, hooked and pointed bills. Accipitrids vary in size in the region from a 28-cm-long (11 in) hawk to 1-m-long (40 in) eagles. Females are usually larger than males, in some species noticeably so. Most raptors are variations of gray, brown, black, and white, usually with brown or black spots, streaks, or bars on various parts of their bodies. The plumages of these birds are actually quite beautiful when viewed close-up, which, unfortunately, is difficult to do. Males and females are usually alike in color pattern. Juvenile raptors often spend several years in *subadult* plumages that differ in pattern from those of adults. Many falcons can be distinguished from hawks by their long, pointed wings, which allow the rapid, acrobatic flight for which these birds are justifiably famous.

Natural History

Ecology and Behavior

Raptors are meat-eaters. Most hunt and eat live prey. They usually hunt alone, although, when mated, the mate is often close by. Hawks, kites, and eagles take mainly vertebrate animals, including some larger items such as monkeys and birds up to the size of small vultures. Prey is snatched with talons first, and then killed and ripped apart with the bill. Some invertebrate prey is also taken. One species, the SNAIL KITE (Plate 29), specializes almost completely on one kind of freshwater marsh snail, which before eating it daintily removes from the shell with its long, pointed bill.

Falcons are best known for their remarkable eyesight and fast, aerial pursuit and capture of flying birds – they are "birdhawks." Thus, the small AMERICAN KESTREL (Plate 29), a telephone wire bird familiar to many people, is a falcon that formerly was known as the Sparrowhawk; the MERLIN, a slightly larger falcon, was known as the Pigeonhawk. Both of these falcons breed in North America, with some populations migrating south to winter in Central and South America. Most people are familiar with stories of PEREGRINE FALCONS diving through the atmosphere (*stooping*, defined as diving vertically from height to gain speed and force) at speeds approaching 320 kph (200 mph) to stun, grab, or knock from the sky an unsuspecting bird. But some falcons eat more rodents than birds, and some even take insects. One species specializes on taking bats on the wing at dawn and dusk. Forest-falcons are just that, falcons of the inner forest. They perch motionless for long periods on tree branches, waiting to ambush prey like birds and lizards. The LAUGHING FALCON (Plate 29) is a bird of open fields and forest edge that specializes on snakes. It perches until it spots a likely candidate for dinner, then swoops down fast and hits powerfully, grabbing the reptile and immediately biting off the head – a smart move because this falcon takes even highly poisonous prey such as coral snakes (p. 106). It then flies off to a high perch to feast on the headless serpent.

Many raptors are territorial, a solitary individual or a breeding pair defending an area for feeding and, during the breeding season, for reproduction. Displays that advertise a territory and may be used in courtship consist of spectacular aerial twists, loops, and other acrobatic maneuvers. Although many raptors are common birds, typically they exist at relatively low densities, as is the case for all *top predators* (a predator at the pinnacle of the food chain, preyed upon by no animal). That is, there usually is only enough food to support one or two of a species in a given area. For example, a typical density for a small raptor species, perhaps one that feeds on mice and small lizards, is one individual per square kilometer. A large eagle that feeds on monkeys may be spaced so that a usual density is one individual per thousand square kilometers.

Breeding

Hawk and eagle nests are constructed of sticks that both sexes place in a tree or on a rock ledge. Some nests are lined with leaves. The female only incubates the one to six eggs (only one or two in the larger species) for 28 to 49 days and gives food to the nestlings. The male frets about and hunts, bringing food to the nest for the female and for her to provide to the nestlings. Both sexes feed the young when they get a bit older; they can fly at 28 to 120 days of age, depending on species size. After fledging, the young remain with the parents for several more weeks or months until they can hunt on their own. Falcon breeding is similar. Falcons nest in vegetation, in a rock cavity, or on a ledge; some make a stick nest, others apparently make no construction. Incubation is from 25 to 35 days, performed only by the female in most of the falcons, but by both sexes in the caracaras. In most falcons, the male hunts for and feeds the female during the egg-laying, incubation, and early nestling periods. Male and female feed nestlings, which fledge after 25 to 49 days in the nest. The parents continue to feed the youngsters for several weeks after fledging until they are proficient hunters.

Ecological Interactions
The hunting behavior of falcons has over evolutionary time shaped the behavior of their prey animals. Falcons hit perched or flying birds with their talons, stunning the prey and sometimes killing it outright. An individual bird caught unawares has little chance of escaping the rapid, acrobatic falcons. But birds in groups have two defenses. First, each individual in a group benefits because the group, with so many eyes and ears, is more likely to spot a falcon at a distance than is a lone individual, thus providing all in the group with opportunities to watch the predator as it approaches and so evade it. This sort of anti-predation advantage may be why some animals stay in groups. Second, some flocks of small birds, such as starlings, which usually fly in loose formations, immediately tighten their formation upon detecting a flying falcon. The effect is to decrease the distance between each bird, so much so that a falcon flying into the group at a fast speed and trying to take an individual risks injuring itself – the "block" of starlings is almost a solid wall of bird. Biologists believe that the flock tightens when a falcon is detected because the behavior reduces the likelihood of an attack.

SNAIL KITES, which have a broad range from the southeastern USA to the middle of South America, and which are fairly common along waterways in parts of southern Mexico, are an endangered species in the USA. There, formerly called the Everglades Kite, they occur only over small bits of Florida. During the 1960s, owing to destruction and draining of their marsh habitats, the entire Florida population was believed to number no more than 60 individuals. Snail kites are extremely specialized in their feeding; in fact, they may be *monophagic* – eating only one thing: freshwater apple snails. This high degree of specialization is one of the kites' biggest problems, because periodic droughts lead to very low food availability and, therefore, to crashes in the kite population.

Lore and Notes
Large raptors have doubtless always attracted people's attention, respect, and awe. Wherever eagles occur, they are chronicled in the history of civilizations. Both the ancient Greeks and Romans associated eagles with their gods. Early Anglo-Saxons were known to hang an eagle on the gate of any city they conquered. Some North American Indian tribes and also Australian Aboriginal peoples deified large hawks or eagles. Several states have used likenesses of eagles as national symbols, among them Turkey, Austria, Germany, Poland, Russia, and Mexico. Eagles are popular symbols on regal coats of arms and one of their kind, a fierce-looking fish-eater, was chosen as the emblem of the USA (although, as most USA schoolchildren know, Benjamin Franklin would have preferred that symbol to be the Wild Turkey.)

People have had a close relationship with falcons for thousands of years. Falconry, in which captive falcons are trained to hunt and kill game at a person's command, is a very old sport, with evidence of it being practiced in China 4000 years ago and in Iran 3700 years ago. One of the oldest known books on a sport is *The Art of Falconry*, written by the King of Sicily in 1248. Falconry reached its zenith during medieval times in Europe, when a nobleman's falcons were apparently among his most valued possessions. The CRESTED CARACARA (Plate 28) is depicted on the national emblem of Mexico.

Status

Several hawks and falcons are considered to be threatened in Mexico. Most of these raptors are rare but have extensive distributions, often ranging from Mexico to northern or central South America; some are more numerous in other regions. The HARPY EAGLE, which ranges from southern Mexico to Argentina, and is perhaps the world's most powerful eagle, is rare over most of its range owing to hunting (CITES Appendix I and USA ESA listed). The APLOMADO FALCON (USA ESA listed), a large falcon which formerly occurred in the southern USA and over most of Mexico, has been eliminated from the USA and all but small bits of Mexico. The CRESTED CARACARA is considered threatened in Florida (USA ESA listed). Trade in essentially all hawks, kites, eagles, and falcons is heavily regulated (most are CITES Appendix II listed). Conservation measures aimed at raptors are bound to be difficult to formulate and enforce because the birds are often persecuted for a number of reasons (hunting, pet and feather trade, ranchers protecting livestock) and they roam very large areas. Also, some breed and winter on different continents, and thus need to be protected in all parts of their ranges, including along migration routes. Further complicating population assessments and conservation proposals, there are still plenty of Neotropical raptor species about which very little is known. For example, for the approximately 80 species of raptors that breed primarily in Central and South America (excluding the vultures), breeding behavior has not been described for 27 species, nests are unknown for 19 species, as is the typical prey taken by six species.

Profiles

Osprey, *Pandion haliatus*, Plate 28d
Crested Caracara, *Polyborus plancus*, Plate 28e
Laughing Falcon, *Herpetotheres cachinnans*, Plate 29a
Collared Forest Falcon, *Micrastur semitorquatus*, Plate 29b
Bat Falcon, *Falco rufigularis*, Plate 29c
American Kestrel, *Falco sparverius*, Plate 29d
Snail Kite, *Rostrhamus sociabilis*, Plate 29e
Gray-headed Kite, *Leptodon cayanensis*, Plate 30a
Plumbeous Kite, *Ictinia plumbea*, Plate 30b
Roadside Hawk, *Buteo magnirostris*, Plate 30c
Gray Hawk, *Buteo nitidus*, Plate 30d
Common Black Hawk, *Buteogallus anthracinus*, Plate 30e
Short-tailed Hawk, *Buteo brachyurus*, Plate 30f

10. Vultures

Birds at the very pinnacle of their profession, eating dead animals, *vultures* are highly conspicuous and among the most frequently seen birds of rural Mexico. That they feast on rotting flesh does not reduce the majesty of these large, soaring birds as they circle for hours high over field and forest. The family of American vultures, Cathartidae, has only seven species, all confined to the New World;

several are abundant animals but one is close to extinction. They range from southern Canada to Tierra del Fuego, with several of the species sporting wide and overlapping distributions. Four species, the KING, BLACK, TURKEY (Plate 31), and LESSER YELLOW-HEADED VULTURES, occur in southern Mexico. The two largest members of the family are known as *condors*, the CALIFORNIA and ANDEAN CONDORS. Vultures are seen often above both open country and forested areas; the BLACK VULTURE, one of the most frequently encountered birds of tropical America, is very common around garbage dumps. Traditional classifications place the New World vulture family within the avian order Falconiformes, with the hawks, eagles, and falcons, but some biologists believe the group is more closely related to the storks.

Mexican vultures are large birds, from 60 to 80 cm (24 to 32 in) long, with wing spans to 1.8 m (6 ft). (The Andean Condor has a wing span of 3 m (10 ft)!). Vultures generally are black or brown, with hooked bills and curious, unfeathered heads whose bare skin is richly colored in red, yellow, or orange. Turkey Vultures, in fact, are named for their red heads, which remind people of turkey heads. The King Vulture, the largest in the region, departs from the standard color scheme, having a white body, black rump and wing feathers, and multi-colored head. Male and female vultures look alike; males are slightly larger than females.

Natural History

Ecology and Behavior
Vultures are carrion eaters of the first order. Most soar during the day in groups, looking for and, in the case of TURKEY VULTURES, sniffing for food. (Turkey Vultures can find carcasses in deep forest and also buried carcasses, strongly implicating smell, as opposed to vision, as the method of discovery.) They can move many miles daily in their search for dead animals. KING and BLACK VULTURES, supplementing their taste for carrion, also occasionally kill animals, usually newborn or those otherwise defenseless. Some of the vultures, especially the Black Vulture, also eat fruit. With their super eyesight, fine-tuned sense of smell, and ability to survey each day great expanses of habitat, vultures are, to paraphrase one biologist, amazingly good at locating dead animals. No mammal species specializes in carrion to the degree that vultures do, the reason being, most likely, that no mammal could search such large areas each day. That is, no mammal that ate only carrion could find enough food to survive.

King Vultures are usually seen in pairs or solitarily, as are LESSER YELLOW-HEADED VULTURES, but the other species are more social, roosting and foraging in groups of various sizes. Black Vultures, in particular, often congregate in large numbers at feeding places, and it is common to find a flock of them at any village dump. At small to medium-sized carcasses, there is a definite pecking order among the vultures: Black Vultures are dominant to Turkey Vultures, and can chase them away; several Black Vultures can even chase away a King Vulture, which is the bigger bird. However, in an area with plenty of food, all three species may feed together in temporary harmony (F. G. Stiles & D. H. Janzen 1983). When threatened, vultures may spit up partially digested carrion, a strong defense against harassment if ever there was one.

Breeding
Vultures are monogamous breeders. Both sexes incubate the one to three eggs, which are placed on the ground in protected places or on the floor of a cave or

tree cavity. Eggs are incubated for 32 to 58 days; both sexes feed the young regurgitated carrion for 2 to 5 months until they can fly. Young vultures at the nest site very rarely become food for other animals, and it has been suggested that the odor of the birds and the site, awash as they are in badly decaying animal flesh, keep predators (and everything else) at bay.

Ecological Interactions

Vultures are abundant and important scavengers around Neotropical towns and villages. Usefully, they help clean up garbage; expert observers have said they "perform at least as much of the Sanitation Department's chores as do its human members" (F. G. Stiles & D. H. Janzen 1983). Scavengers, or *detritus* feeders, are important parts of ecosystems, recycling energy and nutrients back into food webs. Although vultures are an integral part of these systems, most detritus is consumed by small arthropods (insects and the like) and such microorganisms as fungi and bacteria.

BLACK and TURKEY VULTURES roost communally, the two species often together. A common observation has been that once an individual finds a food source, other vultures arrive very rapidly to share the carcass. Biologists strongly suspect that the group roosting and feeding behavior of these birds are related, and that the former increases each individual's food-finding efficiency. In other words, a communal roost serves as an *information center* for finding food. Researchers believe that vultures may locate the probable position of the next day's food while soaring high in the late afternoon, before they return to the roost before sunset. In the morning, the ones that detected potential carcasses set out from the roost to locate the food, with others, less successful the previous day, following in the general direction. Thus, on different days, the birds take turns as leaders and followers, and all benefit from the communal roosting association.

Lore and Notes

Europeans first arriving in the New World thought the region's vultures evil – "the sloth, the filth, and the voraciousness of these birds, almost exceeds credibility," wrote one impressionable Englishman in the late 1700s. But the native peoples, especially in hot, humid areas where rotting of corpses occurs quickly, generally thought well of these carrion-eaters. Ancient Mayans called the KING VULTURE "Oc," including it frequently in their artworks. Vulture feathers were used in Mayan headdresses. Mayans apparently believed that vultures, when near death, changed into armadillos; the proof being that both were "bald."

The Mayan legend of how the vulture came to be black and bald and to feed on carrion goes like this: in the old days, vultures were actually handsome, white birds with feathered heads, which ate only the finest fresh meat; they had an ideal life. One day, the vulture family, out soaring in the sun, spied a feast laid out on banquet tables in a forest clearing. They swooped down and ate the splendid food. Unfortunately for the vultures, the food had been set out by nobles as an offering to the gods. The nobles schemed to punish the unknown culprits. They set out another feast in the clearing and hid behind trees with their witch doctors. When the vultures returned for another meal, the nobles and witch doctors raced out from the trees and threw magic powder on the birds. The vultures, in their panic to escape the people, flew straight up and got too close to the sun, scorching their heads, causing their feathers to fall out. In the clouds, the magic powder turned their white plumage to black. When they returned to earth, the

Great Spirit ruled that for their thievery, from that day forward, vultures would eat only carrion. (A. L. Bowes 1964)

Status

The four Mexican vultures are all somewhat common to very common birds; none are threatened. The only New World vulture in dire trouble is the CALIFORNIA CONDOR (CITES Appendix I and USA ESA listed), which for a while was extinct in the wild. The main causes of the condors' decline in the 20th century were: hunting (they were persecuted especially because ranchers believed they ate newborn cattle and other domestic animals); their ingestion of poisonous lead shot from the carcasses they fed on; and the thinning of their eggshells owing to the accumulation of organochlorine pesticides (DDT) in their bodies. The last eight free-ranging specimens were caught during the mid-1980s in their southern Californian haunts for use in captive breeding programs. The total captive population is now more than 100 individuals, and several have been released back into the wild in California and Arizona.

Profiles

Turkey Vulture, *Cathartes aura*, Plate 31a
Black Vulture, *Coragyps atratus*, Plate 31b
King Vulture, *Sarcoramphus papa*, Plate 31c

11. Pigeons and Doves

The *pigeon* family is a highly successful group, represented, often in large numbers, almost everywhere on dry land, except for Antarctica and some oceanic islands. Their continued ecological success must be viewed as at least somewhat surprising, because pigeons are largely defenseless creatures and quite edible, regarded as a tasty entree by human and an array of non-human predators. The family, Columbidae, includes approximately 250 species, about 20 of which occur in southern Mexico. They inhabit almost all kinds of habitats, from semi-deserts to tropical moist forests, to high-elevation mountainsides. Smaller species generally are called *doves*, larger ones, *pigeons*, but there is a good amount of overlap in name assignments.

All pigeons are generally recognized as such by almost everyone, a legacy of people's familiarity with domestic and feral pigeons. Even small children in zoos, upon encountering an exotic, colorful dove will determine it to be "some kind of pigeon." Pigeons world-wide vary in size from the dimensions of a sparrow to those of a small turkey; Mexico's range from 15 cm-long (6 in) ground-doves to the 38 cm (15 in) BAND-TAILED PIGEON (Plate 32). Doves and pigeons are plump-looking birds with compact bodies, short necks, and small heads. Legs are usually fairly short, except in the ground-dwelling species. Bills are small, straight, and slender. Typically there is a conspicuous patch of naked skin, or *cere*, at the base of the bill, over the nostrils. Although many of the Old World pigeons are easily among the most gaily colored of birds (Asian *fruit doves*, for instance), the New World varieties generally color their soft, dense plumages with understated grays and browns, although a few also have bold patterns of black lines or spots. Many have splotches of iridescence, especially on necks and wings. In the majority of Mexican species, male and female are generally alike in size and color, although females are often a bit duller than the males.

Natural History

Ecology and Behavior

Most of the pigeons are at least partly arboreal, but some spend their time in and around cliffs, and still others are primarily ground-dwellers. They eat seeds, ripe and unripe fruit, berries, and the occasional insect, snail, or other small invertebrate. Even those species that spend a lot of time in trees often forage on the ground, moving along the leaf-strewn forest floor, for example, with the head-bobbing walk characteristic of their kind. Owing to their small, weak bills, they eat only what they can swallow whole; "chewing" is accomplished in the *gizzard*, a muscular portion of the stomach in which food is mashed against small pebbles that are eaten by pigeons expressly for this purpose. Pigeons typically are strong, rapid flyers, which, in essence, along with their cryptic color patterns, provide their only defenses against predation. Most pigeons are gregarious to some degree, staying in groups during the non-breeding portion of the year; some gather into large flocks. Visitors to the Neotropics from North America often are struck by the relative scarcity of sparrows; it is in large part the pigeons of the region that ecologically "replace" sparrows as predominant seed-eaters.

Breeding

Pigeons are monogamous breeders. Some breed solitarily, others in colonies of various sizes. Nests are shallow, open affairs of woven twigs, plant stems, and roots, placed on the ground, on rock ledges, or in shrubs or trees. Reproductive duties are shared by male and female. This includes nest-building, incubating the one or two eggs, and feeding the young, which they do by regurgitating food into the nestlings' mouths. All pigeons, male and female, feed their young *pigeon milk*, a nutritious fluid produced in the *crop*, an enlargement of the esophagus used for food storage. During the first few days of life, nestlings receive 100% pigeon milk but, as they grow older, they are fed an increasing proportion of regurgitated solid food. Incubation time ranges from 11 to 28 days, depending on species size. Nestlings spend from 11 to 36 days in the nest. Parent pigeons of some species give *distraction displays* when potential predators approach their eggs or young; they feign injury as they move away from the nest, luring the predator away.

Ecological Interactions

The great success of the pigeon family – a world-wide distribution, robust populations, the widespread range and enormous numbers of rock doves (wild, domestic pigeons) – is puzzling to ecologists. At first glance, pigeons have little to recommend them as the fierce competitors any hugely successful group needs to be. They have weak bills and therefore are rather defenseless during fights and ineffectual when trying to stave off nest predators. They are hunted by people for food. In several parts of the world they compete for seeds and fruit with parrots, birds with formidable bills, yet pigeons thrive in these regions and have spread to many more that are parrot-less. To what do pigeons owe their success? First, to reproductive advantage. For birds of their sizes, they have relatively short incubation and nestling periods; consequently, nests are exposed to predators for relatively brief periods and, when nests fail, parents have adequate time to nest again before the season ends. Some species breed more than once per year. Also, the ability of both sexes to produce pigeon milk to feed young may be an advantage over having to forage for particular foods for the young. Second, their ability to capitalize on human alterations of the environment points to a high degree of hardiness and adaptability, valuable traits in a world in which people make changes to habitats faster than most organisms can respond with evolutionary changes of their own.

Lore and Notes

Although many pigeons today are very successul animals, some species met extinction within the recent past. There are two particularly famous cases. The DODO was a large, flightless pigeon, the size of turkey, with a large head and strong, robust bill and feet. Dodos lived, until the 17th century, on the island of Mauritius, in the Indian Ocean, east of Madagascar. Reported to be clumsy and stupid (hence the expression, "dumb as a dodo"), but probably just unfamiliar with and unafraid of predatory animals, such as people, they were killed in the thousands by sailors who stopped at the island to stock their ships with food. This caused population numbers to plunge; the birds were then finished off by the pigs, monkeys, and cats introduced by people to the previously predator-free island – animals that ate the Dodos' eggs and young. The only stuffed Dodo in existence was destroyed by fire in Oxford, England, in 1755.

North America's PASSENGER PIGEON, a medium-sized, long-tailed member of the family, suffered extinction because of overhunting and because of its habits of roosting, breeding, and migrating in huge flocks. People were able to kill many thousands of them at a time on the Great Plains in the central part of the USA, shipping the bodies to markets and restaurants in large cities through the mid-1800s. It is estimated that when Europeans first settled in the New World, there were 3 billion Passenger Pigeons, a population size perhaps never equalled by any other bird, and that they may have accounted for up to 25% or more of the birds in what is now the USA. It took only a bit more than 100 years to kill them all; the last one died in the Cincinnati Zoo in 1914.

Throughout recorded history, doves have been one of the most commonly used bird symbols, usually associated with love and other positive human traits. In several ancient civilizations, doves were symbols of fertility, of the soul, of longevity and faithfulness (China); "among the ancient Hebrews the dove was the symbol of purity, of persecuted Israel, and the image of the spirit of God; in the Christian world the white dove was the symbol of the Holy Ghost, of Christ, of the Church, of the Virgin, of the souls of the redeemed, of spirtual love, of innocence, of defenselessness, of charity, of martyrdom, of sorrow, and of the Ascension ..." (B. Rowland 1978)

Status

Several New World pigeons are threatened or endangered, including two from Mexico: the PURPLISH-BACKED (or VERACRUZ) QUAIL-DOVE, which lives only in a small part of the state of Veracruz (and perhaps in sections of Costa Rica and Panama), and the SOCORRO DOVE, which was endemic to small Pacific islands off Mexico's western coast. Socorro Doves were last seen in the wild in the 1950s. It was a victim of human settlement of the islands, and of the cats people brought along. Today, only about 200 of the doves exist, all in captivity. BAND-TAILED PIGEONS (Plate 32), owing to hunting, may eventually be threatened over parts of their range.

Profiles

Band-tailed Pigeon, *Columba fasciata*, Plate 32a
Scaled Pigeon, *Columba speciosa*, Plate 32b
Red-billed Pigeon, *Columba flavirostris*, Plate 32c

Blue Ground-dove, *Claravis pretiosa*, Plate 32d
Common Ground-dove, *Columbina passerina*, Plate 33a
Ruddy Ground-dove, *Columbina talpacoti*, Plate 32e
Inca Dove, *Columbina inca*, Plate 33b
White-tipped Dove, *Leptotila verreauxi*, Plate 33c
White-winged Dove, *Zenaida asiatica*, Plate 33d
White-faced Quail-dove, *Geotrygon albifacies*, Plate 33e

12. Parrots

Everyone knows *parrots* as caged pets, so discovering them for the first time in their natural surroundings is often a strange but somehow familiar experience (like a dog-owner's first sighting of a wild coyote): one has knowledge and expectations of the birds' behavior and antics in captivity, but how do they act in the wild? Along with toucans, parrots are probably the birds most commonly symbolic of the American tropics. The 300+ parrot species that comprise the family Psittacidae (the P is silent; refer to parrots as *psittacids* to impress your friends and tourguides!) are globally distributed across the tropics, with some species extending into subtropical and even temperate zone areas. The family has a particularly diverse and abundant presence in the Neotropical and Australian regions. About 16 species occur in southern Mexico. Ornithologists divide parrots by size: *parrotlets* are small birds (as small as 10 cm, or 4 in) with short tails; *parakeets* are also small, with long or short tails; *parrots* are medium-sized, usually with short tails; and *macaws* are large (up to 1 m, or 40 in) and long-tailed.

Consistent in form and appearance, all parrots are easily recognized as such. They share a group of traits that set them distinctively apart from all other birds. Their typically short neck and compact body yield a form variously described as stocky, chunky, or bulky. All possess a short, hooked bill with a hinge on the upper part that permits great mobility and leverage during feeding. Finally, their legs are short and their feet, with two toes projecting forward and two back, are adapted for powerful grasping and a high degree of dexterity – more so than any other bird. The basic parrot color scheme is green, but some species, such as a few of the macaws, depart from basic in spectacular fashion, with gaudy blues, reds, and yellows. Green parrots feeding quietly amid a tree's high foliage can be difficult to see, even for experienced birdwatchers. Parrots in southern Mexico are most numerous in forested lowland areas. Best views are commonly obtained when flocks fly noisily overhead, when they depart feeding trees, or when a flock is located loafing and squabbling the afternoon away in an isolated, open tree.

Natural History

Ecology and Behavior
Parrots are incredibly noisy, highly social seed and fruit eaters. Some species seem to give their assortment of harsh, screeching squawks during much of the day, whereas others are fairly quiet while feeding. During early mornings and late afternoons, raucous, squawking flocks of parrots characteristically take flight explosively from trees, heading in mornings for feeding areas and later for night roosts, and these are usually the best sighting times. (The typical, if understandable, reaction of a newcomer to the tropical forest after a screeching parrot flock flashes by overhead: "What the *heck* was that?") Parrots are almost always encountered in flocks of four or more, and groups of 50+ smaller parrots are common. Flocks are usually groups of mated pairs and, with brief observation of behavior,

married pairs are often noticeable. Flocks move about seeking fruits and flowers in forests, parkland, and agricultural areas. In flight, parrots are easily identified by their family-specific silhouette: thick bodies and usually long tails, with short, relatively slowly-beating wings. Parrots generally are not considered strong flyers, but are certainly fast over the short run. Most do not need to undertake long-distance flights; they are fairly sedentary in their habits, with some regular movements as they follow the seasonal geographic progressions of fruit ripening and flower blossoming.

Parrots use their special locomotory talent to clamber methodically through trees in search of fruits and flowers, using their powerful feet to grasp branches and their bills as, essentially, a third foot. Just as caged parrots, they will hang at odd angles and even upside down, the better to reach some delicious morsel. Parrot feet also function as hands, delicately manipulating food and bringing it to the bill. Parrots feed mostly on fruits and nuts, buds of leaves and flowers, and on flower parts and nectar. They are usually considered frugivores, but careful study reveals that when they attack fruit, it is usually to get at the seeds within. The powerful bill slices open fruit and crushes seeds. As one bird book colorfully put it, "adapted for opening hard nuts, biting chunks out of fruit, and grinding small seeds into meal, the short, thick, hooked parrot bill combines the destructive powers of an ice pick (the sharp-pointed upper mandible), a chisel (the sharp-edged lower mandible), a file (ridged inner surface of the upper mandible), and a vise" (F. G. Stiles & A. F. Skutch 1989). Thick, muscular parrot tongues are also specialized for feeding, used to scoop out pulp from fruit and nectar from flowers.

Breeding
In most of the southern Mexico parrots, the sexes are very similar or identical in appearance; breeding is monogamous and pairing is often for life. Nesting is carried out during the dry season and, for some, into the early wet season. Most species breed in cavities in dead trees, although a few build nests. Macaw nests are almost always placed 30 m (100 ft) or more above the ground. A female parrot lays two to eight eggs, which she incubates alone for 17 to 35 days while being periodically fed regurgitated food by her mate. The helpless young of small parrots are nest-bound for 3 to 4 weeks, those of the huge macaws, 3 to 4 months. Both parents feed nestlings and fledglings.

Ecological Interactions
Many fruit-eating birds are fruit seed dispersers, but apparently not so parrots. Their strong bills crush seeds, and the contents are digested. For example, in one study in southern Central America, an ORANGED-CHINNED PARAKEET was examined after it fed all morning at a fig tree. It had in its digestive tract about 3500 fig seeds, almost all of which were broken, cracked, or already partially digested. Therefore, the main ecological interaction between parrots and at least some fruit trees is that of seed predator. Because parrots eat fruit and seeds, they are attracted to farms and orchards and in some areas are considered agricultural pests, with implications for their future populations (see below). Macaws and other parrots often congregate at *licks*, exposed riverbank or streamside clay deposits. The clay that is eaten may help detoxify harmful compounds that are consumed in their seed, fruit and leaf diet, or may supply essential minerals that are not provided by a vegetarian diet. Some ecotravel destinations are based on lick locations, for example, in Peru, where visitors reliably find parrots to watch at traditional clay licks.

Lore and Notes

Parrots have been captured for people's pleasure as pets for thousands of years. Greek records from 400 BC describe parrot pets. Ancient Romans wrote of training parrots to speak and even of how they acted when drunk! The fascination stems from the birds' bright coloring, their ability to imitate human speech and other sounds, their individualistic personalities (captive parrots definitely like some people while disliking others), and their long lifespans (up to 80 years in captivity). Likewise, parrots have been hunted and killed for food and to protect crops for thousands of years. Some Peruvian Inca pottery shows scenes of parrots eating corn and being scared away from crops. Historically, people have also killed parrots to protect crops – Charles Darwin noted that in Uruguay in the early 1800s, thousands of parakeets were killed to prevent crop damage. Macaws, the largest parrots, are thought to have been raised in the past for food in the West Indies.

Status

World-wide, about 90 parrot species are considered vulnerable, threatened, or endangered. Only one of the 16 that occur in southern Mexico falls into this group, the YELLOW-HEADED PARROT (Plate 35), which is considered endangered by some authorities. Two other Mexican species, from the northern part of the country, are also endangered: THICK-BILLED (CITES Appendix I and USA ESA listed) and RED-CROWNED PARROTS. The MAROON-FRONTED PARROT, endemic to a small region of northeastern Mexico, is also endangered, but some authorities consider it the same species as the Thick-billed Parrot. Many Mexican and Central American parrot species still enjoy healthy populations and are frequently seen. Unfortunately, however, parrots are subject to three powerful forces that, in combination, take heavy tolls on their numbers: parrots are primarily forest birds, and forests are increasingly under attack by farmers and developers; parrots are considered agricultural pests by farmers and orchard growers owing to their seed and fruit eating, and are persecuted for this reason; and parrots are among the world's most popular cage birds. Among the region's species, SCARLET MACAWS (Plate 34), MEALY PARROTS (Plate 34), and YELLOW-HEADED PARROTS are particularly prized as pets, and nests of these parrots are often robbed of young for local sale as pets or to international dealers (see Close-Up, p. 201). The SCARLET MACAW, in particular, which until recently ranged over a wide swath of southern Mexico (Veracruz, Tabasco, Oaxaca, and Chiapas) now occurs in this region only in eastern Chiapas, on the Guatemalan border. (The species is not considered threatened because it is more numerous in other parts of its range in Central and northern South America.) Without fast, additional protections, many other Mexican parrots will soon be threatened.

Profiles

Scarlet Macaw, *Ara macao*, Plate 34a
Mealy Parrot, *Amazona farinosa*, Plate 34b
Red-lored Parrot, *Amazona autumnalis*, Plate 34c
Yucatán Parrot, *Amazona xantholora*, Plate 34d
White-fronted Parrot, *Amazona albifrons*, Plate 35a
Yellow-headed Parrot, *Amazona oratrix*, Plate 35b
White-crowned Parrot, *Pionus senilis*, Plate 35c
Olive-throated Parakeet, *Aratinga nana*, Plate 35d
Orange-fronted Parakeet, *Aratinga canicularis*, Plate 35e

13. Cuckoos and Anis

Many of the *cuckoos* and *anis* (AH-neez) are physically rather plain but behaviorally rather extraordinary: as a group they employ some of the most bizarre breeding practices known among birds. Cuckoos and anis are both included in the cuckoo family, Cuculidae, which, with a total of about 130 species, enjoys a world-wide distribution in temperate areas and the tropics. Ten species occur in southern Mexico. Cuckoos are mainly shy, solitary birds of forests, woodlands, and dense thickets. Anis are the opposite; gregarious animals that spend their time in small flocks in savannas, brushy scrub, and other open areas, particularly around human habitations. Anis are among those birds that make one wonder where they perched before the advent of fences. Notable *cuculid* relatives of the cuckoos and anis are two species of common ground birds of drier areas of Mexico and southwestern USA, *roadrunners*, of cartoon fame.

Most cuckoos are medium-sized, slender, long-tailed birds. Male and female mostly look alike, attired in plain browns, tans, and grays, often with streaked or spotted patches. Several have alternating white and black bands on their tail undersides. (Many cuckoos of the Old World are more colorful.) They have short legs and bills that curve downwards at the end. Anis are conspicuous medium-sized birds, glossy black all over, with iridescent sheens particularly on the head, neck and breast. Their bills are exceptionally large, with humped or crested upper parts. Roadrunners are largish, streaked brown ground birds with heavy bills, very long legs, and long tails.

Natural History

Ecology and Behavior

Most of the cuckoos are arboreal. They eat insects, apparently having a special fondness for caterpillars. They even safely consume hairy caterpillars, which are avoided by most potential predators because they taste bad or contain sickness-causing noxious compounds. Cuckoos have been observed to snip off one end of the hairy thing, squeeze the body in the bill until the noxious entrails fall out, then swallow the harmless remainder. A few cuckoos, such as the LESSER ROADRUNNER (Plate 36) and rarely seen PHEASANT CUCKOO, are ground-dwellers, eating insects but also vertebrates such as small lizards and snakes. Roadrunners rarely fly, preferring to run along the ground to catch food or to escape from potential predators.

The highly social anis forage in groups, usually on the ground. Frequently they feed around cattle, grabbing the insects that are flushed out of hiding places by the grazing mammals. They eat mostly bugs, but also a bit of fruit. Anis live in groups of 8 to 25 individuals, each group containing two to eight adults and several juveniles. Each group defends a territory from other groups throughout the year. The flock both feeds and breeds within its territory.

Breeding

Cuckoos are known in most parts of the world for being *brood parasites*: they build no nests of their own, and the females lay their eggs in the nests of other species. These other birds often raise the young cuckoos as their own offspring, usually to the significant detriment of their own, often smaller, young. About 50 of the 130 members of the family are brood parasites. Only two of the species in the covered region, STRIPED and PHEASANT CUCKOOS, breed in this manner. The rest of them are somewhat typically monogamous breeders. The male feeds the female in courtship, especially during her egg-laying period. Both sexes build the plain platform nest that is made of twigs and leaves and placed in a tree or shrub. Both sexes incubate the two to six eggs for about 10 days, and both parents feed the young. In several species, the young hop out of the nest before they have been flight-certified, to spend the several days before they can fly flopping around in the vegetation near the nest, being fed by their parents.

Anis, consistent with their highly social ways, are *communal breeders*. In the most extreme form, all the individuals within the group contribute to a single nest, several females laying eggs in it; up to 29 eggs in one nest have been noted. Many individuals help build the stick nest and feed the young. Although at first glance it would seem as if all benefit by having the group breed together, females contributing eggs to a common nest which all build, defend and tend, actually it is the dominant male and female within each group that gain most. Their eggs go in the nest last, on top of the others, which sometimes become buried. Also, some females roll others' eggs out of the nest before they lay their own; thus, it pays to lay last. In some species, several nests are built by pairs within the group's territory.

Lore and Notes

The Mayan folk tale about how the Ani came to have only black feathers is a sad one. It seems the Ani in the past was bright pink. She was very proud of her young nestlings, so when her friend the hawk said he was going off to feed, mother Ani, exaggerating, told him to be sure to avoid the nest of beautiful young birds that was her own. The hawk tried, but when he spied a nest of runty-looking, ugly little birds, he judged that they could not be the fine-looking youngsters described by the Ani, so he ate them. When the Ani found out what her incorrect description had cost her, she put on black feathers to mourn, and all anis ever since have worn black mourning plumage (A. L. Bowes 1964).

Roadrunners in tropical Mexico are considered by some to have great medicinal value; a roadrunner broth, they claim, has the pain-killing power of asprin, particularly for afflictions of the legs.

"Although anis lack beautiful plumage and a melodious voice, they have been amply compensated in other ways. They have an extraordinary afffectionate nature, adaptability which enables them to thrive in a greater range of environments than many other birds, and nesting habits that make them second in interest to none. Few birds crave the close company of their kind more constantly than the anis. I have never seen them quarrel or fight. When one is separated from its flock, it calls and calls until it finds its companions ..." (A. F. Skutch 1983)

Status

None of the cuckoos or anis that occur in southern Mexico are currently considered threatened. Some cuckoos of Central America, in fact, actually are increasing in numbers and expanding their ranges because they do well in the forest edge, thicket, and open areas that increasingly are created through deforestation. Two species of New World cuckoos appear to be threatened: the RUFOUS-BREASTED CUCKOO, endemic to the island of Hispaniola, and the BANDED GROUND-CUCKOO of Ecuador and Colombia. Costa Rica's COCOS CUCKOO is considered vulnerable. The reason is that it is endemic to Cocos Island, which is in the Pacific some 500 miles from the mainland. Although the precise population size is unknown, species confined to a single small place are always vulnerable, because a single catastrophe there could exterminate the species.

Profiles

Squirrel Cuckoo, *Piaya cayana*, Plate 36a
Groove-billed Ani, *Crotophaga sulcirostris*, Plate 36b
Yellow-billed Cuckoo, *Coccyzus americanus*, Plate 36c
Lesser Roadrunner, *Geococcyx velox*, Plate 36d
Lesser Ground Cuckoo, *Morococcyx erythropygus*, Plate 36e

14. Owls

Although some *owls* are common Mexican birds, they are considered here only briefly because most are active only at night and so are rarely seen. But there are a few exceptions. Most owls are members of the family Strigidae, a world-wide group of about 120 species that lacks representation only in Antarctica and remote oceanic islands. Owls are particularly diverse in the tropics and subtropics; southern Mexico has about 20 species. Most people can always identify owls because of several distinctive features. All have large heads with forward-facing eyes, small, hooked bills, plumpish bodies and sharp, hooked claws. Most have short legs and short tails. Owls are clad mostly in mixtures of gray, brown, and black, the result being that they usually are highly camouflaged against a variety of backgrounds. They have very soft feathers. Most are medium-sized birds, but the group includes species that range in length from 15 to 75 cm (6 to 30 in). Males and females generally look alike, although females are a bit larger.

Natural History

Ecology and Behavior

In general, owls occupy a variety of habitats: forests, clearings, fields, grasslands, mountains, marshes. They are considered to be the nocturnal equivalents of the day-active birds-of-prey – the hawks, eagles, and falcons. Most owls hunt at night, taking prey such as small mammals, birds (including smaller owls), and reptiles; smaller owls specialize on insects, earthworms, and other small invertebrates. But some owls hunt at twilight (*crepuscular* activity) and sometimes during the day, including the FERRUGINOUS PYGMY-OWL (Plate 37). Owls hunt by sight and sound. Their vision is very good in low light, the amount given off by moonlight, for instance; and their hearing is remarkable. They can hear sounds that are much lower in sound intensity (softer) than most other birds, and their ears are positioned on their heads asymmetrically, the better for localizing sounds in space. This means that owls in the darkness can, for example, actually hear small rodents moving about on the forest floor, quickly locate the source of the sound, then swoop and grab. Additionally, owing to their soft, loose feathers, owls' flight is essentially silent, permitting prey little chance of hearing their approach. Owls swallow small prey whole, then instead of digesting or passing the hard bits, they regurgitate bones, feathers, and fur in compact *owl pellets*. These are often found beneath trees or rocks where owls perch and they can be interesting to pull apart to see what an owl has been dining on.

Breeding

Most owls are monogamous breeders. They do not build nests themselves, but either take over nests abandoned by other birds or nest in cavities such as tree or rock holes. Incubation of the 1 to 10 or more eggs (often two to four) is usually conducted by the female alone for 4 to 5 weeks, but she is fed by her mate. Upon hatching, the female broods the young while the male hunts and brings meals. Young fledge after 4 to 6 weeks in the nest.

Lore and Notes

The forward-facing eyes of owls are a trait shared with only a few other animals: humans, most other primates, and to a degree, the cats. Eyes arranged in this way allow for almost complete binocular vision (one eye sees the same thing as the other), a prerequisite for good depth perception, which, in turn, is important for quickly judging distances when catching prey. On the other hand, owl eyes cannot move much, so owls swivel their heads to look left or right.

Owls have a reputation for fierce, aggressive defense of their young; many a human who ventured too near an owl nest has been attacked and had damage done! Owls through history have been regarded as symbols of wisdom (for instance, in ancient Greece) or of death, darkness, and supernatural evil (the Mesopotamians and the Romans).

Status

Owls in Mexico are threatened primarily by forest clearing. However, no Mexican species so far are officially listed as threatened. Most currently threatened owls are Old World species. The endangered Northern Spotted Owl, of which so much is heard in the USA, is a *subspecies* of Spotted Owl; because the other subspecies, the Southern Spotted Owl, which occurs in the southwestern USA and Mexico, is still fairly common, the species as a whole is not presently considered threatened.

Profiles

Ferruginous Pygmy-owl, *Glaucidium brasilianum*, Plate 37a
Mottled Owl, *Ciccaba virgata*, Plate 37b
Vermiculated Screech Owl, *Otus guatemalae*, Plate 37c

15. Goatsuckers

Members of the family of birds known as the *goatsuckers*, or Caprimulgidae, do not actually suck the milk of goats, as legend has it, but they do possess one of the most fanciful of bird names. The *caprimulgids* are a group of about 70 species spread over most of the world's land masses, with the exceptions of northern North America and northern Eurasia, southern South America, and New Zealand. Their closest relatives are probably the owls, and some recent classification schemes place the two in the same avian order. Twelve species occur in southern Mexico. One of them, the PAURAQUE (POWR-ah-kay) (Plate 37) is a commonly seen bird over much of the region and probably the most abundant caprimulgid there and in Central America. Pauraques, like most caprimulgids, inhabit open areas such as grassland and farmland, thickets, parkland, and forest edge areas.

Goatsuckers have a very characteristic appearance. In the New World most range from 16 to 32 cm (6 to 12 in) in length. They have long, pointed wings, medium or long tails, and big eyes. Their small, stubby bills enclose big, wide mouths that they open in flight to scoop up flying insects. Many species have bristles around the mouth area. With their short legs and weak feet, they are poor walkers – flying is their usual mode of locomotion. The plumage of these birds is uniformly cryptic: mottled, spotted, and barred mixtures of browns, grays, tans, and black. They often have white patches on their wings or tails that can be seen only in flight.

Natural History

Ecology and Behavior
Most caprimulgids are night-active birds, with some, such as the COMMON NIGHTHAWK (familiar to North Americans) and LESSER NIGHTHAWK (Plate 37), becoming active at twilight (*crepuscular* is the term ecologists use for such a habit). They feed on insects, which they catch on the wing, either with repeated forays from perched locations on the ground or on tree branches, or, as is the case with Common and Lesser Nighthawks, with continuous, often circling flight. Caprimulgids usually gather at night near bright lights to feast on the light-drawn insects. Most of these birds are not seen during the day unless they are accidentally flushed from their roosting spots, which are either on the ground or on tree branches on which they perch sideways. Because of their camouflage coloring, they are almost impossible to see when perched.

Breeding
Goatsuckers breed monogamously. No nest is built. Rather, the female lays her one or two eggs on the ground, perhaps in a small depression in the soil under the branches of a tree, or on a rock or sandbar. Around human developments, they are known for placing their eggs on bare, gravelly rooftops. Either the female alone or both sexes incubate (both in the PAURAQUE) for 18 to 20 days, and both parents feed insects to the young. Goatsuckers engage in *broken-wing displays* if their nest is approached by a predator or a person, which distracts a predator's attention. They flop about on the ground, often with one or both wings held out as if injured, making gargling or hissing sounds, all the while moving away from the nest.

Lore and Notes
Americans tend to call this group either goatsuckers or *nighthawks*, whereas Europeans prefer *nightjars*.

One of the goatsucker family, North America's COMMON POORWILL, may be the only bird known actually to hibernate, as some mammals do, during very cold weather. During their dormant state, poorwills save energy by reducing their metabolic rate and their body temperature, the latter by about 22 °C (40 °F).

Status
None of Mexico's goatsuckers are threatened. In the New World, the WHITE-WINGED NIGHTJAR of Brazil and the USA's PUERTO RICAN NIGHTJAR occur in very limited areas and are critically endangered. The former is known from only a few old specimens and a few modern sightings in central Brazil. Little is known about the bird, but the area of modern sightings falls within a national park, offering some hope for its survival. The Puerto Rican Nightjar (USA ESA listed) occupies dry forest areas of southwestern Puerto Rico; only several hundred pairs remain.

Profiles
Pauraque, *Nyctidromus albicollis*, Plate 37d
Lesser Nighthawk, *Chordeiles acutipennis*, Plate 37e
Mexican Whip-poor-will, *Caprimulgus arizonae*, Plate 37f

16. Swifts and Swallows

Swifts and *swallows*, although not closely related, are remarkably similar in appearance and habit. Most famously, they pursue the same feeding technique – catching insects on the wing during long periods of sustained flight. The swallows (Family Hirundinidae) are a passerine group, 80 species strong, with a world-wide distribution. Twelve species occur regularly in southern Mexico, but several are present only as over-winter migrants from more northerly breeding areas (for instance, PURPLE MARTIN and TREE, BANK, and BARN [Plate 38] SWALLOWS). Swallows are small, streamlined birds, 11.5 to 21.5 cm (4.5 to 8.5 in) in length, with short necks, bills, and legs. They have long, pointed wings and forked tails, plainly adapted for sailing through the air with high maneuverability. Some are covered in shades of blue, green, or violet, but many are gray or brown. The sexes look alike.

Swifts, although superficially resembling swallows, are actually only distantly related; they are not even classified with the passerines. The 80 or so species of swifts (Family Apodidae) are, in fact, most closely related to hummingbirds. About eight species are found in southern Mexico, some fairly rarely. Swifts, like swallows, are slender, streamlined birds, with long, pointed wings. They are 9 to 25 cm (3.5 to 10 in) long and have very short legs, short tails or long, forked tails, and very small bills. Swifts' tails are stiffened to support the birds as they cling to vertical surfaces. The sexes look alike: sooty-gray or brown, with white, grayish or reddish rumps or flanks. Many are glossily iridescent.

Natural History

Ecology and Behavior

Among the birds, swifts and swallows represent pinnacles of flying prowess and aerial insectivory. It seems as if swifts and swallows fly all day, circling low over water or land, or flying in erratic patterns high overhead, snatching insects from the air. Perpetual flight was in the past so much the popular impression of swifts that it was actually thought that they never landed – that they essentially remained flying throughout most of their lives (indeed, it was long ago believed that they lacked feet; hence the family name, Apodidae, literally, *without feet*). They do land, however, although not often. When they do, they use their clawed feet and stiff tail to cling to and brace themselves against vertical structures. They almost never land on the ground, having trouble launching themselves back into the air from horizontal surfaces. A swift spends more time airborne than any other type of bird, regularly flying all night, and even copulating while in the air (a tricky affair, apparently: male and female are partially in freefall during this activity). Swifts are also aptly named, as they are among the fastest flyers on record.

Swallows also take insects on the wing as they fly back and forth over water and open areas. Some also eat berries. Not quite the terra-phobes that swifts are, swallows land more often, resting usually during the hottest parts of the day. Directly after dawn, however, and at dusk, swallows are always airborne.

Breeding

All swallows are monogamous, many species breeding in dense colonies of several to several thousand nesting pairs. Nests are constructed of plant pieces placed in a tree cavity, burrow, or building, or, alternatively, consist of a mud cup attached to a vertical surface such as a cliff. Both sexes or the female alone incubate the three to seven eggs for 13 to 16 days. Both parents feed nestlings, for 18 to 28 days, until they fledge. Swifts are monogamous and most are colonial breeders, but some species nest solitarily. The sexes share breeding chores. Nests consist of plant pieces, twigs, and feathers glued together with the birds' saliva. One to six eggs are incubated for 16 to 28 days, with young fledging at 25 to 65 days of age.

Ecological Interactions

Swallows, small, vulnerable light-weights, are often under competitive pressure for breeding space from other hole-nesting species. Starlings and some sparrows, for instance, sometimes try to take over nests built by swallows and, indeed, such nest usurpation appears to be the chief cause for the serious decline in numbers of the GOLDEN SWALLOW in Jamaica.

Some swallows, such as CLIFF SWALLOWS, locate their colonies near to or actually surrounding the cliff-situated nests of large hawks (predatory birds that take mostly rodents), basking in the protection afforded by having a nest close to a fearsome predator.

Because swifts and swallows depend each day on capturing enough insects, their daily habits are largely tied to the prevailing weather. Flying insects are thick in the atmosphere on warm, sunny days, but relatively scarce on cold, wet ones. Therefore, on good days, swallows, for instance, can catch their fill of bugs in only a few hours of flying, virtually anywhere. But on cool, wet days, they may need to forage all day to find enough food, and they tend to do so over water or low to the ground, where under such conditions bugs are more available.

Lore and Notes

Swallows have a long history of association with people. The ancient Greeks believed swallows to be sacred birds probably because they nested in and flew around the great temples. In the New World, owing to their insect-eating habits, they have been popular with people going back to the time of the ancient Mayan civilization. Mayans, it is believed, respected and welcomed swallows because they reduced insect damage to crops. In fact Cozumel (the word refers to swallows), off the Quintana Roo coast, is the Island of Swallows. Mayan legend has it that swallows, with their graceful fast flight, were specially charged by the Gods: "Years ago when birds were new on Earth ... Cozumel, the swallow, was a dazzling bright bird ... He was so graceful that the Gods were attracted to Cozumel and agreed to make him their chief messenger. In this important position, Swallow was permitted to sit beside the Great Spirit and to carry the King's royal news. The other birds respected Cozumel because of his job ..." (A. L. Bowes 1964)

People's alterations of natural habitats, harmful to so many species, are often helpful to swallows, which adopt buildings, bridges, road culverts, roadbanks, and quarry walls as nesting areas. BARN SWALLOWS (Plate 38) have for the most part given up nesting in anything other than human-crafted structures. The result of this close association is that, going back as far as ancient Rome, swallows have been considered good luck. Superstitions attached to the relationship abound; for example, it is said that the cows of a farmer who destroys a swallow's nest will give bloody milk. Arrival of the first migratory Barn Swallows in Europe is considered a welcoming sign of approaching spring, as is the arrival of CLIFF SWALLOWS at some of California's old Spanish missions.

Status

None of the swifts or swallows that breed or winter in Mexico are threatened. A few Old World species, from Africa and Asia, are known to be quite rare and are considered threatened; for some others, so little is known that we are uncertain of their populations' sizes or vulnerabilities.

Profiles

White-collared Swift, *Streptoprocne zonaris*, Plate 38a
Vaux's Swift, *Chaetura vauxi*, Plate 38b
Chestnut-collared Swift, *Cypseloides rutilus*, Plate 38c
Gray-breasted Martin, *Progne chalybea*, Plate 38d
Barn Swallow, *Hirundo rustica*, Plate 38e
Northern Rough-winged Swallow, *Stelgidopteryx serripennis*, Plate 38f
Mangrove Swallow, *Tachycineta albilinea*, Plate 39a
Cave Swallow, *Hirundo fulva*, Plate 39b

17. Hummingbirds

Hummingbirds are birds of extremes. They are among the most recognized kinds of birds, the smallest of birds, and undoubtedly among the most beautiful, albeit on a minute scale; fittingly, much of their biology is nothing short of amazing. Limited to the New World, the hummingbird family, Trochilidae, contains about 330 species, about 50 of which occur in southern Mexico; only 11 species occur regularly on the Yucatán Peninsula. The variety of forms encompassed by the family, not to mention the brilliant iridescence of most of its members, is indicated in the names attached to some of the different subgroups: in addition to the hummingbirds proper, there are the *emeralds, sapphires, sunangels, sunbeams, comets, metaltails, fairies, woodstars, woodnymphs, pufflegs, sabrewings, thorntails, thornbills,* and *lancebills.* Hummingbirds occupy a broad array of habitat types, from exposed high mountainsides at 4000 m (13,000 ft) to mid-elevation arid areas to sea level tropical forests and mangrove swamps, as long as there are nectar-filled flowers to provide necessary nourishment.

Almost everyone can identify hummingbirds (call them *hummers* to sound like an expert), being familiar with their general appearance and behavior: very small birds, usually gorgeously clad in iridescent metallic greens, reds, violets, and blues, that whiz by us at high speeds, with the smallest among them resembling nothing so much as large flying insects.

Most hummers are in the range of only 6 to 13 cm (2.5 to 5 in) long, although a few of the larger kinds reach 20 cm (8 in), and they tip the scales at an almost imperceptibly low 2 to 9 g (most being 3 to 6 g) – the weight of a large paper-clip! Bill length and shape varies extensively among species, each bill closely adapted to the precise type of flowers from which a species delicately draws its liquid food. Males are usually more colorful than females, and many of them have *gorgets*, bright, glittering throat patches in red, blue, green, or violet. Not all hummers are so vividly outfitted; one group, called *hermits* (because of their solitary ways), are known for dull, greenish-brown and gray plumages. Hummers have tiny legs and feet; in fact, in some classifications they are included with the swifts in the avian order Apodiformes, meaning *those without feet.*

Natural History

Ecology and Behavior

Owing to their many anatomical, behavioral, and ecological specializations, hummingbirds have long attracted the research attention of biologists; the result is that we know quite a bit about them. These highly active, entertaining-to-watch birds are most often studied for one of four aspects of their biology: flying ability, metabolism, feeding ecology, and their aggressive defense of food resources.

(1) Hummers are capable of very rapid, finely controlled, acrobatic flight, more so than any other kind of bird. The bones of their wings have been modified through their evolutionary history to allow for perfect, stationary hovering flight and also for the unique ability to fly *backwards*. Their wings vibrate in a figure eight-like wingstroke at a speed beyond our ability to see each stroke – up to 80 times per second. Because people usually see hummers only during the birds' foraging trips, they often appear never to land, remaining airborne as they zip from flower to flower, hovering briefly to probe and feed at each. But they do perch every now and again, providing opportunities to get good looks at them.

(2) Hummingbirds have very fast metabolisms, a necessary condition for small, warm-blooded animals. To pump enough oxygen and nutrient-delivering blood around their little bodies, their hearts beat up to 10 times faster than human hearts – 600 to 1000 times per minute. To obtain sufficient energy to fuel their high metabolism, hummingbirds must eat many times each day. Quick starvation results from an inability to feed regularly. At night, when they are inactive, they burn much of their available energy reserves and on cold nights, if not for special mechanisms, they would surely starve to death. The chief method to avoid energy depletion on cold nights is to enter into a sleep-like state of *torpor*, during which the body's temperature is lowered to just above that of the outside world, from 17 to 28 °C (30 to 50 °F) below their daytime operating temperatures, saving them enormous amounts of energy. In effect, they put the thermostat down and hibernate overnight.

(3) All hummingbirds are *nectarivores* – they get most of their nourishment from consuming nectar from flowers. They have long, thin bills and specialized tongues to lick nectar from long, thin flower tubes, which they do while hovering. Because nectar is mostly a sugar and water solution, hummingbirds need to obtain additional nutrients, such as proteins, from other sources. Toward this end they also eat the odd insect or spider, which they catch in the air or pluck off spiderwebs. Some ornithologists believe that insects constitute a larger proportion of hummingbird diets than is generally believed, some going so far as to suggest that some species visit flowers more often to catch bugs there than to gather nectar. Recent research in Central America shows that hummers with strongly curved bills (adapted to feed from curved flower nectar tubes) obtain their protein from spiderwebs, hovering while they glean spiders and bugs from the webs; whereas those with straight bills tend to obtain their protein in "flycatching" mode – via aerial capture of flying wasps and flies. The difference makes sense: flycatching with a long, sharply curved bill, we may assume, would be quite difficult.

(4) Many hummers are highly aggressive birds, energetically defending individual flowers or feeding territories from all other hummingbirds, regardless of species. Not all are territorial, however. Some are *trapline* feeders, repeatedly following a regular route around a section of their habitat, checking the same flowers for nectar, which the flowers replenish at intervals. Some traplines cover as

much as a kilometer (0.6 miles) of habitat. Whether birds defend territories depends usually on whether it is an economically attractive option. If the costs of defense (including forcefully evicting intruders), in terms of the amount of energy expended, exceed the amount that can be gained from feeding on the territory, or if nectar-producing flowers are super-abundant in the environment, providing sufficient food for all, then owning and defending a territory is not worthwhile.

Predators on hummingbirds include small, agile hawks and falcons and also frogs and large insects, such as praying mantises, which ambush the small birds as they feed at flowers. Another hazard is large spider webs, from which sometimes they cannot extricate themselves.

Breeding

Hummingbirds are polygamous breeders in which females do almost all the work. In some species, a male in his territory advertises for females by singing squeaky songs. A female enters the territory and, following courtship displays, mates. Afterwards, she leaves the territory to nest on her own. Other species are *lek breeders*. In these systems, males gather at traditional, communal mating sites called leks. For instance, a lek may be in a cleared spot in the forest undergrowth. Each of 3 to 25 males has a small mating territory in the lek, perhaps just a perch on a flower. The males spend hours there each day during the breeding season, advertising for females. Females enter a lek, assess the displaying males, and choose ones to mate with. A male might spend months at the lek, but only have one 15-minute mating interaction with a female; other males may mate with many females in a season. After mating, females leave the lek or territory and build their nests, which are cup-like and made of plant parts, mosses, lichens, feathers, animal hairs, and spider webs. Nests are placed in the small branches of trees, often attached with spider web. The female lays two eggs, incubates them for 15 to 19 days, and feeds regurgitated nectar and insects to her young for 20 to 26 days, until they fledge. Hummers may be found breeding at any time of year, depending on species and geography. Breeding in a particular area often occurs when flowers are most abundant.

Ecological Interactions

The relationship between hummingbirds (nectar consumers) and the flowering plants from which they feed (nectar producers) is mutually beneficial. The birds obtain a high-energy food that is easy to locate and always available because various flowering plant species, as well as groups of flowers on the same plant, open and produce nectar at different times. The flowering plants, in turn, use the tiny birds as other plants use bees – as pollinators. The nectar is produced and released into the part of flower the hummer feeds from for the sole reason of attracting the birds so that they may accidentally rub up against other parts of the flower (*anthers*) that contain pollen grains. These grains are actually reproductive spores that the flower very much "wants" the bird to pick up on its body and transfer to other plants of the same species during its subsequent foraging, thereby achieving for the flower *cross-pollination* – breeding with another member of its species (many plants also have the ability to pollinate themselves). Flowers that are specialized for hummingbird pollination place nectar in long, thin tubes that fit the shape of the birds' bills and also protect the nectar from foraging insects. Hummingbird-pollinated plants often have red, pink, or orange flowers, colors which render them easily detectable to the birds but, owing to peculiarities of their vision, indistinguishable from the background environment to insects. Furthermore, these flowers are often odorless because birds use colorvision and not smell to find them (whereas nectar-eating insects, which the plants want to discourage, often use odor to find flowers).

Interlopers in this mutualistic interaction are a group of pollen-eating mite species. Mites are miniscule arthropods, allied with the spiders and ticks. Some mites may spend their lives on a single plant, feeding and reproducing, but others, perhaps searching for mates or new sites to colonize, try to reach other plants. Walking to another plant for such a small animal is almost out of the question. What to do? The mites jump onto the bills of hummingbirds when the birds visit flowers and become hitchhikers on the bird, usually holing up in their nostrils. The passengers leap off the bird's bill during a subsequent visit to a plant of the same species that they left, necessary because the mites are specialized for certain plants. Recent research suggests that the passenger mites monitor the scents of flowers to identify the correct type, to know when to get off the bus.

Lore and Notes
Hummingbirds, as one might expect, have been the object of considerable myth and legend. The Mayan tale of how hummingbirds became so bright and beautiful is that, way back, the Great Spirit created the hummingbird as a tiny, plain-looking, delicate bird with exceptional flying prowess. The plain hummingbird was nonetheless happy with its existence, and privileged to be the only bird permitted to drink the nectar of pretty flowers. When the drably attired hummer planned her marriage, all the other birds of the kingdom, including the motmot and oriole, donated colorful feathers, so that the hummer's bridal gown was glittering and showy. The Great Spirit, pleased with the hummingbird and her gown, ruled that she could wear it forever (A. L. Bowes 1964).

The Aztecs also took note of hummers, adorning their ceremonial clothes with the birds' gaudy, metallic feathers. Several groups of Indians used these feathers in their wedding ornaments. Hummingbird bodies have a long mythical history in Latin America of being imbued with potent powers as love charms. Having a dead or stuffed hummer in the hand or pocket is thought by some even today to be a sure way to appear irresistible to a member of the opposite sex. In Mexico, even powdered and minced hummingbird is sold for this purpose.

Status
About 25 hummingbird species in Mexico, Central America, or South America are considered threatened, with about half of them now actually endangered. In Mexico, the MEXICAN WOODNYMPH (of the central Pacific area) is considered vulnerable, and the OAXACA HUMMINGBIRD (which occurs only in the southern part of the state), the WHITE-TAILED HUMMINGBIRD (Guerrero and Oaxaca), and SHORT-CRESTED COQUETTE (Guerrero only) are endangered. All hummers are CITES Appendix II listed.

Profiles
Little Hermit, *Phaethornis longuemareus*, Plate 39c
Fork-tailed Emerald, *Chlorostilbon canivetti*, Plate 39d
Green-breasted Mango, *Anthracothorax prevostii*, Plate 39e
White-bellied Emerald, *Amazilia candida*, Plate 40a
Cinnamon Hummingbird, *Amazilia rutila*, Plate 40b
Rufous-tailed Hummingbird, *Amazilia tzacatl*, Plate 40c
Azure-crowned Hummingbird, *Amazilia cyanocephala*, Plate 40d

Buff-bellied Hummingbird, *Amazilia yucatanensis*, Plate 40e
Wedge-tailed Sabrewing, *Campylopterus curvipennis*, Plate 41a
Violet Sabrewing, *Campylopterus hemileucurus*, Plate 41b
Amethyst-throated Hummingbird, *Lampornis amethystinus*, Plate 41c
Striped-tailed Hummingbird, *Eupherusa eximia*, Plate 41d
Berylline Hummingbird, *Amazilia beryllina*, Plate 41e
White-eared Hummingbird, *Basilinna leucotis*, Plate 42a
Blue-throated Hummingbird, *Lampornis clemenciae*, Plate 42b
Magnificent Hummingbird, *Eugenes fulgens*, Plate 42c
Garnet-throated Hummingbird, *Lamprolaima rhami*, Plate 42d
Green Violet-ear, *Colibri thalassinus*, Plate 42e

18. Trogons

Although not as familiar to most people as other gaudy birds such as toucans and parrots, *trogons* are generally regarded by wildlife enthusiasts as among the globe's most visually striking, glamorous of birds; as such, visitors to southern Mexico should try hard not to miss them. The trogon family, Trogonidae, inhabits tropical and semi-tropical regions in the Neotropics, Africa, and southern Asia. It consists of about 40 species of colorful, medium-sized birds with compact bodies, short necks and short "chickenlike" bills. Considering that trogons are distributed over three widely separated geographic regions, it is striking that the family's body plan and plumage pattern are so uniform. Male trogons are rather consistently described as having metallic or glittering green, blue, or violet heads and chests, with deeply contrasting bright red, yellow, or orange underparts. Females are duller in color, usually with brown or gray heads, but share the males' brightly colored breasts and bellies. The characteristic trogon tail is long and squared-off, with horizontal black and white stripes on the underside. Trogons usually sit erect with their distinctive tails pointing straight to the ground.

One trogon stands out from the flock: the regal-looking RESPLENDENT QUETZAL (Plate 43). Famously described as "the most spectacular bird in the New World," the quetzal generally resembles other trogons, but the male's emerald-green head is topped by a ridged crest of green feathers and, truly ostentatiously, long green plumes extend half a meter (18 in) or more past the end of the male's typical trogon tail. Seeing a male quetzal gracefully swooping low through a forest, its long, trailing plumes flashing by, is frequently mentioned by bird lovers as a supreme experience.

Natural History

Ecology and Behavior
Although trogons are distributed throughout many Neotropical forests, they are not limited to warm areas: some species, such as the quetzal and MOUNTAIN TROGON (Plate 43), inhabit cool cloud forests at elevations up to 3000+ m (10,000 ft). One species, the ELEGANT TROGON, ranges northward into southern Arizona (USA). Trogons are generally observed either solitarily or in pairs, and occasionally in small family groups. In spite of their distinctive calls (typically a *cow-cow-cow* call that is one of a tropical forest's characteristic sounds) and brilliant plumages, trogons can be difficult to locate and even to see clearly when spotted perched on a tree branch. This is because, just like green parrots, partly green trogons easily meld into dark green overhead foliage. (Some biologists, in

fact, suspect that the flashy hues of forest birds such as trogons, so glaring and conspicuous when the birds are viewed in the open, might actually appear as dull and inconspicuous to potential predators within the dark confines of closed forests.) Trogon behavior is not much help because typically these birds perch for long periods with little moving or vocalizing, the better presumably to keep them off some predator's dinner plans. Trogons are best seen, therefore, when flying. This often occurs in sudden bursts as they flip off the branches on which a moment before they sat motionless, and sally out in undulatory, short flights to snatch succulent insects. Trogons can thus be considered *sit-and-wait* predators. They also swoop to grab small lizards, frogs, and snails. Trogons are also partial frugivores, taking small fruits from trees while hovering; particular favorites apparently include figs and avocados.

Breeding
Trogons usually nest in cavities in dead trees or in structures high up in trees, such as termite or wasp nests (they have been observed "taking over" a wasp nest, carving a nest hole in it and, adding insult to injury, feasting daily on the wasps during nesting!). Generally the trogon female incubates her two or three eggs overnight and the male does duty during the day. Incubation is 17 to 19 days. Young are tended by both parents; fledging is at 14 to 30 days. Quetzal nest holes generally are about 10 m (33 ft) high in trees. Both sexes share nest duties and apparently divide each day into two shifts each at the nest. Quetzals are known to defend exclusive territories around their nest tree that average about 700 m (2300 ft) in diameter.

Lore and Notes
Most trogon lore concentrates, as might be expected, on the RESPLENDENT QUETZAL, which is thought to have been revered by, or even sacred to, local peoples going back to ancient Mayan and Incan times. One legend of indigenous Guatemalans is that the quetzal received its showy plumage during the European conquest of the Americas: "After a particularly gruesome battle, huge flocks of quetzals (which were then only green) flew down to keep a watch over the dead Mayans, thus staining their breasts red" (C. M. Perrins 1985). Another story is that the Great Spirit of the Mayans decided one day to choose a king of the birds. The various species competed, showing off their beauty, intelligence, strength, knowledge, or artistry. The quetzal initially held back because, although he was ambitious and proud, his plumage was quite plain. An idea came to the quetzal. He convinced his friend, the roadrunner, to lend him some elegant feathers (promising untruthfully to return them later) and, with the new, long, showy tail feathers, was appointed king of the birds (A. L. Bowes 1964). Ancient Mayans probably only royalty and upper classes – used the long quetzal tail plumes for ceremonial head-dresses, feather capes, and artwork. The chief god of the Mayans was *Kukulcan* and the chief god of the Toltecs was *Quetzalcoatl*, both meaning "The Plumed Serpent." The main road though Cancún's hotel zone, you may notice, is Kukulcán Boulevard.

Status
Most of the eight trogon species that occur in southern Mexico are not currently threatened. The RESPLENDENT QUETZAL, however, is considered by some authorities to be endangered (CITES Appendix I and USA ESA listed), primarily owing to destruction of its cloud forest habitat. Large tracts of cloud forest

continue to be cleared for cattle pasture, agriculture, and logging, especially below 2000 m (6500 ft). Another factor threatening quetzal populations is that local people illegally hunt them with rifles, blowguns, and traps for trade in skins, feathers, and live birds. Because of the quetzal's special prominence as the national bird of Guatemala (depicted on its state seal and on its currency) and as a particularly gorgeous poster animal for conservation efforts, several special reserves have been established for the bird in Guatemala, and quetzals are specially protected in some of Costa Rica's national parks and reserves. A system of cloud forest reserves stretching from Mexico to Panama is required for complete protection of the quetzal and other higher-elevation Central American trogons. The El Triunfo Biosphere Reserve in southern Chiapas, for instance, protects extensive quetzal habitat. One other Mexican trogon, the EARED TROGON, native to the northern part of the country, is seriously threatened.

Profiles

Resplendent Quetzal, *Pharomachrus mocinno*, Plate 43a
Slaty-tailed Trogon, *Trogon massena*, Plate 43b
Violaceous Trogon, *Trogon violaceus*, Plate 43c
Black-headed Trogon, *Trogon melanocephalus*, Plate 43d
Mountain Trogon, *Trogon mexicanus*, Plate 43e

19. Kingfishers

Kingfishers are handsome, bright birds most often encountered along rivers and streams or along the seashore. Classified with the motmots in the Order Coraciiformes, the approximately 90 kingfisher species are grouped in three families and range throughout the tropics, with a few of them pushing deeply northward and southward into temperate areas. Only six kingfisher species, all in the Family Cerylidae, reside in the New World; five are found in southern Mexico (one is migratory from North America). They differ in size (see below), from 12 to 40 cm (5 to 16 in), but all are of a similar form: large heads with very long, robust, straight bills, short necks, short legs, and, for some, highly noticeable crests. The kingfisher color scheme in the New World is also fairly standardized: dark green or blue-gray above, white and/or chestnut-orange below.

Natural History

Ecology and Behavior

New World kingfishers, as the name suggests, are all mainly fish eaters (that is, they are *piscivores*). Usually seen hunting alone, they sit quietly, attentively on a low perch – a tree branch over the water, a bridge spanning a stream – while scanning the water below. When they locate suitable prey, they swoop and dive, plunging head-first into the water (to depths of 60 cm, or 24 in) to seize it. If successful, they quickly emerge from the water, return to the perch, beat the fish against the perch to stun it, then swallow it whole, head first. Thus, kingfishers are sit-and-wait predators of the waterways. They will also, when they see movement below the water, hover over a particular spot before diving in. BELTED KINGFISHERS (Plate 44), migrants that breed to the north and only winter in the covered region, commonly fly out 500 m (¼ mile) or more from a lake's shore to hover 3 to 15 m (10 to 50 ft) above the water, searching for fish. Kingfisher diets are occasionally supplemented with tadpoles and insects. Kingfishers fly fast and purposefully, usually in straight and level flight, from one perch to another; often they are seen only as flashes of blue or green darting along waterways.

Kingfishers are highly territorial, aggressively defending their territories from other members of their species with noisy, chattering vocalizations, chasing, and fighting. They inhabit mostly lowland forests and waterways, but some range up to elevations of 2500 m (8000 ft).

Breeding
Kingfishers are monogamous breeders that nest in holes. Both members of the pair help defend the territory in which the nest is located, and both take turns digging the 0.75 to 1.5 m (2 to 5 ft) long nest burrow into the soft earth of a river or stream bank. Both parents incubate the three to eight eggs, up to 24 hours at a stretch, for a total of 19 to 26 days. The young are fed increasingly large fish by both parents until they fledge at 25 to 38 days old. Fledglings continue to be fed by the parents for up to 10 weeks. At some point after they are independent, the parents expel the young from the territory. Many young kingfishers apparently die during their first attempts at diving for food. Some have been seen first "practicing" predation by capturing floating leaves and sticks.

Ecological Interactions
Mexican kingfishers can be arranged quite nicely by size order, from the largest, the RINGED KINGFISHER (Plate 44), to the smallest, the 12.5 cm (5 in) long AMERICAN PYGMY KINGFISHER (Plate 44). The size order is not accidental. Ecologists believe that such graded variation in size allows all the different kingfisher species to coexist in the same places because, although they are so alike in habits and, especially, in the kinds of foods they eat, they actually avoid serious competition by specializing on eating fish of different sizes. Ringed Kingfishers eat large fish, BELTEDS eat slightly smaller fish, and Pygmys eat tiny fish, tadpoles, and insects. Intriguingly, the larger the species of kingfisher, the higher is the average perch height above the water from which it hunts – from 1.4 m (4 ft) for the smallest species to about 10 m (32 ft) for the largest.

Lore and Notes
Kingfishers are the subject of a particularly rich mythology, a sign of the bird's conspicuousness and its association with water throughout history. In some parts of the world, kingfishers are associated with the biblical Great Flood. It is said that survivors of the flood had no fire and so the kingfisher was chosen to steal fire from the gods. The bird was successful, but during the theft, burned his chest, resulting in the chestnut-orange coloring we see today. According to the ancient Greeks, Zeus was jealous of Alcyone's power over the wind and waves and so killed her husband by destroying his ship with thunder and lightning. "In her grief, Alcyone threw herself into the sea to join her husband, and they both turned immediately into kingfishers. The power that sailors attributed to Alcyone was passed on to the Halcyon Bird, the kingfisher, which was credited with protecting sailors and calming storms" (D. Boag 1982). Halcyon birds were thought to nest 7 days before and 7 days after the winter solstice and these days of peace and calm, necessary to rear young, were referred to as *halcyon* days.

Status
All New World kingfishers are moderately to very abundant; none are considered threatened. Some breed quite well in the vicinity of human settlements. Eleven

species of kingfishers currently reside on lists of vulnerable or threatened animals, but they are Old World in their distributions, most from Polynesia and the Philippines.

Profiles

Belted Kingfisher, *Ceryle alcyon*, Plate 44a
Ringed Kingfisher, *Ceryle torquata*, Plate 44b
Amazon Kingfisher, *Chloroceryle amazona*, Plate 44c
Green Kingfisher, *Chloroceryle americana*, Plate 44d
American Pygmy Kingfisher, *Chloroceryle aene*, Plate 44e

20. Motmots

Motmots – beautiful kingfisher relatives with several distinctive features and a ridiculous name. The name probably originates with the BLUE-CROWNED MOT-MOT (Plate 45), whose common call is *whoot-whoot, whoot-whoot*. Family Momotidae includes only nine species, all limited geographically to the Neotropics. Motmots are mainly residents of low-altitude forests, but occur in other habitats also, such as orchards, tree-lined plantations, suburban parks, and some dryer scrub areas; they are common inhabitants of many archaeological ruin sites. Motmots are colorful, slender, small to medium-sized birds, 18 to 48 cm (7 to 19 in) long. They have fairly long, broad, bills that are downcurved at the end; the bills have serrated edges, adapted to grab and hold their animal prey. The most peculiar motmot feature, however, is the tail. In most motmots, two central feathers of the tail grow much longer than others. Soon, feather barbs near the end of these two feathers drop off, either from the bird's preening or from brushing against tree branches, resulting in short lengths of barbless vane and, below this area, in what is commonly described as a *racquet head* appearance of the feather ends. Male and female motmots are alike in size and coloring. If you get a sufficiently close look, you will discover that motmots are among the most handsome, visually stunning of Mexican birds, with their bodies of blended shades of green, black masks and, in some, brilliant blue head patches.

Natural History

Ecology and Behavior

Motmots are predators on insects (particularly beetles, butterflies, dragonflies, and cicadas), spiders, and small frogs, lizards and snakes, which they snatch in the air, off leaves, and from the ground. Typically they perch quietly on tree branches or on telephone wires or fences, sometimes idly swinging their long tails back and forth, until they spy a suitable meal. They then dart quickly, seize the prey, and ferry it back to the perch for munching. If the item is large or struggling – a big beetle or a lizard – the hungry motmot will hold it tight in its serrated bill and whack it noisily against the perch before swallowing it. Motmots are also frugivores, eating small fruits, up to the size of plums, which they collect from trees while hovering. Motmots are never observed in flocks. They are seen either solitarily or in pairs, and may remain in pairs throughout the year (although the sexes may separate during the day to feed). An unusual feature of their behavior is that motmots are usually active well into the twilight, going to sleep later than do most birds.

Breeding

Some courtship activities have been observed, with male and female motmots calling back and forth high up in the trees, and sometimes holding bits of green leaves in their bills. Motmots are burrow nesters, like their kingfisher cousins. Both male and female dig the burrow, often placed in the vertical bank of a river or roadside. Tunnels up to 4 m (13 ft) long have been uncovered, but most are in the order of 1.5 m (5 ft). Both parents incubate the two to four eggs. Young are fed and brooded by male and female for 24 to 30 days, at the end of which the juvenile motmots are well feathered and able to fly from the burrow entrance.

Lore and Notes

One Mayan legend explains both how motmot tails came to have racquet ends and why motmots are burrow nesters. The motmot, with his brilliant coloring and long tail, considered himself a bit above other, less regal-looking birds. When a large storm was brewing and all the birds set to work to prepare to weather the storm and survive, building dams and storing fruits and seeds, the motmot, too pretty and important to be bothered with work, hid in the brush and went to sleep. He did not notice that his long tail stuck out into a trail, where working birds frequently stepped on it, causing barbs to fall out. The storm never materialized and, later, as all the birds gathered to preen in relief, they laughed at the motmot's ruined tail. The embarrassed motmot fled into the dark forest, dug a burrow, and became an underground recluse (A. L. Bowes 1964).

"It is a paradox that one of the most beautiful birds is hatched and reared in a foul hole in a bank, only to emerge at last with its lovely plumage undefiled. It is still more strange that the TURQUOISE-BROWED MOTMOT (Plate 45) acquires its colors in the earth, for they are not brightly glittering like gems and metals but as soft and delicately blended as the rainbow and the sunset sky" (A. F. Skutch 1983).

Status

Several motmots, including the TODY MOTMOT (Plate 45), are rare over parts of their ranges, but are not considered threatened owing to their greater abundance in other parts of their ranges. The KEEL-BILLED MOTMOT is on some lists of threatened species, chiefly because its extremely patchy distribution is not well understood. That is, Keel-billeds have an extensive range within Central America but they are common only in small, isolated areas within that range, in southern Mexico, Belize, Guatemala, Nicaragua, and Costa Rica.

Profiles

Tody Motmot, *Hylomanes momotula*, Plate 45a
Blue-crowned Motmot, *Momotus momota*, Plate 45b
Turquoise-browed Motmot, *Eumomota superciliosa*, Plate 45c
Russet-crowned Motmot, *Momotus mexicanus*, Plate 45d

21. Woodpeckers

We are all familiar with *woodpeckers*, at least in name and in their cartoon incarnations. These are industrious, highly specialized birds of the forest – where there are trees in the world, there are woodpeckers (excepting only Australia, the polar regions, and some islands). The group, encompassing 200+ species, from the 9-cm (3.5-in) *piculets* to large woodpeckers up to 50 cm (20 in) long, is contained in the Family Picidae, placed, along with the toucans, in the Order Piciformes. Sixteen species of various sizes occur in southern Mexico (one, the YELLOW-BELLIED SAPSUCKER, is a North American migrant, present only during winter) and they occupy diverse habitats and employ various feeding methods. Small and medium-sized birds, woodpeckers have strong, straight, chisel-like bills, very long tongues that are barbed and often sticky-coated, and toes that spread widely, firmly anchoring the birds to tree trunks and branches. They come in various shades of olive-green, brown, and black and white, usually with small but conspicuous head or neck patches of red or yellow. Some have red or brown crests. A few woodpeckers are actually quite showy, and some have striking black and white stripes on chest or back. The sexes usually look alike. Because of the tapping sounds they produce as they butt their bills against trees and wooden structures, and owing to their characteristic stance – braced upright on vertical tree trunks – woodpeckers often attract our notice and so are frequently observed forest-dwellers.

Natural History

Ecology and Behavior

Woodpeckers are associated with trees and are adapted to cling to a tree's bark and to move lightly over its surface, searching for insects; they also drill holes in bark and wood into which they insert their long tongues, probing for hidden bugs. They usually move up tree trunks in short steps, using their stiff tail as a prop, or third foot. Woodpeckers eat many kinds of insects and their larvae that they locate on or in trees. They also are not above a bit of flycatching, taking insects on the wing, and many supplement their diets with fruits, nuts, and nectar. Members of the genus *Melanerpes*, in particular, which includes the GOLDEN-FRONTED and ACORN WOODPECKERS (Plates 46, 47), also eat a lot of fruit. A few species have ants as a dietary staple. The *sapsuckers*, a type of woodpecker, use their bills to drill small holes in trees that fill with sap, which is then eaten. Some woodpeckers also forage on the ground.

Woodpeckers are monogamous, and some live in large family groups. Tropical woodpeckers usually remain paired throughout the year. They sleep and nest in cavities that they excavate in trees. Woodpeckers hit trees with their bills for three very different reasons: for drilling bark to get at insect food; for excavating holes for roosting and nesting; and for *drumming*, sending signals to other wood-peckers. Thus, one never knows upon first hearing the characteristic drumming sound whether the bird in question is feeding, signalling, or carving a new home. Woodpeckers typically weave up and down as they fly (*undulatory* flight), a behavior suited to make it more difficult for predators to track the birds' movements.

Breeding

A mated male and female woodpecker carve a nesting hole in a tree. Sometimes they line the cavity with wood chips. Both sexes incubate the two to four eggs for 11 to 18 days, males typically taking the entire night shift. Young are fed by both parents for 20 to 35 days until they fledge. Juveniles probably remain with the parents for several months more, or longer in those species in which families of up to 20 individuals associate throughout the year.

Ecological Interactions

The different woodpecker species that live in one region usually are graded in size. There are small, medium (for instance, GOLDEN-FRONTED WOODPECKER, Plate 46), and large (PALE-BILLED WOODPECKER, Plate 47) species. The size differences permit coexistence in the same area of such ecologically similar species because birds of various sizes specialize on different foods, thereby reducing negative effects of competition. When there are two or more woodpecker species of about the same size inhabiting the same place, it always appears to be the case that the potential competitors forage in different ways, or for different items – again, eliminating competition that could drive one of the species to extinction in the region over which they overlap.

Woodpeckers as a group have both beneficial and harmful effects on forests. On the one hand, they damage living, dying, and dead trees with their excavations and drilling, but on the other, they consume great quantities of insects, such as *tree borers*, that can themselves significantly damage forests. Because other birds use tree holes for roosting and/or nesting, but do not necessarily or cannot dig holes themselves, sometimes the carpenter-like woodpeckers end up doing the work for them. Many species occupy deserted woodpecker holes. More sinisterly, some birds "parasitize" the woodpecker's work by stealing holes. For instance, COLLARED ARACARIS (Plate 48), medium-sized toucans, have been observed evicting PALE-BILLED WOODPECKERS from their nest holes.

Lore and Notes

In ancient Roman mythology, Saturn's son, Picus, was a god of the forests. The sorceress Circe, attracted to the handsome Picus, courted him, but was rejected. In her wrath, she transformed Picus into a woodpecker, providing the basis for the woodpecker family's name, Picidae. The Mayans of ancient America thought that the woodpecker was a lucky bird, possessor of a lucky green stone that it kept under its wing. The legend was that the luck would be transferred to any person who could find a woodpecker hole and cover it. The bird would excavate a new hole, but that one, too, should be covered. After nine excavations, the woodpecker would drop the charm, allowing the person to claim it. During the height of the Mayan civilization, presumably, there were quite a few highly frustrated woodpeckers.

Woodpeckers damage trees and buildings and also eat fruit from gardens and orchards (especially cherries, apples, pears, and raspberries) and so in some parts of the tropics are considered significant pests and treated as such. The HISPAN-IOLAN WOODPECKER was routinely killed in the Dominican Republic because of its habits of carving holes in royal palms and boring into woody cacao pods to eat the inner pulp and insects.

Status

Two of the largest woodpeckers, which survived until several decades ago, are now probably extinct: northern Mexico's IMPERIAL WOODPECKER, which, at 58 cm (22 in) long, was the largest member of the family; and the USA's IVORY-BILLED WOODPECKER, a victim of the destruction of its old growth river forest habitat. None of Mexico's other woodpeckers are presently threatened, but several species are noticeably declining as forests continue to be cut. Several woodpeckers appear on lists of threatened animals, including the HELMETED WOODPECKER of South America and the RED-COCKADED WOODPECKER (listed USA ESA endangered) of the southeastern USA.

Profiles

Golden-fronted Woodpecker, *Melanerpes aurifrons*, Plate 46a
Golden-olive Woodpecker, *Piculus rubiginosus*, Plate 46b

Yucatán Woodpecker, *Centurus pygmaeus*, Plate 46c
Ladder-backed Woodpecker, *Picoides scalaris*, Plate 46d
Chestnut-colored Woodpecker, *Celeus castaneus*, Plate 47a
Acorn Woodpecker, *Melanerpes formicivorus*, Plate 47b
Lineated Woodpecker, *Dryocopus lineatus*, Plate 47c
Pale-billed Woodpecker, *Campephilus guatemalensis*, Plate 47d
Gray-breasted Woodpecker, *Centurus hypopolius*, Plate 47e

22. Toucans

Spectacular. No other word fits them – *toucans* are spectacular animals. Their shape, brilliant coloring, and tropical quintessence make them one of the most popular "poster animals" for the tropical forests of the Americas and one most visitors want to see. It's hardly surprising, therefore, that the logos of several conservation organizations and tour companies feature toucans. The toucan family, Ramphastidae, is classified with the woodpeckers, and contains about 40 species – the toucans and the usually smaller *toucanets* and *aracaris* (AH-rah-SAH-reez); all are restricted to the American tropics. Three species occur in southern Mexico.

The first sighting of toucans in the wild is always exhilarating – the large size of the bird, the bright colors, the enormous, almost cartoonish bill. Toucans are usually first noticed flying from treetop to treetop in small groups. Your eyes immediately lock onto the flight silhouette; something is different here! As one observer put it, it looks as if the bird is following its own bill in flight (J. C. Kricher 1989). The effect of the bill seeming to lead the bird is that toucans appear unbalanced while flying. The bird's most distinguishing feature, the colorful, disproportionately large bill, is actually light, mostly hollow, and used for cutting down and manipulating the diet staple, tree fruit.

Natural History

Ecology and Behavior

Toucans are gregarious forest birds, usually observed in flocks of 3 to 12. They follow each other in *strings* from one tree to another, usually staying in the high canopy (a toucan only occasionally flies down to feed at shrubs, or to pluck a snake or a lizard from the forest floor). The birds are playful, grasping each other's bills in apparent contests, and tossing fruit to each other. Toucans are primarily fruit-eaters, preferring the darkest, so ripest, fruit. Their long bill allows them to perch on heavier, stable branches and reach a distance for hanging fruits. They snip the fruit off, hold it at the tip of the bill, and then, with a forward flip of the head, toss the fruit into the air and into their throats. (It seems, we humans think, an inefficient eating method, but the toucans do quite nicely with it.) Toucans also increase their protein intake by consuming the occasional insect, spider, or small reptile, or even bird eggs or nestlings. Sometimes individual fruit trees are defended by a mated toucan pair from other toucans or from other frugivorous birds – defended by threat displays and even, against other toucans, by bill clashes.

Breeding

Toucans nest (and some sleep) in tree cavities, either natural ones or those hollowed out by woodpeckers, in either live or dead trees. Nests can be any height above the ground, up to 30 m (100 ft) or more. Both sexes incubate and feed the two to four young. Toucans are apparently monogamous. Some species, such as the COLLARED ARACARI (Plate 48), seem to breed cooperatively; that is, other family members, in addition to the mother and father, help raise the young in a single nest.

Ecological Interactions
Small fruit seeds pass unharmed through toucan digestive tracts and large seeds are regurgitated, also unharmed. Thus, these frugivores aid in the dispersal of tree seeds, and, together with other fruit-eaters, are responsible for the positions of some forest trees. In other words, many forest trees grow not where a parent tree drops its seeds, but where frugivorous birds do so.

Lore and Notes
According to stories and legends, some groups of South American Indians that existed at least until the European conquests (for instance, the U'wa, or Tunebo, of Venezuela and Colombia) recognized frugivorous birds such as toucans as beneficial dispersers of fruit seeds. To reward the animals, the people left fruit and nuts for them whenever crops were harvested.

Toucans are commonly known in many areas of the Neotropics as "Dios te de," (God gives it to you), apparently because the three-syllable call of the large CHESTNUT-MANDIBLED TOUCAN, which occurs from Honduras south to Colombia, sounds like this expression.

"... the toucan traces an undulatory course from treetop to treetop. The sudden opening of the wings imparts ... its peculiar character, in keeping with the whole aspect and behavior of the bird – who is not so much grotesque and ungainly as unexpected, an artist's fantasy come to life in flesh and feathers. Clumsy in appearance, something of an avian clown, the toucan is sufficiently agile to meet all the demands of its arboreal life ..." (A. F. Skutch 1983).

Status
Toucans are common residents in the various regions in which they occur, except where there is extensive deforestation. None of the family are currently threatened in Mexico. Some toucans, e.g., the CHESTNUT-MANDIBLED, have suffered substantial population declines in heavily deforested areas of Central America, for instance, in some regions of Panama. Also, some toucan species may be scarce locally due to hunting. Several toucans, including the KEEL-BILLED (Plate 48), are listed by CITES but, rather than being immediately threatened, they are listed because they are considered "look-alikes" of threatened species, and so they need to be monitored during international trade.

Profiles
Collared Aracari, *Pteroglossus torquatus*, Plate 48a
Keel-billed Toucan, *Ramphastos sulfuratus*, Plate 48b
Emerald Toucanet, *Aulacorhynchus prasinus*, Plate 48c

All of the bird families considered below are of *passerine*, or *perching*, birds, contained within the Order Passeriformes (see p. 122).

23. Woodcreepers

Woodcreepers are small and medium-sized brown birds that pursue a mostly arboreal lifestyle. The family, Dendrocolaptidae, consists of about 50 species, all tree-climbing birds of the Neotropics. They are common in forests at low and moderate elevations, but also at the forest edge and in semi-open areas. In contrast to many other groups, the woodcreepers, including the 12 species that occur in southern Mexico, are fairly uniform in size, plumage colors, and natural history. Most are slender birds, 20 to 36 cm (8 to 14 in) in length. The sexes look alike, with plumages mostly of various shades of brown, chestnut, or tan. Many have white patches of varying dimension on breast, head, or back; and most have some spotting, streaking, or banding, particularly on the chest. Woodcreepers resemble woodpeckers to some extent, with longish bills (some strongly curved downwards) and stiff tails that they use to brace themselves against tree trunks. Owing to their physical similarities, the various woodcreepers are often difficult to tell apart, even for experienced birdwatchers. Body size, bill size and shape, and type of streaking are used to distinguish members of the group.

Natural History

Ecology and Behavior
Woodcreepers feed by moving upwards on tree trunks and also horizontally along branches, peering under bark and into moss clumps and epiphytes, using their long bills to probe and snatch prey in tight nooks and crannies. Unlike woodpeckers, they do not dig holes in search of prey. The foraging technique is quite standardized: a woodcreeper flies to the base of a tree and then spirals up the trunk, using its stiff, spiny tail as a third foot to brace itself in a vertical posture against the tree; it checks for prey as it climbs. At the top, the bird flies down to the base of the next tree, and repeats the process. The group is primarily *insectivorous*, but also takes spiders as well as small lizards and amphibians. Many woodcreepers are frequent participants, with antbirds, tanagers, and motmots, among others, in *mixed-species flocks* (see p. 178) that follow swarms of army ants, taking prey that rush out from hiding places to avoid the voracious ants. Woodcreepers are most often observed singly or in pairs, but occasionally in small family groups. They roost in tree crevices or holes.

Breeding
Most woodcreepers practice standard monogamy, with the sexes equally sharing nesting chores. However, in some, apparently no real pair-bonds are established and, after mating, females nest alone. Nests are usually tree crevices or holes, but sometimes are established in arboreal termite nests. Parents line nests with wood chips. The two or three eggs are incubated for 17 to 21 days and young fledge 18 to 24 days after hatching.

Ecological Interactions
Quite often, a large number of woodcreeper species coexist in the same local area. The food items they take and their foraging methods are quite similar – which ordinarily should engender much harmful competition among the birds and lead to elimination of the weaker competitors. However, the species apparently avoid much direct competition by having sufficiently different body sizes and bill sizes and shapes to support specialization on slightly different prey and foraging on slightly different parts of trees.

Lore and Notes
Some woodcreepers have reputations for being extremely aggressive toward other species, for instance, for harassing and evicting roosting or nesting woodpeckers from tree cavities.

Status

Many woodcreepers are quite abundant in southern Mexico. A few species are uncommon over certain regions, but are not threatened because they are more abundant in other parts of their ranges. One Brazilian species with low population numbers, the MOUSTACHED WOODCREEPER, is considered vulnerable to threat.

Profiles

Tawny-winged Woodcreeper, *Dendrocincla anabatina*, Plate 49a
Olivaceous Woodcreeper, *Sittasomus griseicapillus*, Plate 49b
Ivory-billed Woodcreeper, *Xiphorhynchus flavigaster*, Plate 49c
Spot-crowned Woodcreeper, *Lepidocolaptes affinis*, Plate 49d

24. Antbirds

Antbirds are small and medium-sized, rather drably attired inhabitants of the lower parts of the forest that have intriguing feeding behavior; unfortunately, owing to their usual behavior of hopping about along the dark forest floor, they are difficult to observe. Antbirds (Family Formicariidae) are active passerines, about 251 species strong, which are confined chiefly to the warm forests and thickets of the Neotropics. The family name refers to the feeding behavior of some of the species, which regularly follow ant swarms, snatching small creatures that leave their hiding places to avoid the predatory ants. Nine species occur in southern Mexico.

Antbirds range in size from 8 to 36 cm (3 to 14 in). The smallest are the *antwrens* and *antvireos*, whereas the ones known formally as *antbirds* are mid-sized, and the largest are *antshrikes* and *antthrushes*. In fine detail, these birds are quite varied in appearance and some are boldly patterned, but males mostly appear in understated shades of dark gray, brown, or black, with varying amounts of white on backs, shoulders, or wings. A few species are black and white striped. Female plumage is likewise dull, generally olive, brown, or chestnut. Both sexes in some species have red eyes surrounded by patches of bare skin that are bright blue or other colors.

Natural History

Ecology and Behavior

Antbirds are mostly found at the lower levels of the forest or on the forest floor – they are shade dwellers. Most species practice *insectivory*, although some of the larger ones also eat fruit or small lizards, snakes, and frogs. Some are ant-followers, feeding in *mixed-species flocks* that follow army ants (see below), but others are ground and foliage *gleaners*, many rummaging around on the ground through the leaf litter, tossing dead leaves aside with their bills as they search for insects. When those that follow ants for a living breed, they temporarily cease ant following and establish and defend breeding territories. In some species, family groups remain together, male offspring staying with the parents, even after acquiring mates.

Breeding

Many antbirds appear to mate for life. Courtship feeding occurs in some of these birds, males passing food to females prior to mating. Many antbirds build cup nests out of pieces of plants that they weave together. Nests are usually placed in a fork of branches low in a tree or shrub. Some nest in tree cavities. Male and female share nest-building duty, as well as incubation of the two to three eggs and feeding insects to the young. Incubation is 14 to 20 days and young remain in the nest for 9 to 18 days.

Ecological Interactions

Antbirds are participants in two related ecological interactions that facilitate feeding. Some antbirds are "professional" *ant followers* (that is, they do it all the time), such as the BICOLORED ANTBIRD of southern Central America, and some follow army ant swarms occasionally. Because these birds usually refrain from eating the ants – a high formic acid content makes these insects unpalatable to most – but simply follow swarms of ants, using them to scare food out into the open as hunters use beaters to flush animals from hiding spots, this interaction between birds and bugs is a *commensalism*: one population benefits and the other, the ants, is essentially unaffected by it. Army ant swarms, consisting of from 50,000 to a million tiny carnivores, generally advance across the forest floor with a front that is 3 to 15 m (10 to 51 ft) wide, driving out all animals they encounter; the ants get some and the birds get many of the others.

A second phenomenon, *mixed species flocks* of foragers that roam large territories within the forest, is suspected of being a strong *mutualistic* interaction, in which all participants benefit. Interestingly, we are not sure in this case precisely what those benefits are, but many bird species throughout the tropics participate in such feeding flocks, so benefits to all there must be. Some of the flocks follow army ants, but others do not. Antbirds are regular members of these feeding flocks, with some South American species even appearing to be regular flock "leaders." Other birds join the feeding assemblages occasionally, or for parts of a day – tanagers (frequently), motmots, cuckoos, woodcreepers, ovenbirds. Sometimes upwards of 50 different species are observed moving in a single flock. Individuals tend to remain with the same flock for a year or more, breeding within the flock's territory. These flocks move through forests at a reported typical speed of about 1 km (0.6 miles) every 3 hours, although not in a straight line. A common experience of visitors to tropical forests is that, as they walk along a trail, very few birds are noticed. Suddenly, from nowhere, a large flock appears, birds of many species are everywhere, calling, foraging, fluttering around, flying, moving; then, in a few moments, just as suddenly, they are gone, passed on, and the quiet calm returns.

As you might expect, to avoid direct competition, the various species within a flock will specialize on feeding in particular places with respect to the flock; in fact, on close examination, the feeding flocks appear to be tightly structured. Some species always forage on or near the ground, others on trees at 2, 5 or 10 m (6, 15, 30 ft) up. Some explore live leaves for bugs, others specialize on dead leaves. Some might make constant, short sallies, flycatching. One species might always be at the flock center, another typically at the flock periphery. Some birds concentrate on catching prey that flushes in response to the flock's whirlwind arrival, while others search the leaf litter. Notice the advantage of this arrangement for an individual bird, as opposed to its joining a single-species flock: because feeding behavior is usually the same for all members of a species, in a single-species flock an individual needs to compete continually for the same food in the same part of the forest using the same feeding method.

What are the possible benefits of foraging in large flocks of diverse species? First, there is always safety in numbers – more birds means more eyes and ears available to detect dangerous predators, such as falcons or snakes. Also, some of the flock members (often shrike-tanagers and antshrikes), in shifts, seem to serve as *sentinels,* feeding less, keeping alert for danger, and giving alarm calls when predators are detected. It has been noted that some of the flock participants that feed on the forest floor will not even put their head down into the leaf litter unless sentinels are posted. Further, should a predator strike, the likelihood of any one individual flock member being taken is small. Second, there must be some feeding advantage that results from the moving, mixed-species flock – probably the propensity of the rapidly moving flock to flush insects from hiding places, thus making them easier to find for all.

Lore and Notes
The strange compound names of these birds, such as antwrens, antshrikes, and antthrushes, apparently arose because the naturalists from outside of the Neotropics who named them could not ascertain local names or believed that local people had no names for these species. Although the birds do not resemble, for example, wrens, shrikes, or thrushes, they were so designated owing to their relative sizes.

Status
None of the antbirds of southern Mexico are presently considered threatened, and many are very common, if difficult to see, residents of various forest areas. About 30 species within the family Formicariidae, mostly Brazilian, are considered vulnerable or already threatened; at least four species in Brazil are endangered, including the BLACK-HOODED ANTWREN and the FRINGE-BACKED FIRE-EYE.

Profiles
Barred Antshrike, *Thamnophilus doliatus*, Plate 50a
Black-faced Antthrush, *Formicarius analis*, Plate 50b
Dot-winged Antwren, *Microrhopias quixensis*, Plate 50c
Dusky Antbird, *Cercomacra tyrannina*, Plate 50d
Scaled Antpitta, *Grallaria guatimalensis*, Plate 50e

25. Manakins

The *manakins*, Family Pipridae, are a Neotropical group of about 60 species of small, compact, stocky passerine birds, 9 to 19 cm (3.5 to 7.5 in) long, with short tails and bills, and two attention-grabbing features: brightly colored plumages and perhaps the most elaborate courtship displays among birds. Some male manakins are outstandingly beautiful, predominantly glossy black but with brilliant patches of bright orange-red, yellow, or blue on their heads and/or throats. Some have deep blue on their undersides and/or backs. The exotic appearance of male manakins is sometimes enhanced by long, streamer-like tails, up to twice the length of the body, produced by the elongation of two of the central tail feathers. Females, in contrast, are duller and less ornate, usually shades of yellowish olive-green or gray. To accompany the bird's courtship displays, the wing feathers of some species, when moved in certain ways, make whirring or snapping sounds. Only three species occur in southern Mexico.

Natural History

Ecology and Behavior

Manakins are highly active forest birds, chiefly of warmer, lowland areas, although some range up into cloud forests. Residents of the forest understory, they eat mostly small fruits, which they pluck from bushes and trees while in flight, and they also take insects from the foliage. Manakins are fairly social animals when it comes to feeding and other daily activities, but males and females do not pair. They employ a non-monogamous mating system and, in fact, most of our knowledge about manakin behavior concerns their breeding behavior – how females choose males with which to mate and, in particular, male courting techniques. To use the ornithological jargon, manakins are *promiscuous* breeders. No pair-bonds are formed between males and females. Males mate with more than one female and females probably do the same. After mating, females build nests and rear young by themselves. Males, singly or in pairs, during the breeding season stake out display sites on tree branches, in bushes, or on cleared patches of the forest floor, and then spend considerable amounts of time giving lively vocal and visual displays, trying to attract females. An area that contains several of these performance sites is called a *lek*, and thus manakins, along with other birds such as some grouse, some cotingas, and some hummingbirds, are *lekking* breeders.

At the lek, male manakins *dance*, performing elaborate, repetitive, amazingly rapid and acrobatic movements, sometimes making short up and down flights, sometimes rapid slides, twists, and turn-arounds, sometimes hanging upside down on a tree branch while turning rapidly from side to side and making snapping sounds with their wings. The details of a male's dance are *species specific*, that is, different species dance in different ways. Females, attracted to leks by the sounds of male displays and by their memories of lek locations – the same traditional forest sites are used from one year to the next – examine the energetically performing males with a critical eye and then choose the ones they want to mate with, sometimes making the rounds several times before deciding. In a few species, such as the LONG-TAILED MANAKIN (Plate 51), two and sometimes three males (*duos* and *trios*) join together in a coordinated dance on the same perch. Long-tails in their dance alternate *leapfrog hops* with bouts of slow, *butterfly* flight, and the males jointly give a synchronous call that sounds like "toledo" up to 19 times per minute. A duo may give up to 5100 "toledos" in a single day, apparently alerting passing females that the males are ready to display and mate. In these curious cases, one male is dominant, one subordinate, and only the dominant of the pair eventually gets to mate with interested females. Why the subordinate male appears to help the dominant one obtain matings (Are they closely related? Do subordinate males stand to inherit display sites when the dominants die? Do subordinates achieve "stolen" matings with females when the dominants are temporarily distracted?), why the manakins dance at all, and on what basis females choose particular males to become the fathers of their young, are all areas of continuing scientific inquiry.

Breeding

Male manakins take no part in nesting. The female builds a shallow cup nest that she weaves into a fork of tree branches, 1 to 15 m (3 to 51 ft) off the ground. She incubates the one to two eggs for 17 to 20 days, and rears the nestlings herself, bringing them fruit and insects, for 13 to 20 days.

Ecological Interactions

Manakins, like most birds that use open, cup-like nests, often suffer very high rates of nest destruction. In one small study, only about 7% of eggs survived the incubation stage and hatched. Most nests were lost to predators, for which the suspect list is quite lengthy: ground-dwelling as well as arboreal snakes, birds such as motmots, puffbirds, toucans, and magpie-jays, large arboreal lizards, and mammals such as opossums, monkeys, kinkajous, and coati.

Largely frugivorous, manakins are important seed dispersers of the fruit tree species from which they feed (see p. 182).

Lore and Notes

Colorful manakin feathers were often used by the indigenous peoples of Central and South America for ornamental purposes, especially for clothing and masks used during dances and solemn festivals.

Status

None of Mexico's manakins are currently considered threatened. Several South American species possibly face trouble, including the GOLDEN-CROWNED MAN-AKIN of Brazil and the BLACK-CAPPED MANAKIN of Argentina and Brazil, both of which are considered vulnerable to threat.

Profiles

White-collared Manakin, *Manacus candei*, Plate 51a
Red-capped Manakin, *Pipra mentalis*, Plate 51b
Long-tailed Manakin, *Chiroxiphia linearis*, Plate 51c

26. Cotingas

Owing to their variety of shapes, sizes, ecologies, and breeding systems, as well as to their flashy coloring, *cotingas* are usually considered to be among the Neotropic's glamour birds. The family, Cotingidae, is closely allied with the *manakins* and contains about 65 species that live primarily in lowland tropical forests. All observers of these birds stress the group's diversity. The cotingas include tiny, warbler-sized birds and large, crow-sized birds (in fact, the group contains both some of the smallest and largest passerine birds); fruit-and-insect eaters but also some that eat only fruit (which among birds is uncommon); species in which the sexes look alike and many in which males are spectacularly attired in bright spectral colors but females are plain; territorial species that breed monogamously and *lekking* species that breed *promiscuously* (see below); and, without doubt, some of the strangest looking birds of Neotropical forests.

The classification of the cotingas is controversial. If two closely related groups, the *tityras* and *becards*, are included with them, there are a total of about nine species that occur in southern Mexico. Perhaps the only generalizations that apply to all cotingas is that they have short legs and relatively short, rather wide bills, the better to swallow fruits. Males of some of the group are quite ornate, with patches of gaudy plumage in unusual colors. For instance, some of the typical cotingas are lustrous blue and deep purple, and some are all white; others are wholly black, or green and yellow, or largely red or orange or gray. Unusual-looking cotingas include two from southern Central America: the male THREE-WATTLED BELLBIRD, which is brown with a shiny white head and attached to its bill area are three hanging, worm-like, darkly colored wattles; and the BARE-NECKED UMBRELLABIRD, which has an umbrella-shaped black crest.

Natural History

Ecology and Behavior

Cotingas primarily inhabit the high canopy of the forest. They are fruit specialists, a feature of their natural history that has engendered much study. They eat small and medium-sized fruits that they take off trees, often while hovering. Some cotingas, such as the *pihas* and *fruitcrows*, supplement the heavily frugivorous diet with insects taken from the treetop foliage, but others, particularly the *bellbirds*, feed exclusively on fruit – which has both benefits and problems (see below).

Some cotingas pair up, defend territories, and breed conventionally in apparent monogamy. But others, such as umbrellabirds and bellbirds, are *lekking* species, in which males individually stake out display trees and repeatedly perform vocal and visual displays to attract females. Females enter display areas (*leks*), assess the jumping and calling males, and choose the ones they wish to mate with. In this type of breeding, females leave after mating and then nest and rear young alone.

Breeding

Cotinga nests, usually placed in trees or bushes, are generally small, open, and inconspicuous, some nest cups being made of loosely arranged twigs, some of mud, and some of pieces of plants. Many species lay only a single egg, some one or two eggs. Incubation is 17 to 28 days and the nestling period is 21 to 46 days, both stages quite long for passerine birds.

Ecological Interactions

A key feature of tropical forests, and of the animal communities that inhabit them, is the large number of birds (cotingas, finches, manakins, parrots, orioles, tanagers, toucans, and trogons make up a partial list) that rely on fruit as a diet staple. Frugivory (fruit-eating) represents a trade, each participant – the fruit-bearing tree and the fruit-eating bird – offering the other something of great value (and therefore it is a kind of *mutualism* – p. 66). The complex web of relationships between avian fruit-eaters and fruit-producing trees is particularly interesting because it nicely demonstrates ecological interactions between plants and animals. Birds benefit greatly from the association because, for several reasons, fruit is an excellent type of food on which to specialize. It is conspicuous, abundant, and, in the tropics, usually available year-round. Also, unlike some other food types, like insects, fruit is easy to find and devour – it rarely hides or resists being eaten. But fruit is not ideal. Problems involved with eating mostly or only fruit include the fact that fruit is relatively low in protein so, for example, nestlings fed only fruit grow very slowly (as is the case in some cotingas). Trees benefit from the relationship by having birds eat, transport, and then drop their seeds far away, thereby achieving efficient, successful reproduction – something well worth their investment in fruit production. Seeds dropped right near a parent tree must compete with the parent for sun and soil nutrients, and most often do not survive. Seeds dropped at a distance from a parent tree have a much greater chance of survival. Parent trees, therefore, make use of birds as winged, animate, seed *dispersal agents*. Because of the cotingas' feeding specialization on fruit, they are considered to be major dispersers of tree seeds. Owing to the cotingas' high-canopy habits, it is often difficult to determine the precise fruits they go after. They are believed to feed heavily at palms, laurels, and incense trees, and also at, among others, members of the blackberry/raspberry family.

Status

In addition to several threatened and endangered South American species, four Central American cotingas are considered threatened: the BARE-NECKED UMBRELLABIRD, THREE-WATTLED BELLBIRD, YELLOW-BILLED COTINGA, and TURQUOISE COTINGA. The umbrellabird, for instance, is sparsely distributed in several localized areas scattered throughout its range (Costa Rica and Panama) and, often, only a few individuals can be found where it does occur. Leks that have been discovered have contained only three to six displaying males. The prime threat to umbrellabirds and other cotingas is deforestation; a particular problem for the umbrellabird, an *altitudinal migrant*, is that to live and reproduce it requires both highland forest breeding habitat and lowland forest wintering habitat. Thus, for the bird to survive, both types of forests in a single area, together with connecting forest corridors, must be preserved.

Profiles

Rufous Piha, *Lipaugus unirufus*, Plate 51d
Lovely Cotinga, *Cotinga amabilis*, Plate 51e
Masked Tityra, *Tityra semifasciata*, Plate 52a
Rose-throated Becard, *Pachyramphus aglaiae*, Plate 52b

27. American Flycatchers

The American *flycatchers* comprise a huge group of passerine birds that is broadly distributed over most habitats from Alaska and northern Canada to the southern tip of South America. The flycatcher family, Tyrannidae, is among the most diverse of avian groups. With about 380 species, flycatchers usually contribute a hefty percentage of the avian biodiversity in every locale. For instance, it has been calculated that flycatchers make up fully one-tenth of the land bird species in South America, and perhaps one-quarter of Argentinian species. The group is represented in southern Mexico by a healthy contingent of about 65 species.

Flycatchers range in length from 6.5 to 30 cm (2.5 to 12 in). At the smallest extreme are some of the world's tiniest birds, weighing, it is difficult to believe, only some 7 g (¼ oz). Bills are usually broad and flat, the better to snatch flying bugs from the air. Tail length is variable, but some species have very long, forked tails, which probably aid the birds in their rapid, acrobatic, insect-catching maneuvers. Most flycatchers are dully turned out in shades of gray, brown, and olive-green; many species have some yellow in their plumage, and a relatively few are quite flashily attired in, for example, bright expanses of red or vermilion. One set of frequently seen flycatchers, the best known of which is the GREAT KISKADEE (Plate 53), share a common, bright color scheme, with yellow chests and bellies, and black and white striped heads. A great many of the smaller, drabber flycatchers, clad in olives and browns, are extremely difficult to tell apart in the field, even for experienced birdwatchers. Flycatcher sexes are usually similar in size and coloring.

Natural History

Ecology and Behavior

Flycatchers are common over a large array of different habitat types, from high elevation mountainsides, to lowland moist forests, to treeless plains and grasslands, to marshes and mangrove swamps; they are especially prevalent in rainforests. As their name implies, most flycatchers are *insectivores*, obtaining most of their food by employing the classic flycatching technique: they perch motionless on tree or shrub branches or on fences or telephone wires, then dart out in short, swift flights to snatch from the air insects foolhardy enough to enter their field of vision; they then return time and again to the same perch to repeat the process. Many flycatchers also take insects from foliage as they fly through vegetation, and many supplement their diets with berries and seeds. Some of the larger flycatchers will also take small frogs and lizards, and some, such as the GREAT KISKADEE, consider small fish and tadpoles delicacies to be plucked from shallow edges of lakes and rivers. A few species have ceded flycatching to their relatives and now eat only fruit. Almost all of the relatively few flycatchers that have been studied inhabit exclusive territories that mated pairs defend for all or part of the year.

Breeding

Flycatchers are mainly monogamous. Some forest-dwelling species, however, breed *promiscuously*: groups of males call and display repeatedly at traditional courting sites called *leks*, attracting females that approach for mating but then depart to nest and raise young by themselves. Many flycatchers are known for spectacular courtship displays, males showing off to females by engaging in aerial acrobatics, including flips and somersaults. In monogamous species, males may help the females build nests. Some build cup nests, roofed nests, or globular hanging nests placed in trees or shrubs, others construct mud nests that they attach to vertical surfaces such as rock walls, and some nest in holes in trees or rocks. Tropical flycatchers generally lay two eggs that are incubated by the female only for 12 to 23 days; nestlings fledge when 14 to 28 days old.

Ecological Interactions

Some flycatchers show marked alterations in their lifestyles as seasons, locations, and feeding opportunities change. Such ongoing capacity for versatile behavior in response to changing environments is considered a chief underlying cause of the group's great ecological success. An excellent example is the EASTERN KINGBIRD'S drastic changes in behavior between summer and winter. Breeding during summer in North America, these flycatchers are extremely aggressive in defending their territories from birds and other animals, and they feed exclusively at that time on insects. But a change comes over the birds during the winter, as they idle away the months in South America's Amazon Basin. There, Eastern Kingbirds congregate in large, non-territorial flocks with apparently nomadic existences, and they eat mostly fruit.

Some small flycatchers, known as *tody flycatchers*, construct large, hanging, woven, or *felted*, nests that take up to a month or more to build. These nests tend to hang from slender vines or weak tree branches, which provides a degree of safety from climbing nest predators such as snakes and small mammals. Often, however, such efforts are ineffective – nest predation rates are quite high. In response, some of the tody flycatchers purposefully build their nests near to colonies of stinging bees, apparently seeking additional protection from predators.

Lore and Notes

Of all the groups of birds, it is probably among the flycatchers that the most undiscovered species remain. This distinction is owing to the group's great diversity, its penetration into essentially all terrestrial habitats, and the inconspicuousness of many of its members. In fact, as people reach previously inaccessible locations – hidden valleys, cloud-draped mountain plateaus – in the remotest parts of South America, previously unknown flycatchers are indeed sighted. One new species was first identified in 1976 in northern Peru, and another, weighing only 7 g (¼ oz), was found in southern Peru in 1981. Two more species were first described in the scientific literature in 1997.

Status

The BELTED FLYCATCHER, endemic to Chiapas and southern Guatemala, is considered by some authorities to be threatened, as is the WILLOW FLYCATCHER (USA ESA listed) in northern Mexico and southwestern USA. No other Mexican species are known to be threatened. Probably the only flycatchers in Central America that are near-threatened are the TAWNY-CHESTED FLYCATCHER of Costa Rica and Nicaragua, and perhaps the COCOS FLYCATCHER, which is endemic to Costa Rica's Cocos Island. At least six South American species are presently endangered, including the very rare ASH-BREASTED TIT-TYRANT, which is confined to small, wooded patches of Peru and adjacent Bolivia.

Profiles

Vermilion Flycatcher, *Pyrocephalus rubinus*, Plate 52c
Scissor-tailed Flycatcher, *Tyrannus forficatus*, Plate 52d
Dusky Flycatcher, *Empidonax oberholseri*, Plate 52e
Hammond's Flycatcher, *Empidonax hammondii*, Plate 52f
Tropical Kingbird, *Tyrannus melancholicus*, Plate 53a
Great Kiskadee, *Pitangus sulphuratus*, Plate 53b
Boat-billed Flycatcher, *Megarhynchus pitangua*, Plate 53c
Social Flycatcher, *Myiozetetes similis*, Plate 53d
Sulphur-bellied Flycatcher, *Myiodynastes luteiventris*, Plate 53e
Cassin's Kingbird, *Tyrannus vociferans*, Plate 53f
Brown-crested Flycatcher, *Myiarchus tyrannulus*, Plate 54a
Dusky-capped Flycatcher, *Myiarchus tuberculifer*, Plate 54b
Yucatán Flycatcher, *Myiarchus yucatanensis*, Plate 54c
Tropical Pewee, *Contopus cinereus*, Plate 54d
Tufted Flycatcher, *Mitrephanes phaeocercus*, Plate 54e
Greater Pewee, *Contopus pertinax*, Plate 54f

28. Wrens

Wrens are small brownish passerines with an active, snappish manner and, characteristically, upraised tails. The approximately 60 wren species comprise the family Troglodytidae, a group for the most part confined in distribution to the Western Hemisphere. About 25 species occur in southern Mexico. Among other traits, wrens are renowned for their singing ability, vocal duets, and nesting behavior. They range in length from 10 to 20 cm (4 to 8 in) and usually appear mainly in shades of brown or reddish brown, with smaller bits of gray, tan, black, and white. Some of these birds are tiny, weighing in at less than 15 g, or half an ounce. Wings and tails are frequently embellished with finely barred patterns. Wrens have rather broad, short wings and owing to this, are considered poor flyers. The sexes look alike. Wren's tails may be the group's most distinguishing feature, much of the time being held stiffly upright, at military attention. Tails are waved back and forth during both courtship and aggressive displays.

Natural History

Ecology and Behavior

Wrens are cryptically colored and fairly secretive in their habits as they flip, flutter, hop, and poke around the low levels of the forest, and through thickets, grasslands, and marshes, searching for insects. They are completely insectivorous or nearly so. Often spending the year living in pairs, they defend territories in which during the breeding season they will nest. Some of the larger wrens, such as the BANDED-BACKED WREN (Plate 55), spend their days in pairs or small family flocks, and, owing to their size, are a bit bolder in their movements. After using their nests for breeding, wrens will use them as roosting places – or "dormitories," as one researcher puts it. The vocalizations of wrens have been studied extensively. A pair will call back and forth as they lose sight of each other while foraging in thickets, keeping in contact. In some species, mated pairs sing some of the bird world's most complex duets, male and female rapidly alternating in giving parts of one song (as we think of it), so rapidly and expertly that it actually sounds as if one individual utters the entire sequence. Such duets probably function as "keep-out" signals, warning away from the pair's territory other members of the species, and in maintaining the pair-bond between mated birds. Other wrens, such as North America's WINTER WREN, have amazingly complex songs, long trains of notes in varied sequences up to 10 seconds or more in duration; researchers place these vocalizations at the very pinnacle of bird song complexity.

Breeding

Wrens of southern Mexico are mainly monogamous, but some breed *cooperatively*, with members of the small family group helping out at the single nest of the parents. Nests, generally of woven grass, are placed in vegetation or in tree cavities. They are small but elaborate nests, roofed, with inconspicuous side entrances. Intriguingly, in some species the male builds many more nests on his territory than his mate (or mates, in polygynous species) can use, apparently as a courtship signal, perhaps as an inducement for a female to stay and mate. Only the female incubates the two to five eggs, for 13 to 19 days. Sometimes she is fed by the male during this period. Nestlings are fed by both parents for 14 to 19 days, until fledging.

Ecological Interactions

Some wrens enjoy a *commensal* relationship with ants. They often place their nests in cactus or acacia (pp. 27, 66) plants that serve as residences for ant colonies. The ants do not bother the birds and vice versa, yet the birds' nests bask in the ants' formidable protective powers against such possible egg and nestling predators as snakes and small mammals.

Lore and Notes

The HOUSE WREN (Plate 55) has all but abandoned breeding in its forgotten natural haunts and now, quite profitably, associates itself with people. These birds root about near and in human settlements, looking for insects. Nests are often placed in crannies and crevices within buildings or other structures. (Many wrens nest in naturally occurring cavities, hence the family name, Troglodytidae, or *cave dweller*.) The bird has been so successful living with people that it is quite common throughout its range, from Canada to the southern tip of South America. The House Wren on Cozumel – the island's only wren – is considered by most to be a separate species, called the COZUMEL WREN.

From European antiquity come tales of the wren being considered among bird-dom's royalty, perhaps even the very King of Birds. One story of how the wren was crowned, likely one of Aesop's famous fables, is that in a contest to prove royal worth, the wily wren flew higher than any other bird by hiding on the back of the ascending eagle, then flying higher itself when the eagle tired.

Status

Most of the wrens that occur in southern Mexico are fairly or very abundant. However, three Mexican species are considered vulnerable to threat, mostly owing to small and/or declining populations: the SUMICHRAST'S SLENDER-BILLED WREN (which occurs only in extreme northern Oaxaca and central Veracruz), NAVA'S SLENDER-BILLED WREN (occurs only in a small region where Veracruz, Oaxaca, and Chiapas meet), and an island species, the CLARION WREN. Three wren species are endangered: Cuba's ZAPATA WREN and Colombia's APOLINAR'S and NICEFORO'S WRENS.

Profiles

Spot-breasted Wren, *Thryothorus maculipectus*, Plate 55a
White-breasted Wood-wren, *Henicorhina leucosticta*, Plate 55b
Southern House Wren, *Troglodytes aedon*, Plate 55c
Banded-backed Wren, *Campylorhynchus zonatus*, Plate 55d
White-bellied Wren, *Uropsila leucogastra*, Plate 55e
Canyon Wren, *Catherpes mexicanus*, Plate 55f

29. Thrushes and Mockingbirds

The more than 300 species of *thrushes* inhabit most terrestrial regions of the world and include some of the most familiar park and garden songbirds. The family, Turdidae, has few defining, common features that set all its members apart from other groups, as perhaps could be expected; so large an assemblage of species is sure to include a significant amount of variation in appearance, ecology, and behavior. Thrushes as a group are tremendously successful birds, especially when they have adapted to living near humans and benefiting from their environmental modifications. Most obviously, on five continents, a thrush is among the most common and recognizable garden birds, including North America's AMERICAN ROBIN, Europe's REDWING and BLACKBIRD, and Mexico's CLAY-COLORED ROBIN (Plate 56). Thrushes of the Western Hemisphere are slender-billed birds that range from 12.5 to 31 cm (5 to 12 in) in length. Generally they are not brightly colored; instead, they come in drab browns, grays, brown-reds, olive, and black and white. The sexes are very similar in appearance. During their first year of life, young thrushes are clad in distinctively spotted plumages.

Family Mimidae is a New World group that consists of about 30 species of mid-sized, often slender songbirds, 20 to 32 cm (8 to 13 in) long, with characteristically long tails; they are known variously as *mockingbirds, thrashers,* and *catbirds*. Most are brown or gray with lighter chests, which are often streaked or spotted; within a species, the sexes generally look alike.

Natural History

Ecology and Behavior

Among the thrushes are species that employ a variety of feeding methods and that take several different food types. Many eat fruits, some are primarily insectivorous, and most are at least moderately omnivorous. Although arboreal birds, many thrushes frequently forage on the ground for insects, other arthropods, and, a particular favorite, delicious earthworms. The thrushes associated with gardens and lawns in southern Mexico usually forage like the familiar thrushes from North America or Europe – they hop and walk along the ground, stopping at intervals, cocking their heads to peer downwards. These birds are residents of many kinds of habitats – forest edge, clearings, and other open areas such as shrub areas and grasslands, gardens, parks, suburban lawns, and agricultural areas. Many thrushes are quite social, spending their time during the non-breeding season in flocks of the same species, feeding and roosting together. Some of the tropical thrushes make seasonal migrations from higher to lower elevations, following abundant food supplies.

Mockingbirds and catbirds are mostly birds of the ground, shrubs, and low trees. They forage on the ground in open areas and gardens for insects and other small invertebrate animals, and also take some fruit. These birds, as a group, are known for their virtuoso singing performances, their highly intriguing ability to closely mimic the songs of other species, and their aggressive territoriality during breeding seasons – many a person who wandered innocently across a mockingbird territory during nesting has been hit on the head by the swooping mockers.

Breeding

Thrushes breed monogamously, male and female together defending exclusive territories during the breeding season; pairs may associate year round. Nests, usually built by the female and placed in tree branches, shrubs, or crevices, are cup-shaped, made of grass, moss, and like materials, and often lined with mud. Two to six (usually two or three) eggs are incubated by the female only for 12 to 14 days. Young are fed by both parents for 12 to 16 days prior to their fledging. Likewise, catbirds and mockingbirds are monogamous. Cup nests are built of sticks and leaves by both sexes or by the female alone; she also incubates the two to five eggs for about 12 days. Young are fed in the nest by both sexes for 11 to 14 days until they fledge.

Lore and Notes

English colonists in the New World gave the AMERICAN ROBIN, a thrush, its name because it resembled England's common ROBIN – both birds have reddish breasts. The New World bird, however, is more closely related to Europe's BLACK-BIRD, also a common garden bird and a true thrush. Not content with incorrectly labeling birds that were new to them with English names, British settlers around the world, homesick, it is thought, imported birds from the British Isles to their new domains so that familiar birds would surround their new homes. The effects on native birds were often disastrous, and European thrushes such as the SONG THRUSH and BLACKBIRD, among many other birds, are now naturalized, dominant citizens of, for instance, Australia and New Zealand.

Status

None of the approximately 20 species of thrushes that occur in southern Mexico are considered threatened and many are very common birds. One of Mexico's mockers, the SOCORRO MOCKINGBIRD, endemic to one of Mexico's Pacific islands, is endangered; probably only a few hundred individuals survived as of the early 1990s. The BLACK CATBIRD (Plate 56), endemic to the Yucatán Peninsula, is apparently rare over much of its range, and is considered threatened on some of Belize's offshore islands.

Profiles

Wood Thrush, *Hylocichla mustelina*, Plate 56a
Clay-colored Robin, *Turdus grayi*, Plate 56b
White-throated Thrush, *Turdus assimilis*, Plate 56c
Tropical Mockingbird, *Mimus gilvus*, Plate 56d
Black Catbird, *Melanoptila glabrirostris*, Plate 56e
Blue Mockingbird, *Melanotis caerulescens*, Plate 56f
Blue-and-white Mockingbird, *Melanotis hypoleucus*, Plate 56

30. Jays

Jays are members of the Corvidae, a passerine family of 100 or so species that occurs just about everywhere in the world – or, as ecologists would say, *corvid* distribution is *cosmopolitan*. The group includes the *crows, ravens,* and *magpies.* Twelve species occur in southern Mexico. Although on many continents birds of open habitats, Neotropical jays are primarily woodland or forest birds. Jays, aside from being strikingly handsome birds, are known for their versatility, adaptability, and for their seeming intelligence; in several ways, the group is considered by ornithologists to be the most highly developed of birds. They are also usually quite noisy.

Members of the family range in length from 20 to 71 cm (8 to 28 in), many near the higher end – large for passerine birds. Corvids have robust, fairly long bills and strong legs and feet. Many corvids (crows, ravens, rooks, jackdaws) are all or mostly black, but the jays are different, being attired in bright blues, purples, greens, yellows, and white. American jays tend to be blue, and many, such as STELLER'S JAY (Plate 57), have conspicuous crests. Mexico's BROWN JAY (Plate 57) is an exception among Neotropical jays, being plain brown and white and crest-less; owing to this, even researchers who work with the bird affectionately refer to it as the most homely of the group. In corvids, the sexes generally look alike. The COMMON RAVEN, all black, is the largest passerine bird; it ranges over most of Mexico save for the eastern coast and Yucatán region.

Natural History

Ecology and Behavior

Jays eat a large variety of foods (and try to eat many others) and so are considered omnivores. They feed on the ground, but also in trees, taking birds' eggs and nestlings, carrion, insects (including some in flight), and fruits and nuts. Bright and versatile, they are quick to take advantage of new food sources and to find food in agricultural and other human-altered environments. Jays use their feet to hold food down while tearing it with their bills. Hiding food for later consumption, *caching,* is practiced widely by the group.

Corvids are usually quite social, Mexican jays being no exception. BROWN JAYS, for instance, remain all year in small groups of relatives, 5 to 10 individuals strong, that forage together within a restricted area, or *home range,* and at the appropriate time, breed together on a group-defended territory. They inhabit more open wooded areas, forest edges, fields, plantations, and habitat along waterways and near human settlements. Jays are raucous and noisy, giving varieties of harsh, grating, loud calls (some that sound like *jay*) as the foraging flock straggles from tree to tree.

Breeding

YUCATÁN, GREEN, BROWN and some populations of SCRUB JAYS (Plate 57) breed cooperatively. Generally the oldest pair in the group breeds and the other members serve only as *helpers*, assisting in nest construction and feeding the young. Courtship feeding is common, the male feeding the female before and during incubation, which she performs alone. Bulky open nests, constructed primarily of twigs, are placed in trees or rock crevices. Two to seven eggs are incubated for 16 to 21 days, the young then being fed in the nest by parents and helpers for 20 to 24 days.

Ecological Interactions

Jays and other corvids are often scavengers. Other carrion-eating birds exist, such as vultures, but jays and crows and their relatives contribute a good deal to breaking down dead animals so that the nutrients bound up in them are recycled into food webs. (And where there are roads, civic-minded corvids assist highway departments in keeping them clear of automobile-killed animals.) Corvids' omnivory also drives them to be predators on bird nests – generally on species that are smaller than they are, of which there are many. Jays, crows and magpies tear up nests and eat eggs and nestlings. They are considered to be responsible for a significant amount of the nest predation on many songbird species, particularly those with open-cup nests. Owing to their seed-caching behavior, jays are important to trees as dispersal agents. In the USA, for example, the BLUE JAY'S acorn-burying habit must surely result in the maintenance and spread of oak forests.

Lore and Notes

Considered by many to be among the most intelligent of birds, and by ornithologists as among the most highly evolved, corvids have a folklore rife with tales of crows, ravens and magpies as symbols of ill-omen. This undoubtedly traces to the group's frequently all-black plumage and habit of eating carrion, both sinister traits. Ravens, in particular, have long been associated in many Northern cultures with evil or death, although these large, powerful birds also figure more benignly in Nordic and Middle Eastern mythology. Several groups of indigenous peoples of northwestern North America consider the COMMON RAVEN sacred and sometimes, indeed, as a god. As for jays themselves, their loud, chattering and scolding voices together with their physical beauty, have sometimes earned for these birds derogatory reputations. In fact, in Shakespearian England, "jay" was a term for a dishonest or loose woman.

Status

Most corvids are common or very common birds. Some adjust well to people's activities, indeed often expanding their ranges when they can feed on agricultural crops. Two smaller blue jays that occur only over small parts of southern Mexico, DWARF (central Veracruz and northern Oaxaca) and WHITE-THROATED JAYS (Guerrero and Oaxaca), are considered endangered. The only other corvids in real trouble are two island species, the HAWAIIAN CROW and the MARIANA CROW, both of which are critically endangered. SCRUB JAYS in Florida (USA), members of the same species as the MEXICAN SCRUB JAY, are considered threatened (USA ESA listed).

Profiles

Brown Jay, *Cyanocorax morio*, Plate 57a
Green Jay, *Cyanocorax yncas*, Plate 57b

Yucatán Jay, *Cyanocorax yucatanicus*, Plate 57c
White-throated Magpie-jay, *Calocitta formosa*, Plate 57d
Steller's Jay, *Cyanocitta stelleri*, Plate 57e
Mexican Scrub Jay, *Aphelocoma coerulescens*, Plate 57f

31. Warblers and Other Small, Flitting Birds

Birdwatchers and perhaps other users of this book realize that there are a large number of tiny birds that flit about trees, shrubs, thickets, and grasslands, birds that go by such names as *vireo, warbler, gnatcatcher, yellowthroat, redstart*. For good reasons, these birds are treated here only lightly. Owing to their sizes and agile natures, they are often difficult to identify, even for journeymen birdwatchers. Although they are beautiful little birds, even experienced birders sometimes despair of trying to differentiate the various species. Also, American warblers are common birds of North America, and so are not in any way exotic to travellers from that continent. In fact, many of the warblers seen in southern Mexico, particularly from November through March, are breeders from North America, escaping the northern cold to winter in the tropics. Warblers are also more important birds in North than in Central America, in the sense that they are often so diverse and numerous in many northern forest habitats that, as a group, they make up more of the birdlife than all other birds combined. But for our purposes here, brief descriptions of a few of these birds will suffice. When one's interest is sufficiently aroused by these tiny birds to warrant further exploration, it is time to consider oneself a birdwatcher and to invest in a professional field guide!

Warblers, gnatcatchers, and the BANANAQUIT (Plate 59) are very small birds that flit and move jauntily about in trees, shrubs, and gardens, for the most part searching for or pursuing their dietary staple, insects. American warblers (Family Parulidae), also known as *wood warblers*, are a group of approximately 110 species, with wide distributions over the New World's forests, fields, marshes, and gardens. About 60 species occur regularly in southern Mexico, many of them seasonal migrants from North American breeding sites. Warblers are brightly colored, predominantly yellow or greenish, often mixed with varying amounts of gray, black and white; a few have even more color, with patches of red, orange or blue. Gnatcatchers are members of a large, primarily Old World family, Sylviidae. There are 11 gnatcatcher species, tiny gray birds that flit about in wooded and forest edge areas; three species occur in southern Mexico. The Bananaquit, which has been variously lumped into the warbler group, considered a honeycreeper (p. 196), and made the only member of its own family, Coerebidae (its classification is controversial), is a tiny yellow and olive/grayish bird with a broad Neotropical distribution, from southern Mexico and the Caribbean to northern Argentina.

Natural History

Ecology and Behavior

These birds are commonly found in a variety of natural habitats and in gardens and plantations. They forage in lively fashion mainly for insects and spiders; many also pierce berries to drink juice and some partake of nectar from flowers. Many warblers typically join *mixed species feeding flocks* with other small songbirds such as honeycreepers and tanagers. Warblers generally are territorial birds: either during the breeding season (in migratory species) or year-round (in nonmigratory, purely tropical species) a male and a female defend a piece of real estate from other members of the species. Some tropical warblers and gnatcatchers remain paired throughout the year. For the most part, warblers that remain all year in the tropics reside at middle and high elevations, whereas migratory warblers are found at a variety of elevations. Unusual among birds, BANANAQUITS build not only breeding nests, but also lighter, domed *dormitory* nests, which they sleep in individually.

Breeding

Warblers, gnatcatchers, and the BANANAQUIT are monogamous, but partners do not necessarily make equal contributions to breeding efforts. Warblers build open cup or roofed nests in trees or bushes, but sometimes on the ground. Often the female alone builds the nest and incubates the two to three eggs for 14 to 17 days; the male may feed his incubating mate. Young fledge after 8 to 15 days in the nest. Gnatcatchers build cup nests several meters up in trees, weaving together vegetation, moss, and spider web. Both sexes build the nest, incubate the two to three eggs for 13 to 15 days, and feed chicks, which fledge 11 to 14 days after hatching. Both Bananaquit sexes build the round, domed, breeding nest. Only the female incubates the two to three eggs, for 12 to 13 days. Both parents feed the chicks in the nest for 17 to 19 days by regurgitating food to them.

Ecological Interactions

For many years, North American scientists interested in warblers and many other songbirds concentrated their research on the birds' ecology and behavior during breeding, essentially ignoring the fact that the birds spent half of each year wintering in the tropics, many of them in southern Mexico and Central America. Now, with the realization that the birds' biology during the non-breeding season is also important for understanding their lives, their ecology and behavior during the winter have become areas of intense interest. Being addressed in research studies are such questions as: Are species that are territorial during breeding also territorial on their wintering grounds, and if so, in what way? Do individual birds return to the same spot in the tropics each year in winter as they do for nesting during the North American spring? Do migratory birds compete for food on their wintering grounds with those species that remain all year in the tropics? Are the highly colored plumages of many migratory birds more of use in signalling to other birds on the breeding or wintering grounds?

Status

Only one warbler that occurs in Mexico is now endangered, the GOLDEN-CHEEKED WARBLER (USA ESA listed), which breeds in central Texas (USA) but migrates through and visits Mexico in winter. Two other notable warblers in trouble are KIRTLAND'S WARBLER (considered vulnerable to threat, but USA ESA listed), which breeds in the USA (Michigan) and winters in the Bahamas, and BACHMAN'S WARBLER (critically endangered; USA ESA listed), which breeds in the southeastern USA and winters in Cuba. Kirtland's Warbler, which nests only in stands of young Jack-pine trees, has been the victim of its own specialization on one type of breeding habitat combined with a shrinking availability of that habitat, by destruction of its wintering habitat, and by BROWN-HEADED COW-BIRDS, which lay their eggs in the nests of warblers and other species, reducing their reproductive success (p. 195). Several other American warbler species are probably now at risk, but there is at present insufficient information about their populations to judge their statuses with any certainty.

Profiles

Mangrove Warbler, *Dendroica petechia*, Plate 58a
Magnolia Warbler, *Dendroica magnolia*, Plate 58b
Rufous-capped Warbler, *Basileuterus rufifrons*, Plate 58c
Common Yellowthroat, *Geothlypis trichas*, Plate 58d
Golden-browed Warbler, *Basileuterus belli*, Plate 58e
American Redstart, *Setophaga ruticilla*, Plate 59a
Painted Redstart, *Myioborus pictus*, Plate 59b
Bananaquit, *Coereba flaveola*, Plate 59c
Blue-gray Gnatcatcher, *Polioptila caerulea*, Plate 59d
Slate-throated Redstart, *Myioborus miniatus*, Plate 59e

32. Blackbirds and Orioles

Diversity is the key to comprehending the American orioles and blackbirds. The passerine family Icteridae includes about 95 species, which partition neatly into very different groups called *blackbirds, caciques (kah-SEE-kays), cowbirds, grackles, meadowlarks, orioles,* and *oropendolas;* they vary extensively in size, coloring, ecology, and behavior. These *icterids* are highly successful and conspicuous birds throughout their range, which encompasses all of North, Central, and South America. Distinguishing this varied assemblage from other birds are a jaunty deportment and a particular feeding method not widely used by other birds, known as *gaping* – a bird places its closed bill into crevices or under leaves, rocks or other objects, then forces the bill open, exposing the previously hidden space to its prying eyes and hunger. Most icterids are tropical in distribution and about 25 species, a few of which are seasonal migrants from the north, occur in southern Mexico. The icterid group inhabits marshes and almost all types of terrestrial habitats, and occupies warm lowland areas, middle elevations, as well as colder, mountainous regions. Many of these birds have adapted well to human settlements and are common denizens of gardens, parks, and urban and agricultural areas. The wide ranges of sizes, shapes, colors, mating systems, and breeding behaviors of these birds attract frequent interest from avian researchers.

Grackles, common in city areas, are primarily black birds with slender bills and, usually, long tails. Blackbirds are often marsh-dwellers (the term "blackbirds" is also sometimes used as a synonym for the entire icterid group, as in "the New World Blackbirds"). Orioles are small, bright, often exquisitely marked birds in yellow or orange mixed with black and white, whose preferred habitat is forest. Meadowlarks are yellow, black, and brown grassland birds. Oropendolas are spectacular, larger birds of tropical forests and woodlands that often breed in colonies. Caciques, which also breed in colonies, are smaller, sleeker black birds, frequently with red or yellow rumps and yellow bills. Finally, cowbirds, usually quite inconspicuous in various shades of brown and black, have a dark family secret – they are *brood parasites.*

Icterids range in length from 15 to 56 cm (6 to 22 in) – medium to fairly large birds. Bills are usually sharply pointed and conical. Black is the predominant plumage color in the group, but many combine it with bright reds, yellows, or oranges. In some species, the sexes are alike (particularly in the tropical species), but in others, females look very different from males, often more cryptically outfitted in browns, grays, or streaked plumage. Pronounced size differences between the sexes, females being smaller, are common; male oropendolas, for instance, may weigh twice as much as females. Bills and eyes are sometimes brightly colored.

Natural History

Ecology and Behavior

Icterids occur in all sorts of habitat types – woodlands, thickets, grassland, marshes, forest edges, and the higher levels of closed forests – but they are especially prevalent in more open areas. Their regular occupation of marshes has always been viewed as interesting, as they are not obviously adapted for living in aquatic environments – they do not have webbed feet, for example, nor are they able to float or dive. They eat a wide variety of foods including insects and other small animals, fruit and seeds. Some are fairly omnivorous, as befitting birds that frequently become scavengers in urban and suburban settings. A common feature of the group is that seed-eaters (*granivores*) during the non-breeding periods become insect-eaters during breeding, and feed insects to their young. Gaping for food is frequent and will be seen repeatedly if one observes these birds for any length of time. Orioles and caciques join in *mixed-species foraging flocks*; in a single fruit tree one may see two or more oriole species feeding with several species of tanagers, honeycreepers, and others. Caciques often associate with oropendolas, fruitcrows, and other birds in foraging flocks. Outside of the breeding season, icterids, particularly the blackbirds and grackles, typically gather in large, sometimes enormous, flocks that can cause damage to roosting areas and agricultural crops.

Breeding

Icterid species pursue a variety of breeding strategies. Some, such as the orioles, breed in classically monogamous pairs, male and female defending a large territory in which the hanging pouch nest is situated. But others, including many caciques and the oropendolas, nest in colonies. The members of an oropendola colony weave large, bag-like or pouch-like nests that hang from the ends of tree branches, many on the same tree. In a rare form of non-monogamous breeding, 3 to 10 male MONTEZUMA'S OROPENDOLAS (Plate 60) establish a colony in a tree (often an isolated one) and defend a group of 10 to 30 females that will mate and nest in the colony. The males engage in fighting and aggressive displays, competing among themselves to mate with the females. Detailed observations show that the most dominant males (the *alpha* animals) in each colony, usually heavier males, obtain up to 90% of all matings, and therefore are the fathers of most of a colony's young. Caciques, also with pouch-like nests, breed either solitarily in the forest or in colonies. In one study it was noted that each cacique in a colony tries to locate its nest toward the center of the colony, presumably because there is less chance of suffering nest predation at the colony's center. Perhaps most intriguing to scientists that study mating systems is that some very closely related icterid species have very different mating systems and breeding behaviors.

Icterid nests range from hanging pouches woven from grasses and other plant materials to open cups lined with mud to roofed nests built on the ground, hidden in meadow grass. Nests are almost always built by females. The female also incubates the two to three eggs, for 11 to 14 days, while the male guards the nest. Nestlings are fed for 10 to 30 days either by both parents (monogamous species) or primarily by the female (polygamous species).

Most of the cowbirds are brood parasites, building no nests themselves. Rather, females, after mating with one or more males, lay their eggs, up to 14 or more per season, in the nests of other species – other icterids as well as other birds – and let *host* species raise their young (see below).

Ecological Interactions

As are many *cuckoo* species, especially those of the Old World, five of the six species of cowbirds are brood parasites – including the three that occur in southern Mexico: BROWN-HEADED, GIANT and BRONZED COWBIRDS (Plate 60). In these species, a female, the parasite, lays eggs in the nests of other species, the hosts, and her young are raised by the foster parents. Some of the cowbirds specialize on icterid hosts – the Giant Cowbird parasitizes only caciques and oropendolas. Some host species have evolved the abilities to recognize cowbird eggs and eject them from their nests, but others have not. The cowbirds benefit from the interaction by being freed from defending a nesting territory and from nest-building and tending chores – what must amount to significant savings of energy and also decreased exposure to predators. The host species suffer reproductive harm because a female cowbird often ejects a host egg when she lays her own (when the nest is left unguarded). Also, more often than not, the cowbird's young are larger than the host's own, and are thus able to out-compete them for food brought to the nest by the adult birds. The host's own young often starve or are significantly weakened. Because of these harmful effects, the very successful cowbirds are believed to be responsible for severe population declines in North America of several species of small passerine birds. Because one population in these interactions benefits and one is harmed, the relationships between cowbirds and their hosts is *parasitic* – social parasitism in this case. How can brood parasitic behavior arise? Evolutionary biologists posit that one way would be if, long ago, some female cowbirds that built nests had their nests destroyed mid-way through their laying period. With an egg to lay but no nest in which to place it, females in this situation may have deposited the eggs in the nests of other species, which subsequently raised the cowbird young.

Breeding colonies of caciques and oropendolas are often located in trees that contain or are near large bee or wasp nests. The wasps or bees swarm in large numbers around the birds' nests. Apparently the birds benefit from this close association because the aggressiveness the stinging insects show toward animals that try to raid the birds' hanging nests offers a measure of protection.

Lore and Notes

Cacique is an interesting name for a bird: in Spanish it means "chief" or "boss;" in Mexico, it also has the suggestion of "tyrant;" in Chile and some other parts of South America, it means "one who leads an easy life."

According to Mayan legend, *grackles*, perhaps because of their self-important, strutting ways, were known as the guardians of the court of the Great Spirit God.

Status

The icterid group includes some of the most abundant birds of the Western Hemisphere, such as the RED-WINGED BLACKBIRD (Plate 61) and COMMON GRACKLE. Some of the Neotropical orioles, such as the YELLOW-TAILED ORIOLE (Plate 61) have been severely reduced in numbers because they are hunted as prized cage birds. None of the Mexican blackbirds and orioles are considered threatened, but several are now endangered elsewhere: Puerto Rico's YELLOW-

SHOULDERED BLACKBIRD (USA ESA listed), Martinique's MARTINIQUE ORIOLE, Brazil's FORBE'S BLACKBIRD, and a few others in South America are endangered from combinations of habitat destruction, brood parasitism, and the pet trade.

Profiles

Great-tailed Grackle, *Quiscalus mexicanus*, Plate 60a
Montezuma's Oropendola, *Psarocolius montezuma*, Plate 60b
Chestnut-headed Oropendola, *Psarocolius wagleri*, Plate 60c
Bronzed Cowbird, *Molothrus aeneus*, Plate 60d
Melodious Blackbird, *Dives dives*, Plate 60e
Yellow-billed Cacique, *Amblycercus holosericeus*, Plate 61a
Red-winged Blackbird, *Agelaius phoeniceus*, Plate 61b
Black-cowled Oriole, *Icterus dominicensis*, Plate 61c
Yellow-backed Oriole, *Icterus chrysater*, Plate 61d
Yellow-tailed Oriole, *Icterus mesomelas*, Plate 61e
Black-vented Oriole, *Icterus wagleri*, Plate 62a
Audubon's Oriole, *Icterus graduacauda*, Plate 62b
Hooded Oriole, *Icterus cucullatus*, Plate 62c
Orange Oriole, *Icterus auratus*, Plate 62d
Altamira Oriole, *Icterus gularis*, Plate 62e
Streaked-backed Oriole, *Icterus pustulatus*, Plate 62f

33. Tanagers

Tanagers comprise a large New World group of beautifully colored, small passerine birds, most of which are limited to tropical areas. They are among the tropics' most common and visible birds, primarily owing to their habit of associating in *mixed-species flocks* that gather in the open, often near human habitation, to feed in fruit trees, and they are a treat to watch. All told, there are some 230 species of tanagers (Family Thraupidae), the group including the typical *tanagers*, the *honeycreepers*, and the *euphonias*; about 25 species occur in southern Mexico. Some of the tanagers migrate north or south to breed in temperate areas of North and South America (four breed in the USA, including the WESTERN TANAGER, among the most wildly colored of North American birds). Tanagers inhabit all forested and shrubby areas of the American tropics, over a wide range of elevations, and are particularly numerous in wet forests and forest edge areas. Not devotees of the dark forest interior, they prefer the lighter, upper levels of the forest canopy and more open areas; some prefer low, brushy habitat.

Tanagers vary in length from 9 to 28 cm (3.5 to 11 in), with most concentrated near the smaller end of the range. They are compact birds with fairly short, thick bills and short to medium-long tails. Tanagers' outstanding physical attribute is their bright coloring – they are strikingly marked with patches of color that traverse the entire spectrum, rendering the group among the most fabulously attired of birds. It has been said of the typical tanagers (genus *Tangara*) that they must "exhaust the color patterns possible on sparrow-sized birds" (F. G. Stiles and A. F. Skutch 1989). Yellows, reds, blues and greens predominate, although a relatively few species buck the trend and appear in plain blacks, browns, or grays. The sexes usually look alike. Euphonias are small, stout tanagers, whose appearances (all species being slightly different) revolve around a common theme: blue-black above, with yellow foreheads, breasts, and bellies. Honeycreepers are also usually brilliantly colored.

Natural History

Ecology and Behavior

Most tanager species associate in mixed-species tanager flocks usually together with other types of birds; finding five or more tanager species in a single group is common. A mixed flock will settle in a tree full of ripe fruit and enjoy a meal. These flocks move through forests or more open areas, searching for fruit-laden trees. Although tanagers mostly eat fruit, some also take insects from foliage or even out of the air. And although most species are arboreal, a few are specialized ground foragers, taking seeds and bugs. Tanagers usually go after small fruits that can be swallowed whole, such as berries, plucking the fruit while perched. After plucking it, a tanager rotates the fruit a bit in its bill, then mashes it and swallows. (Ecologists divide frugivorous birds into *mashers*, such as tanagers, and *gulpers*, such as trogons and toucans, which swallow fruit whole and intact.) One explanation is that mashing permits the bird to enjoy the sweet juice prior to swallowing the rest of the fruit. This fits with the idea that mashers select fruit based partially on taste, whereas gulpers, which swallow intact fruit, do not (D. J. Levey *et al.* 1994).

Some tanagers, such as the *ant-tanagers*, are frequent members of mixed-species flocks (along with antbirds, woodcreepers and others) that spend their days following army ant swarms, feeding on insects that rush from cover at the approach of the devastating ants (see p. 178). Euphonias specialize on mistletoe berries, but eat other fruits and some insects as well. The honeycreepers are tanagers that are specialized for nectar feeding, their bills and tongues modified to punch holes in flower bottoms and suck out nectar; they also feed on some fruits and insects.

Breeding

Most tanagers appear to breed monogamously, although a number of bigamists have been noted (BLUE-GRAY and SCARLET-RUMPED TANAGERS, Plates 64, 65, among them). Breeding is usually concentrated during the transition from dry to wet season, when fruit and insects are most plentiful. In many species, male and female stay paired throughout the year. Males of many species give food to females in *nuptial feeding*, and during courtship displays make sure that potential mates see their brightly colored patches. Either the female alone or the pair builds a cup nest in a tree or shrub. Two eggs are incubated by the female only for 12 to 18 days and young are fed by both parents for 12 to 18 days prior to their fledging. A pair of tiny euphonias build a nest with a roof and a side entrance, often within a bromeliad plant.

Ecological Interactions

Tanagers, as mashing frugivores, sometimes drop the largest seeds from the fruits they consume before swallowing but, nonetheless, many seeds are ingested; consequently, these birds are active seed dispersers. Some ecologists believe tanagers to be among the most common dispersers of tropical trees and shrubs, that is, they are responsible for dropping the seeds that grow into the trees and shrubs that populate the areas they inhabit. Euphonias, for example, are crucial for the mistletoe life cycle because, after eating the berries, they deposit their seed-bearing droppings on tree branches, where the seeds germinate, the mistletoe plants starting out there as epiphytes.

Some tanagers and other species of Neotropical songbirds are considered *sedentary*. That is, they do not migrate or change location much at all. They breed in one place and then stay more or less in the same area during non-breeding portions of the year. From a conservation viewpoint, sedentary habits are sometimes considered a negative quality. If a species lives in only one habitat type – needs that habitat to survive – then it has little built-in flexibility in its habitat requirements. And should that one habitat be severely reduced, say lowland tropical forest in southern Mexico, then the species may be doomed. But if a species is not sedentary, that is, if it lives during the year in different places and so different habitat types, it has some flexibility in its habitat requirements – which can be both good and bad for conservation. On the one hand, if one habitat is temporarily unavailable, for instance during some seasons of the year, the birds may be able to survive by moving to the other type. On the other hand, to protect a species using two habitats, for instance a high-elevation site during breeding and a low-elevation site during winter, both kinds of habitats must be preserved. Careful observation of some birds previously thought sedentary indicates that many may actually make some movements during the year. For instance, recent research on COMMON BUSH-TANAGERS (Plate 64) near Oaxaca found that although they mostly remain at high elevations, they occasionally move down in winter to low-elevation forests.

Status

The only endangered Mexican tanager is the AZURE-RUMPED (or CABANIS) TANAGER, which occurs at middle and high elevations in limited regions of southern Chiapas and southern Guatemala. It is a fairly uncommon bird with small, declining populations. The only other endangered tanager in the region is Costa Rica's BLACK-CHEEKED ANT-TANAGER. Fortunately, despite being among tropical America's most beautiful birds, tanagers are not favorites of the international pet trade, probably because they have never been popularized as cage birds outside their native regions. However, several of the euphonias, such as the BLUE-HOODED EUPHONIA, are increasingly scarce and the reason may be that, although they are not hunted for the international trade, they *are* prized as cage birds within Central American countries. Several of South America's tanagers, especially in Brazil, are considered threatened or endangered, primarily owing to habitat loss.

Profiles

Red-legged Honeycreeper, *Cyanerpes cyaneus*, Plate 63a
Scrub Euphonia, *Euphonia affinis*, Plate 63b
Yellow-throated Euphonia, *Euphonia hirundinacea*, Plate 63c
Blue-crowned Chlorophonia, *Clorophonia occipitalis*, Plate 63d
Common Bush-tanager, *Chlorospingus ophthalmicus*, Plate 64a
Gray-headed Tanager, *Eucometis penicillata*, Plate 64b
Yellow-winged Tanager, *Thraupis abbas*, Plate 64c
Blue-gray Tanager, *Thraupis episcopus*, Plate 64d
Red-crowned Ant-tanager, *Habia rubica*, Plate 65a
Red-throated Ant-tanager, *Habia fuscicauda*, Plate 65b
Crimson-collared Tanager, *Ramphocelus sanguinolentus*, Plate 65c
Scarlet-rumped Tanager, *Ramphocelus passerinii*, Plate 65d
Hepatic Tanager, *Piranga flava*, Plate 66a
Summer Tanager, *Piranga rubra*, Plate 66b
Flame-colored Tanager, *Piranga bidentata*, Plate 66c
Rose-throated Tanager, *Piranga roseogularis*, Plate 66d

34. Sparrows and Grosbeaks

The *sparrows* and *grosbeaks* are large, diverse groups, totaling about 320 species, that include some of Mexico's most common and visible passerine birds. The groups' classification is continually revised, but here we can consider them to be separate families: the sparrows, *seedeaters, towhees,* and *grassquits* in Family Ember-izidae, and the grosbeaks, *saltators,* and *buntings* in Family Cardinalidae. The groups are almost *cosmopolitan* in distribution in the New World, meaning representatives occur just about everywhere, in all kinds of habitats and climates, from Alaska and northern Canada south to Tierra del Fuego. In fact, one species, the SNOW BUNTING, a small black and white bird, breeds farther north than any other land bird, in northern Alaska, Canada, and Greenland.

Sparrows and grosbeaks are generally small birds, 9 to 22 cm (3.5 to 9 in) in length, with relatively short, thick, conical bills that are specialized to crush and open seeds. In some species, the upper and lower halves of the bill can be moved from side to side, the better to manipulate small seeds. Sparrows have relatively large feet, which they use in scratching the ground to find seeds. Coloring varies greatly within the group but the plumage of most is dull brown or grayish, with many sporting streaked backs. The sexes generally look alike.

Natural History

Ecology and Behavior
Sparrows and grosbeaks are mostly seed eaters, although many are considered almost omnivorous and even those that specialize on seeds for much of the year often feed insects to their young. Some species also eat fruit. These birds mainly inhabit open areas such as grassland, parkland, brushy areas, and forest edge. They are birds of thickets, bushes, and grasses, foraging mostly on the ground or at low levels in bushes or trees. Because many species spend large amounts of time in thickets and brushy areas, they can be quite inconspicuous. Towhees, for instance, birds of thickets and the forest floor, specialized to use their feet to scratch through the leaf litter in search of food, are often heard but not seen.

Most species are strongly territorial, a mated pair aggressively excluding other members of the species from sharply defined areas. In the typical sparrows, pairs often stay together all year; other species within the group often travel in small family groups. Sometimes, territories are defended all year round and almost all available habitat in a region is divided into territories. The result is that those individuals that do not own territories must live furtively on defended territories, always trying to avoid the dominant territory owner, retreating when chased, and waiting for the day when the owner is injured or dies and the territory can be taken over. Only when one of these *floaters* ascends in the hierarchy to territory ownership status can he begin to breed. In species that have this kind of territorial system, the floater individuals that live secretly on other individual's territories, waiting and watching, were termed by their discoverer an avian *underworld*, and the name has stuck.

Breeding
Most sparrows and grosbeaks are monogamous breeders. The female of the pair usually builds a cup-shaped or, more often in the tropics, a domed nest, from grasses, fine roots and perhaps mosses and lichens. Nests are concealed on the ground or low in a shrub or tree. The female alone incubates two to four eggs, for

12 to 14 days. Both male and female feed nestlings, which fledge after 10 to 15 days in the nest.

Ecological Interactions

HOUSE SPARROWS, which occur throughout Mexico except for the Yucatán region, are small gray, brown, and black birds that currently enjoy an almost worldwide distribution, living very successfully in close association with humans. Nests are often placed in buildings. Formally, however, this bird is not a New World sparrow at all, but a member of an Old World family, the Passeridae; until recently, it was restricted to the Eastern Hemisphere. How and why these sparrows arrived in the West, and the unintended consequences of their arrival, is a cautionary tale of human interference in the natural distribution of animals. European settlers brought House Sparrows and other garden birds, such as starlings, to North America and released them, so that the animals around their new homes in the New World would resemble the animals they remembered from their old homes, an ocean away. A small number of House Sparrows released on the East Coast of the USA in the 1800s spread to the north, west, and south, and, after rapidly colonizing all of North America, the species is still spreading. It reached Central America in the mid-20th century and is now successfully ensconced there in most urban and suburban areas, where it competes for food with native sparrows. Species such as the House Sparrow that, owing to people's machinations, are now distributed outside of their natural ranges are said to be *introduced*, as opposed to naturally occurring species, which are termed *indigenous* or *native*. A species that spreads far beyond its original distribution, whether aided by people or not, is called, in ecological terms, an *invader*.

Status

Many species in this group are among the most abundant and frequently observed birds in the areas that they inhabit. A few sparrows of Central and South America are threatened or endangered, including three in Mexico (mostly owing to very small population sizes): the WORTHEN'S and SIERRA MADRE SPARROWS of central Mexico, and the GUADALUPE JUNCO, endemic to Isla Guadalupe off the Pacific coast of Baja California.

Profiles

Grayish Saltator, *Saltator coerulescens*, Plate 67a
Black-headed Saltator, *Saltator atriceps*, Plate 67b
Northern Cardinal, *Cardinalis cardinalis*, Plate 67c
Yellow-eyed Junco, *Junco phaeonotus*, Plate 67d
White-throated Towhee, *Pipilo albicollis*, Plate 67e
Collared Towhee, *Pipilo ocai*, Plate 67f
Green-backed Sparrow, *Arremonops chloronotus*, Plate 68a
Orange-billed Sparrow, *Arremon aurantiirostris*, Plate 68b
Rusty Sparrow, *Aimophila rufescens*, Plate 68c
White-collared Seedeater, *Sporophila torqueola*, Plate 68d
Bridled Sparrow, *Aimophila mystacalis*, Plate 68e
Blue-black Grassquit, *Volatinia jacarina*, Plate 69a
Yellow-faced Grassquit, *Tiaris olivacea*, Plate 69b
Blue-black Grosbeak, *Cyanocompsa cyanoides*, Plate 69c
Blue Bunting, *Cyanocompsa parellina*, Plate 69d
Blue Grosbeak, *Passerina caerulea*, Plate 69e

Environmental Close-Up 3
Illegal Traffic in Rare Animals:
The Case of the Mexican Parrot

Because most peoples' conception of pets ranges narrowly among the more common denizens of pet stores (cat, dog, "parakeet" – or budgerigar or "budgie," – hamster, gerbil, a few species of common tropical fish), they are usually unaware of, and almost always surprised when informed about, the vast illicit traffic in exotic wild pets. The international trade in wild animals is a huge business – an enterprise worth many billions of US dollars. Some of it is legal, but a significant portion of it, perhaps a third, worth (in 1997) as much as an estimated US$10 billion annually, consists of illegal sale of rare and endangered species. The capture and sale of these animals is considered to be one of the prime factors driving many of the species toward extinction. The problem is world-wide and it involves all types of animals. Perhaps 50,000 primates, 4 million birds, 10 million reptiles, and 350 million tropical fish are captured and sold each year (many of the reptiles are sold not as pets but as dried skins). Most of the trade in live animals is to feed the ever-growing demand for exotic pets.

The trade flourishes, even though for the most part illegal, because of the great demand for these animals, because of the relatively light penalties for wildlife trafficking and smuggling, and because of the huge profits available to the traffickers. For instance, some land tortoises of Madagascar, bought for a few dollars from local gatherers and smuggled to Europe or the USA, fetch US$10,000 each. An African Gray Parrot bought for $8 on a street in Nigeria, sells for $5000 in the USA. And a Scarlet Macaw (Plate 34), purchased for $25 from a local trapper in Chiapas, can bring $3000+ in the USA. There is something particularly insidious about trade in rare species: the greater the degree of rarity of an animal, the higher the demand by collectors, and so the more effort is taken by traffickers to capture and smuggle the few remaining individuals of the species. In other words, the more endangered a species is, the more collectors wish to possess it, and the more harm is done by taking additional individuals from their natural habitats.

Another aspect of this illicit trade is the amazing waste it entails. Because animals must be hidden in boxes, crates, or more exotic contrivances, and smuggled across international borders, a huge proportion of the animals die in transit. The estimates are that between 60% and 80% of smuggled animals die before reaching their final destinations. In one gruesome discovery, for example, customs agents found 1500 dead baby crocodiles in an illegal shipment of 2000. For birds, which are in many ways delicate creatures, probably only about 10% survive capture and shipment. In addition to the loss/death of individual animals targeted by trappers and traffickers, often other members of the species are killed or injured during capture – adult female primates, for example, are often killed to separate them from their young, which are then captured and sold. Furthermore, in the case of some species such as parrots, young birds are obtained by chopping down the trees containing nests, and the adult parents often will not attempt to breed again that year – further reducing population sizes.

The big markets for exotic pets, as you might expect, are the richer countries of the world – the USA, Japan, and those of Europe. The main suppliers are the nations of South America, Africa, and East Asia. Of particular concern in Mexico is the illicit traffic in native parrots to the USA. According to Traffic USA, an organization that monitors the wildlife trade, parrots probably make up the largest segment of the illicit trade in wild birds, comprising perhaps 25% of the industry during the 1980s. Parrots are in demand by collectors and pet-owners for the obvious reasons – they are large and colorful and many can be trained to imitate human speech. Until recently, parrots from Mexico, Central America, and South America moved freely into the USA, supplying the demand. Starting in the mid-1970s, and continuing into the 1980s, first one, then a growing list of Central and South American countries banned parrot exports, cutting supplies and raising demand for parrots, thereby stimulating smuggling.

Mexico traditionally was a major source of parrots sold in the USA. But Mexico, too, banned the export of parrots, in 1982 (but not the capture and sale of parrots for within-country use – doubtless owing to the tradition of owning caged birds in Mexico and because of the large number of people dependent for their livelihoods on parrot trapping and sales), creating a black market in the birds and leading to smuggling. Because the activity is illegal, it's difficult to define the extent of the trade – how many and which types of animals are caught and smuggled per year, how they are shipped, where they eventually wind up. Some of the birds are flown to the USA, others come via boat, but most, it is believed, come overland and pass through checkpoints on the USA/Mexico border. Statistics on about 600 seizures per year along the Texas/Mexico border, compiled by Traffic USA, indicate the most commonly smuggled parrots during the early 1990s were Yellow-naped, Red-crowned, and Yellow-headed (Plate 35) Parrots and the Green Parakeet. The Red-crowned and Yellow-headed Parrots are considered by conservation organizations to be threatened. Many experts believe that, due to illegal hunting for the pet trade, Yellow-headed Parrots and Scarlet Macaws will be extinct in Mexico within 20 years. (Macaws, possibly because they are already so rare in the wild, are infrequently seized at border crossings.) Seized birds are initially put in quarantine to make sure they are healthy, then given to zoos or sold at auction to cover quarantine expenses.

The economics of smuggling go something like this: a poacher in the outback, a *pajarero*, may get from a regional dealer perhaps $20 for a young Scarlet Macaw. (Usually young birds are most highly prized and hence, most valuable: they are smaller and so easier to transport, and, at their final destinations, are most easily trained to imitate speech; but adult parrots that have already been trained are also very valuable.) The same bird is sold in a wildlife market in Mexico City for $450. And after arrival at a USA petshop, it retails for $3000+. Similar figures for a Yellow-headed Parrot are $10, $130, and $1000+. As for the actual tranportation over the border, a variety of methods are used. Some parrots are put in boxes and floated across the Río Grande by raft. Some are hidden in cars and trucks. And others are simply strapped to a smuggler's body (a "mule") and walked across. "Mules" supposedly get as little as $5 to $10 per smuggled bird; when caught, they are usually released without being prosecuted.

What's being done and what can be done to reduce this trade in Mexican parrots? Probably only eliminating the demand for parrots in the USA and other places can stop illegal harvesting and shipping, and without a massive popular education campaign, that is unlikely to happen. The only other method to reduce the flow of parrots is stronger policing and greater penalties for traffickers who are caught. The number of border inspectors specially trained to deal with wildlife is growing. In 1991, when Mexico finally signed the CITES agreements (p. 71), the entire country had only about 25 inspectors to enforce all of its wildlife laws. Since 1996, Mexico has had professional wildlife inspectors at all of its international border crossings, including, expecially, those with the USA. On the USA side, the need to reduce the trade in rare wildlife is increasingly gaining attention. Because of environmental measures enacted with NAFTA (North American Free Trade Agreement), more money was made available by the US Congress to hire wildlife inspectors. Finally, the Wild Bird Conservation Act, passed by the US Congress in 1992, seeks to enhance bird conservation and reduce smuggling in a number of ways, including allowing the importing of approved species of wild birds caught under strict rules and under the auspices of the Mexican government and US Fish and Wildlife Service, and of captive-raised birds from approved bird-breeding facilities (thereby, hopefully, reducing the demand for illegal birds).

An unintended consequence of the cross-border parrot trade: sometimes birds escape during transportation or after assuming the life of caged pets, then survive in the "wilds" of suburban USA, forming flocks and even breeding in warmer areas, for instance, in the Los Angeles area. My sister, who lives in West Los Angeles, years ago complained to me about the "noisy parrots" that lived in a tree on her city street, waking her at dawn each day with their squawking. Sure enough, when I visited, I spotted a small group of Yellow-headed Parrots cavorting just outside her windows. In 1997, the free, non-pet population of Yellow-headed Parrots in the greater LA area was thought to number less than 100 individuals; Red-crowned Parrots, endemic to northeastern Mexico, were estimated to number about 1100.

Chapter 9

Mammals

Introduction

Leafing through this book, the reader will have noticed the profiles of many more birds than mammals. This may at first seem discriminatory, especially when it is recalled that many people themselves are mammals and, owing to that direct kinship, are probably keenly interested and motivated to see and learn about mammals. Are not mammals as good as birds? Why not include more of them? There are several reasons for the discrepancy – good biological reasons. One is that, even though the tropics generally have more species of mammals than temperate or arctic regions, the total number of mammal species worldwide, and the number in any region, is less than the number of birds. In fact, there are in total only about 4500 mammal species, compared with 9000 birds, and the relative difference is reflected in the fauna of southern Mexico. But the more compelling reason not to include more mammals in a book on commonly sighted wildlife, is that, even in regions sporting high degrees of mammalian diversity, mammals are relatively rarely seen – especially by short-term visitors. Most mammals lack that basic protection from predators that birds possess, the power of flight. Consequently, mammals being considered delicious fare by any number of predatory beasts (eaten in good numbers by reptiles, birds, other mammals, and even the odd amphibian), most are active nocturnally, or, if day active, are highly secretive. Birds often show themselves with abandon, mammals do not. Exceptions are those mammals that are beyond the pale of predation – huge mammals and fierce ones. But there are no elephants in Mexico, nor prides of lions. Another exception is monkeys. They are fairly large and primarily arboreal, which keeps them safe from a number of kinds of predators, and thus permits them to be noisy and conspicuous.

A final reason for not including more mammals in the book is that of the about 450 mammal species found in Mexico, almost 80% of them are bats and rodents, for the most part small nocturnal animals that, even if spotted, are very difficult for anyone other than experts to identify to species.

General Characteristics

If birds are feathered vertebrates, mammals are hairy ones. The group first arose, so fossils tell us, approximately 245 million years ago, splitting off from the primitive reptiles during the late Triassic Period of the Mesozoic Era, before the birds did the same. Four main traits distinguish mammals and confer upon them great advantage over other types of animals that allowed them in the past to prosper and spread and continue to this day to benefit them: hair on their bodies which insulates them from cold and otherwise protects from environmental stresses; milk production for the young, freeing mothers from having to search for specific foods for their offspring; the bearing of live young instead of eggs, allowing breeding females to be mobile and hence safer than if they had to sit on eggs for several weeks; and advanced brains, with obvious enhancing effects on many aspects of animal lives.

Classification

Mammals are quite variable in size and form, many being highly adapted – changed through evolution – to specialized habitats and lifestyles, for example, bats specialized to fly, marine mammals specialized for their aquatic world. The smallest mammals are the *shrews*, tiny insect eaters that weigh as little as 2.5 g ($\frac{1}{10}$ oz). The largest are the *whales*, weighing in at up to 160,000 kg (350,000 lb, half the weight of a loaded Boeing 747) – as far as anyone knows, the largest animals ever.

Mammals are divided into three major groups, primarily according to reproductive methods. The *monotremes* are an ancient group that actually lays eggs and still retains some other reptile-like characteristics. Only three species survive, the platypus and two species of spiny anteaters; they are fairly common inhabitants of Australia and New Guinea. The *marsupials* give birth to live young that are relatively undeveloped. When born, the young crawl along mom's fur into her *pouch*, where they find milk supplies and finish their development. There are about 240 marsupial species, including kangaroos, koalas, wombats, and opossums; they are limited in distribution to Australia and the Neotropics (the industrious but road-accident-prone VIRGINIA OPOSSUM also inhabits much of northern Mexico and the USA). The majority of mammal species are *eutherians*, or *true* mammals. These animals are distinguished from the other groups by having a *placenta*, which connects a mother to her developing offspring, allowing for long internal development. This trait, which allows embryos to develop to a fairly mature form in safety, and for the female to be mobile until birth, has allowed the true mammals to be rather successful, becoming, in effect, the dominant vertebrates on land for millions of years. The true mammals include those with which most people are intimately familiar: rodents, rabbits, cats, dogs, bats, primates, elephants, horses, whales – everything from house mice to ecotravellers.

The 4500 species of living mammals are divided into about 20 orders and 115 families. Approximately 450 species occur in Mexico, more than half of them within the region covered by this book.

Features of Tropical Mammals

There are several important features of tropical mammals and their habitats that differentiate them from temperate zone mammals. First, tropical mammals face different environmental stresses than do temperate zone mammals, and respond to stresses in different ways. Many temperate zone mammals, of course, must endure cold winters, snow, and low winter food supplies. Many of them respond with *hibernation*, staying more or less dormant for several months until conditions improve. Tropical mammals do not encounter extreme cold or snow, but they face dry seasons, up to 5 months long, that sometimes severely reduce food supplies. But for some surprising reasons, they cannot alleviate this stress by hibernating, waiting for the rainy season to arrive and increased food supplies. When a mammal in Canada or Alaska hibernates, so too do most of its predators. This is not the case in the tropics. A mammal sleeping away the dry season in a burrow would be easy prey to snakes and other predators. Moreover, a big danger to sleeping mammals would be ... army ants! These voracious insects are very common in the tropics and would quickly eat a sleeping mouse or squirrel. Also, external parasites, such as ticks and mites, which are inactive in extreme cold, would continue to be very active on sleeping tropical mammals, sucking blood and doing considerable damage. Last, the great energy reserves needed to be able to sleep for an extended period through warm weather may be more than any mammal can physically accumulate. Therefore, tropical mammals need to stay active throughout the year. One way they counter the dry season's reduction in their normal foods is to switch food types seasonally. For instance, some rodents that eat mostly insects during the rainy season switch to seeds during the dry; some bats that feed on insects switch to dry-season fruits (D. H. Janzen and D. E. Wilson 1983).

The abundance of tropical fruit brings up another interesting difference between temperate and tropical mammals: a surprising number of tropical mammals eat a lot of fruit, even among the *carnivore* group, which, as its name implies, should be eating meat. Most carnivores in southern Mexico, including the dogs, cats, raccoons, and weasels, are known to eat fruit and some seem to prefer it. Upon reflection, it makes sense that these mammals should consume fruit. Fruit is very abundant in the tropics, available all year, and, at least when it is ripe, easily digested by mammalian digestive systems. A consequence of such *frugivory* (fruit-eating) is that many mammals have become, together with frugivorous birds, major dispersal agents of fruit seeds, which they spit out or which travel unharmed through their digestive tracts. These mammals, therefore, spread fruit tree seeds as they travel, seeds that eventually germinate and become trees. Some biologists believe that, even though the carnivores plainly are specialized for hunting down, killing, and eating animal prey, it is possible that fruit was always a major part of their diet (D. H. Janzen and D. E. Wilson 1983).

Finally, there are some differences in the *kinds* of mammals inhabiting tropical and temperate regions. For instance, bears are nowhere to be found in southern Mexico (in fact, there is only a single Neotropical bear species, distributed sparsely from Panama to northern South America), nor are most social rodents like beavers and prairie dogs. Rabbits are few in number of species (only one occurs in the Yucatán) and usually in their abundances, also. On the other hand, some groups occur solely in the tropics or do fabulously well there. There are about 50 species of New World monkeys, all of which occur in tropical areas (but only three occur in Mexico). Arboreal mammals such as monkeys and procyonids (raccoon relatives) are plentiful in tropical forests probably because there is a rich, resource-filled, dense canopy to occupy and feed in. Also, the closed canopy blocks light to the ground, which only allows an undergrowth that is sparse and poor in resources, and consequently permits few opportunities for mammals to live and feed there. Bats thrive in the tropics, being very successful both in terms of number of species and in their abundances. For instance, eight families of bats probably occur in Chiapas, including more than 90 species; only four families and 40 species occur in the entire USA. While all the North American bats are insect eaters, the Neotropical bats are quite varied in lifestyle, among them being fruit-eaters, nectar-eaters, and even a few that consume animals or their blood (D. H. Janzen and D. E. Wilson 1983).

The social and breeding behaviors of various mammals are quite diverse. Some are predominantly solitary animals, males and females coming together occasionally only to mate. Others live in family groups. Some are rigorously territorial, others are not. Details on social and breeding behavior are provided within the individual family descriptions that follow.

Seeing Mammals in Tropical Mexico

No doubt about it, mammals are tough. One can go for weeks in southern Mexico and, if in the wrong places at the wrong times, see very few of them. A lot of luck is involved – a deer, a small herd of peccaries, a porcupine happens to cross the trail a bit ahead of you. Or you will be out birdwatching along a gravel road in the pre-dawn grayness when (as happened once to me) a small anteater, a NORTHERN TAMANDUA (Plate 74), will stroll slowly out of the forest, almost bump into your legs, detour around you, and walk back into the forest. I offer three pieces of mammal-spotting advice. First, if you have time and are a patient sort, stake out a likely looking spot near a stream or watering hole, be quiet, and wait to see what approaches. Second, try taking strolls very early in the morning; at this time, many nocturnal mammals are quickly scurrying to their day shelters. Third, although only for the stout-hearted, try searching with a flashlight at night around field stations, campgrounds, or forest lodges. After scanning the ground (for safety's sake as well as for mammals), shine the light toward the middle regions of trees, and look for bright, shiny eyes reflecting the light. You will certainly stumble across some kind of mammal or another; then it is simply a matter of whether you scare them more than they scare you.

For the region covered by this book, mammal diversity increases from north to south: the fewest species, about 100, occur over the northern tip of the Yucatán Peninsula; about 120 occur in the middle of the peninsula, and the most, between 120 and 160 species, occur near the Guatemalan border and, especially, in Chiapas and eastern Oaxaca.

Family Profiles

1. Opossums

Marsupials are an ancient group, preceding in evolution the development of the *true*, or *placental*, mammals, which eventually replaced the less-advanced marsupials over most parts of the terrestrial world. Marsupials alive today in the Australian and Neotropical regions therefore are remnants of an earlier time when the group's distribution spanned the Earth. Of the eight living families of marsupials, only three occur in the New World, and only one, the *opossums*, occurs in Mexico. That family, Didelphidae, is distributed widely over the northern Neotropics (with one member, the VIRGINIA OPOSSUM, reaching deeply northwards into the USA). About seven species represent the family in southern Mexico. They are a diverse group, occupying essentially all of the region's habitats except high mountain areas. Some, such as the COMMON OPOSSUM (Plate 70), are abundant and frequently sighted, while others are rarely seen.

All opossums are basically alike in body plan, although species vary considerably in size. Their general appearance probably has not changed much during the past 40 to 65 million years. As one nature writer put it, the opossums scurrying around today are much the same as the ones dinosaurs encountered (J. C. Kricher 1989). Basically, these mammals look like rats, albeit in the case of some, such as the Common Opossum, like large rats. Their distinguishing features are a long, hairless tail, which is *prehensile* (that is, opossums can wrap it around a tree branch and hang from it), and large, hairless ears. Females in most of the opossums have pouches for their young on their abdomens, but a few groups of species do not. Opossum hindfeet have five digits each, one digit acting as an opposable thumb. The hindfeet of the WATER OPOSSUM (Plate 70) are webbed. The Virginia Opossum, one of the region's largest, grows to nearly a meter (3 ft) in length, if the long tail is included. A large adult of the species can weigh up to 4 kg (9 lb), but most weigh 1 to 2.5 kg (2 to 5 lb). Similar in size and general appearance is the Common Opossum. (The two large opossums, Virginia and Common, occur together over much of southern Mexico and Central America; the Virginia ranges from there north to the USA and the Common ranges from there south to South America.) Some of the species are much smaller, the MEXICAN MOUSE OPOSSUM (Plate 70), for instance, being only 25 to 30 cm (10 to 12 in) in length, tail included, and weighing only 100 g (4 oz). Opossums come in a narrow range of colors – shades of gray, brown, and black. Male and female opossums generally look alike, but males are usually larger than females of the same age.

Natural History

Ecology and Behavior

Most opossums are night-active *omnivores*, although some also can be seen during the day. Their reputation is that they will eat, or at least try to eat, almost anything they stumble across or can catch; mostly they take fruit, eggs, and invertebrate and small vertebrate animals. A COMMON OPOSSUM forages mainly at night, often along ponds and streams, sometimes covering more than a kilometer (half-mile) per night within its home range, the area within which it lives and seeks food. Opossums that have been studied are not territorial – they do not defend part or all of their home ranges from others of their species. Some opossums forage mainly on the ground, but most are good climbers and are able to forage also in trees and shrubs; and some species are chiefly arboreal. After a night's foraging, an opossum spends the daylight hours in a cave, a rock crevice, or a cavity in a tree or log. Most opossums are unsociable animals, usually observed singly. The exception is during the breeding season, when males seek and court females, and two or more may be seen together.

Predators on opossums include owls, snakes, and carnivorous mammals. Some opossums apparently are somewhat immune to the venom of many poisonous snakes. The response of the COMMON OPOSSUM to threat by a predator is to hiss, growl, snap its mouth, move its body from side to side, and finally, to lunge and bite. They often try to climb to escape. The VIRGINIA OPOSSUM is famous for faking death ("playing possum") when threatened, but that behavior is rare or absent in the Common Opossum and others.

Breeding

Female opossums give birth only 12 to 14 days after mating. The young that leave the reproductive tract are only about 1 cm ($\frac{1}{2}$ in) long and weigh less than 0.5 g. These tiny opossums, barely embryos, climb unassisted along the mother's fur and into her pouch. There they grasp a nipple in their mouth. The nipple swells, essentially attaching the young; they remain there, attached, for about 2 months. Usually more young are born (up to 20) than make it to the pouch and attach correctly. In studies, six young, on average, are found in females' pouches (which have up to 13 nipples). Following the pouch phase, the female continues to nurse her young for another month or more, often in a nest she constructs of leaves and grass in a tree cavity or burrow.

Ecological Interactions

The VIRGINIA and COMMON OPOSSUMS have what can be considered commensal relationships (p. 66) with people. Throughout southern Mexico, populations of these mammals are concentrated around human settlements, particularly around garbage dumps, where they feed. They also partake of fruit crops and attack farmyard birds. Consequently, opossums are more likely to be seen near towns or villages than in uninhabited areas. Of course, these opossums pay a price for the easy food – their pictures are commonly found in the dictionary under "roadkill."

Lore and Notes

COMMON OPOSSUMS are known as foul-smelling beasts. Their reputation probably stems from the fact that they apparently enjoy rolling about in fresh animal droppings. Also, when handled, they employ some unattractive defense mechanisms – tending to squirt urine and defecate.

Status

None of the region's opossum species are now threatened or endangered, although one, the CENTRAL AMERICAN WOOLLY OPOSSUM, is considered vulnerable to threat, and another, the WATER OPOSSUM, is fairly rare. VIRGINIA OPOSSUM is hunted for food in some areas. COMMON and VIRGINIA OPOSSUMS are sometimes killed intentionally near human settlements to protect fruit crops and poultry, and unintentionally but abundantly, by cars.

Profiles

Common Opossum, *Didelphis marsupialis*, Plate 70a
Central American Woolly Opossum, *Caluromys derbianus*, Plate 70b
Water Opossum, *Chironectes minimus*, Plate 70c
Gray Four-eyed Opossum, *Philander opossum*, Plate 70d
Mexican Mouse Opossum, *Marmosa mexicana*, Plate 70e

2. Bats

Of all the kinds of mammals, perhaps we can comprehend the lives, can mentally put ourselves into the skins, of non-human primates (monkeys and apes) best, and of *bats*, least. That is, monkeys, owing to their similarities to people, are somewhat known to us, familiar in a way, but bats are at the other extreme: foreign, exotic, mysterious. The reasons for their foreignness are several. Bats, like birds, engage in sustained, powered flight – the only mammals to do so ("rats with wings," in the memorable phrasing of a female acquaintance). Bats are active purely at night – every single species. Bats navigate the night atmosphere chiefly by "sonar," or *echolocation*: not by sight or smell but by broadcasting ultrasonic sounds – extremely high-pitched chirps and clicks – and then gaining information about their environment by "reading" the echos. Although foreign to people's primate sensibilities, bats, precisely because their lives are so very different from our own, are increasingly of interest to us. In the past, of course, bats' exotic behavior, particularly their nocturnal habits, engendered in most societies not ecological curiosity but fear and superstition.

Bats are flying mammals that occupy the night. They are widely distributed, inhabiting most of the world's tropical and temperate regions, excepting some oceanic islands. With a total of about 980 species, bats are second in diversity among mammals only to rodents. Ecologically, they can be thought of as nighttime equivalents of birds, which dominate the daytime skies. Bats of the Neotropics, although often hard to see and, in most cases, difficult for anyone other than experts to identify, are tremendously important mammals, and that is why they are treated here. Their diversity and numbers tell the story: 39% of all Neotropical mammal species are bats, and there are usually as many species of bats in a Neotropical forest than of all other mammal species combined. Researchers estimate that most of the mammalian biomass (the total amount of living tissue, by weight) in any given Neotropical region resides in bats. Of the more than 200 species of mammals in southern Mexico, about half are bats. For instance, 91 of the 185 mammals that occur in Chiapas are bats. Mexico has 14 endemic bat species. Some of the region's more common and more interesting bats are profiled here.

Bats have true wings, consisting of thin, strong, highly elastic membranes that extend from the sides of the body and legs to cover and be supported by the elongated fingers of the arms. (The name of the order of bats, Chiroptera, refers to the wings: *chiro*, meaning hand, and *ptera*, wing.) Other distinctive anatomical features include bodies covered with silky, longish hair; toes with sharp, curved claws that allow the bats to hang upside down and are used by some to catch food; scent glands that produce strong, musky odors; and, in many, very odd-shaped folds of skin on their noses (noseleaves) and prominent ears that aid in echolocation. Like birds, bats' bodies have been modified through evolution to conform to the needs of energy-demanding flight: they have relatively large hearts, low body weights, and fast metabolisms.

Because few people get near enough to resting or flying bats to examine them closely, I describe the forms of individual species only in the plate section. It will suffice to say that the bats under our consideration here range from tiny, 5-g bats with 5-cm (2-in) wingspans (the HAIRY-LEGGED BAT; Plate 72) to one of the New World's larger bats, which weighs up to 95 g ($\frac{1}{5}$ lb) and has a wingspan greater than 60 cm (2 ft) (the WOOLLY FALSE VAMPIRE BAT; Plate 71). Females in most species are larger than males, although there are exceptions, such as the NECTAR BAT (Plate 71), in which males are larger.

Natural History

Ecology and Behavior

Neotropical bats are renowned for their insect-eating ways. Indeed, most species specialize on insects. They use their sonar not just to navigate the night but to detect insects, which they catch on the wing, pick off leaves, or scoop off the ground. Bats use several methods to catch flying insects. Small insects may be captured directly in the mouth; some bats use their wings as nets and spoons to trap insects and pull them to their mouth; and others scoop bugs into the fold of skin membrane that connects their tail and legs, then somersault in mid-air to move the catch to their mouth. Small bugs are eaten immediately on the wing, while larger ones, such as large beetles, are taken to a perch and dismembered. Not all species, however, are insectivores. Neotropical bats have also expanded ecologically into a variety of other feeding *niches*: some specialize in eating fruit, feeding on nectar and pollen at flowers, preying on vertebrates such as frogs or birds, eating fish, or even, in the case of the COMMON VAMPIRE BAT (Plate 72), sipping blood.

Bats spend the daylight hours in day roosts, usually tree cavities, shady sides of trees, caves, rock crevices, or, these days, in buildings or under bridges. Some bats make their own individual roosting sites in trees by biting leaves so that they fold over, making small *tents* that shelter them from predators as well as from the elements. More than one species of bat may inhabit the same roost, although some species will associate only with their own kind. For most species, the normal resting position in a roost is hanging by their feet, head downwards, which makes taking flight as easy as letting go and spreading their wings. Many bats leave roosts around dusk, then move to foraging sites at various distances from the roost. Night activity patterns vary, perhaps serving to reduce food competition among species. Some tend to fly and forage intensely in the early evening, become less active in the middle of the night, then resume intense foraging near dawn; others are relatively inactive early in the evening, but more active later on. Bats do not fly continuously after leaving their day roosts, but group together at a *night roost*, a tree for instance, where they rest and bring food. Fruit-eaters do not rest in the tree at which they have discovered ripe fruit, where predators might find them, but make several trips per night from the fruit tree to their night roost. Bats are highly social animals, roosting and often foraging in groups.

Most of the species decribed below are common representatives of groups of bats, each of which differs anatomically, ecologically, and behaviorally.

FISHING BAT (Plate 71). Relatively large bats, Fishing Bats roost in hollow trees and buildings near fresh or salt water. They have very large hind feet and claws that they use to pull fish, crustaceans, and insects from the water's surface. These bats fly low over still water, using their sonar to detect the ripples of a fish just beneath or breaking the water's surface. Grabbing the fish with their claws, they then move it to their mouth, land, hang upside down, and feast.

WOOLLY FALSE VAMPIRE BAT (Plate 71). These are fairly large bats that inhabit forest regions, particularly evergreen rainforests, from southern Mexico to Brazil. They roost in small groups in caves or hollow trees and logs. They take insects, some fruit, and, being one of only a few carnivorous New World bats, vertebrates such as lizards, birds, and, in particular, small rodents such as mice. The bat's name refers to the dense fur covering its body, and to the fact that, in the past, large bats such as this one were often mistaken for vampire bats.

BROAD-EARED FREE-TAILED BAT (Plate 71). These are medium-sized bats with large ears and a conspicuous thick naked tail that projects beyond the end of the membrane that connects the rear legs. They occur over a wide range of habitat types and are often seen at dusk flying over settled areas and over streams and ponds. Broad-eared bats are insect-eaters, known to be specialists on moths. Roosting, in groups up to 50+ animals, is in caves, rocks, trees, and buildings. Some groups of free-tailed bats, for instance, the BRAZILIAN FREE-TAILED BAT, are known for roosting in huge cave colonies of millions of individuals – possibly the largest aggregations of warm-blooded animals.

JAMAICAN FRUIT-EATING BAT (Plate 72). A medium-sized fruit-eater of wet and dry forests, these bats also take insects and pollen from flowers. They pluck fruit and carry it to a night roost 25 to 200 m (80 to 650 ft) away to eat it. Observers estimate that nightly each bat carries away from trees more than its own weight in fruit. Jamaican Fruit Bats roost in caves, hollow trees, or in foliage. Breeding is apparently polygynous (a single male mates with several females), because small roosts are always found to contain one male plus several females (up to 11) and their dependent young.

COMMON VAMPIRE BAT (Plate 72). Vampire bats are the only mammals that feed exclusively on blood; the only true mammal parasites. Day roosts are in hollow trees and caves. At night vampires fly out, using both vision (they have larger eyes and better vision than most bats) and sonar to find victims, usually large mammals. They not only fly well, they are also agile walkers, runners, and hoppers, of great assistance in perching on, feeding on, and avoiding swats by their prey. They use their sharp incisor teeth to bite the awake or sleeping animal, often on the neck, and remove a tiny piece of flesh. An anti-clotting agent in the bat's saliva keeps the small wound oozing blood. The vampire laps up the oozing blood – it does not suck. The feeding is reported to be painless (we won't ask how researchers know this, but, with a shudder, we can guess). Because blood is such a nutritious food, these bats need only about 15 ml ($\frac{1}{2}$ oz) a day. Vampires breed at any time of the year; young are fed blood from the mother's mouth for several months until they can get their own.

NECTAR BAT (Plate 71). These are small bats that can hover for a few seconds at flowers to take pollen or nectar; however, most of their omnivorous diet consists of fruit and insects. They roost in large groups in both dry and wet forest habitats. Young use their teeth to cling to their mothers' fur after birth, being carried along during foraging trips; pups can fly on their own at about a month old.

SHORT-TAILED FRUIT BAT (Plate 71). These are small, very common bats that live in large groups, up to several hundred, usually in caves or tree cavities. They are primarily fruit-eaters, but also seasonally visit flowers for nectar. Usually they pick fruit from a tree, then return to a night roost to consume it. After giving birth, females carry their young for a week or two during their nightly foraging; older young are left in the day roost. Because of their abundance and frugivory, these bats are important dispersers of tree seeds in Neotropical forests.

HAIRY-LEGGED BAT (Plate 72). This species is a common representative in southern Mexico of the genus *Myotis*, the *little brown bats*, which are distributed widely over the Neotropics and, indeed, much of the world. *Myotis* bats roost in large groups in hollow trees, rock crevices, and buildings; males usually roost separately from females and their young. At sunset they leave the roost in search of flying insects, and return just before dawn. Young are carried by the mother for a few days after birth, but are then left behind with other young in the roost when the female leaves to forage. Pups can fly at about 3 weeks of age, are weaned at 5 to 6 weeks, and are reproductively mature at only 4 months.

BLACK MASTIFF BAT (Plate 72). These smallish bats are common inhabitants of low-elevation rainforests from Mexico to northern Argentina. They roost, often in groups of 50 to 100, in hollow trees and rocky areas. They are also common in towns and villages, where they roost in buildings. Food consists mainly of insects.

MEXICAN FUNNEL-EARED BAT (Plate 72). These are small, light bats with longish, slender wings, useful for slow, fluttering flight in open areas. They usually cruise low to the ground and use their tail membrane, which stretches between the rear legs, to catch small bugs. Funnel-eared Bats occur over a wide range of habitats, and are common in and around human settlements. They mostly roost in caves, often in large numbers.

Breeding

Bat mating systems are diverse, various species employing monogamy (one male and one female breed together), polygyny (one male and several females), and promiscuity (males and females both mate with more than one individual); the breeding behavior of many species has yet to be studied in detail. Some Neotropical bats breed at particular times of the year, but others have no regular breeding seasons. Most bats produce a single pup at a time.

Ecological Interactions

Bats are beneficial to forests and to people in a number of ways. Many Neotropical plants have bats, instead of bees or birds, as their main pollinators. These species generally have flowers that open at night and are white, making them easy for bats to find. They also give off a pungent aroma that bats can home in on. Nectar-feeding bats use long tongues to poke into flowers to feed on nectar – a sugary solution – and pollen. As a bat brushes against a flower, pollen adheres to its body, and is then carried to other plants, where it falls and leads to cross-pollination. Fruit-eating bats, owing to their high numbers, are important seed dispersers, helping to regenerate forests by transporting and dropping fruit seeds onto the forest floor. Also, particularly helpful to humans, bats each night consume enormous numbers of annoying insects.

Bats eat a variety of vertebrate animals; unfortunately for some of them they play right into the bat's hands ... uh, feet. Some bats that specialize on eating frogs, it has been discovered, can home in on the calls that male frogs give to attract mates. These frogs are truly in a bind: if they call, they may attract a deadly predator; if they do not, they will lack for female company. However, some types of bat prey have developed anti-bat tactics. Several groups of moth species, for instance, can sense the ultrasonic chirps of some echolocating insectivorous bats; when they do, they react immediately by flying erratically or diving down into vegetation, decreasing the success of the foraging bats. Some moths even make their own clicking sounds, which apparently confuse the bats, causing them to break off approaches. The interaction of bats and their prey animals is an active field of animal behavior research because the predators and the prey have both developed varieties of tactics to try to outmaneuver or outwit the other.

Relatively little is known about which predators prey on bats. The list, however, includes birds of prey (owls, hawks), snakes, other mammals such as opossums, cats, and (yes) people, and even other bats. Some monkeys actually hunt tent-roosting bats that they find in tree leaves. Tiny bats, such as the 5-g ($\frac{1}{6}$ oz) HAIRY-LEGGED BAT, are even captured by large spiders and cockroaches. Bats, logically, are usually captured in or near their roosts, where predators can reliably find and corner them. One strong indication that predation is a real concern to bats is that many species reduce their flying in bright moonlight. Bats showing this "lunar phobia" include the JAMAICAN FRUIT-EATING BAT and SHORT-TAILED FRUIT BAT.

Lore and Notes

Bats have frightened people for a long time. The result, of course, is that there is a large body of folklore that portrays bats as evil, associated with or incarnations of death, devils, witches, or vampires. Undeniably, it was bats' alien lives – their activity in the darkness, flying ability, and strange form – and people's ignorance of bats that were the sources of these myriad superstitions. Many cultures, worldwide, have evil bat legends, from Australia to Japan and the Philippines, to Europe, the Middle East, and Central and South America.

Many ancient legends tell of how bats came to be creatures of the night. But the association of bats with vampires – blood-sucking monsters – may have originated in recent times with Bram Stoker, the English author who in 1897 published *Dracula* (the title character, a vampire, could metamorphose into a bat). Vampire bats are native only to the Neotropics. Stoker may have heard stories of their blood-lapping ways from travellers, and for his book, melded the behavior of these bats with legends of vampires from India and from Slavic Gypsy culture. Although not all New World cultures imparted evil reputations to bats, it is not surprising, given the presence of vampire bats, that some did. The Mayans, for instance, associated bats with darkness and death; there was a "bat world," a part of the underworld ruled by a bat god, through which dead people had to pass. Some groups in what is now Guatemala apparently worshipped and greatly feared a bat-god named Camazotz.

Speaking of vampire bats, these bats presumably are much more numerous today than in the distant past because they now have domesticated animals as prey. Before the introduction of domesticated animals to the Neotropics, vampires would have had to seek blood meals exclusively from mammals such as deer and peccaries; now they have, over large parts of their range, herds of large domesticated animals to feed on. In fact, examinations of blood meals reveal that vampire bats in settled areas feed almost exclusively on ranch and farm animals – cattle, horses, poultry, etc. By the way, vampire bats rarely attack people, although it is not unheard of. In some regions, they may transmit rabies.

Status

Determining the statuses of bat populations is difficult because of their nocturnal behavior and habit of roosting in places that are hard to census. With some exceptions, all that is known for most Neotropical species is that they are common or not common, widely or narrowly distributed. Some species are known from only a few museum specimens, or from their discovery in a single cave, but that does not mean that there are not healthy but largely hidden wild populations. All of the bats profiled here are common. At least two Mexican species are currently endangered: the MEXICAN LONG-NOSED BAT (USA ESA listed), a relative of the WOOLLY FALSE VAMPIRE BAT, and *Myotis findleyi*, a Mexican endemic in the same family as the HAIRY-LEGGED BAT. Recently Mexican biologists noted that there are several endemic bats with limited distributions that occur wholly outside of any of Mexico's protected areas – suggesting that they could be threatened by development in the future.

Because many forest bats roost in hollow trees, deforestation is obviously a primary threat. Further, many bat populations in temperate regions in Europe and the USA are known to be declining and under continued threat by a number of agricultural, forestry, and architectural practices (about five bat species or subspecies on the USA mainland are endangered, USA ESA listed). Traditional roost sites have been lost on large scales by mining and quarrying, by the destruction of old buildings, and by changing architectural styles that eliminate many building overhangs, church belfries, etc. Many forestry practices advocate the removal of hollow, dead trees, which frequently provide bats with roosting space. Additionally, farm pesticides are ingested by insects, which are then eaten by bats, leading to death or reduced reproductive success.

Profiles

Fishing Bat, *Noctilio leporinus*, Plate 71a
Woolly False Vampire Bat, *Chrotopterus auritus*, Plate 71b
Short-tailed Fruit Bat, *Carollia perspicillata*, Plate 71c
Nectar Bat, *Glossophaga soricina*, Plate 71d
Broad-eared Free-tailed Bat, *Nyctinomops laticaudatus*, Plate 71e
Black Mastiff Bat, *Molossus ater*, Plate 72a
Hairy-legged Bat, *Myotis keaysi*, Plate 72b
Jamaican Fruit-eating Bat, *Artibeus jamaicensis*, Plate 72c
Common Vampire Bat, *Desmodus rotundus*, Plate 72d
Mexican Funnel-eared Bat, *Natalus stramineus*, Plate 72e

3. Primates

People's reactions to *monkeys* are interesting. All people, it seems, find monkeys striking, even transfixing, when first encountered, but then responses diverge. Many people adore the little primates and can watch them for hours, whether it be in the wild or at zoos. But others, myself included, find them a bit, for want of a better word, unalluring; we are slightly uncomfortable around them. What is so intriguing is that it is probably the same characteristic of monkeys that both so attracts and repels people, and that is their quasi-humanness. Whether or not we acknowledge it consciously, it is doubtless this trait that is the source of all the attention and importance attached to monkeys and apes. They look like us, and, truth be told, they act like us, in a startlingly large number of ways. Aristotle, 2300 years ago, noted similarities between human and non-human primates, and Linnaeus, the Swedish originator of our current system for classifying plants and animals, working more than 100 years pre-Darwin, classed people together in the same group with monkeys. Therefore, even before Darwin's ideas provided a possible mechanism for people and monkeys to be distantly related, we strongly suspected there was a link; the resemblance was too close to be accidental. Given this bond between people and other primates, it is not surprising that visitors to parts of the world that support non-human primates are eager to see them and very curious about their lives. Three monkey species occur in Mexico, all within the region covered by this book.

Primates are distinguished by several anatomical and ecological traits. They are primarily arboreal animals. Most are fairly large, very smart, and highly social – they live in permanent social groups. Most have five very flexible fingers and toes per limb. Primates' eyes are in the front of the skull, facing forward (eyes in the front instead of on the sides of the head are required for binocular vision and good depth perception, without which swinging about in trees would be an extremely hazardous and problematic affair), and primates have, for their sizes, relatively large brains. Female primates give birth usually to a single, very helpless infant.

Primates are distributed mainly throughout the globe's tropical areas and many subtropical ones, save for the Australian region. They are divided into four groups: (1) *Prosimians* include several families of primitive primates from the Old World. They look the least like people, are mainly small and nocturnal, and include lemurs, lorises, galago (bushbaby), and tarsiers. (2) *Old World Monkeys* (family Cercopithecidae) include baboons, mandrills, and various monkeys such as rhesus and proboscis monkeys. (3) *New World Monkeys* (family Cebidae) include many kinds of monkeys and the tiny marmosets. (4) The *Hominoidea* contains the gibbons, orangutans, chimpanzees, gorillas, and ecotravellers.

New World monkeys, in general, have short muzzles and flat, unfurred faces, short necks, long limbs, and long tails that are often used as fifth limbs for climbing about in trees. They are day-active animals that spend most of their time in trees, usually coming to the ground only to cross treeless space that they cannot traverse within the forest canopy. About 40 species of New World Monkeys are distributed from southern Mexico to northern Argentina. Several species of *howler monkeys*, named for the tremendous roaring calls they give at dawn and dusk, occur from Central America to northern South America. The MANTLED HOWLER MONKEY (Plate 73) ranges over parts of southern Mexico (but not the Yucatán region) and most of Central America. They are large, weighing between 4 and 7 kg (9 to 16 lb); howlers, in fact, are among the New World's largest monkeys. These are all black except for a noticeable fringe of brown or blonde hair on their sides (the "mantle"). Males are larger than females. The YUCATÁN BLACK HOWLER (Plate 73), which occurs only in the Yucatán region, is very similar to the Mantled Howler, except that it is all black with longish hair. *Spider monkeys*, named for their long, slender limbs, range from southern Mexico to Brazil's lower Amazon region. The CENTRAL AMERICAN SPIDER MONKEY (Plate 73), like the howlers, is also large, and weighs up to 8 kg (17 lb). They have strong, *prehensile* tails and long, thin limbs. Coat color is variable – they are black, brown, or reddish, with lighter underparts and a light, unfurred facial mask and muzzle.

Natural History

Ecology and Behavior

Howler Monkeys. Howlers inhabit a variety of forest habitat types, but apparently prefer lowland wet evergreen forests. They are highly arboreal and rarely come to the ground; typically they spend most of their time in the upper reaches of the forest. In contrast to many other New World monkeys, howlers are relatively slow-moving and more deliberate in their canopy travels. They eat fruit and a lot of leafy material; in fact, in one study, leaves comprised 64% of their diet, fruit and flowers, 31%. Owing to their specialization on a super-abundant food resource – leaves – the home range of a troop need not be very large. Howlers are frequently inconspicuous because they are slow-moving and often quiet; people on trails may pass directly below howlers without noticing them. They are most assuredly not inconspicuous, however, when the males let loose with their incredible roaring vocalizations (females also roar occasionally). Their very loud, piercing choruses of roars, at dawn, during late afternoon, and, frequently, during heavy rain, are a characteristic and wonderful part of the rainforest environment. These vocalizations are probably used by the howlers to communicate with other troops, to advertise their locations and to defend them; although troops of these monkeys do not maintain exclusive territories, they do appear to defend current feeding sites. The males' howling can be heard easily at 3 km (1.8 miles) away in a forest or 5 km (3 miles) away across water. Whereas MANTLED HOWLERS have a fairly broad distribution, the YUCATÁN BLACK HOWLER occurs only in the eastern and southern parts of Mexico's Yucatán and in Belize and northern Guatemala. They live in lowland areas, often along rivers and streams. In some areas, troops of these monkeys that were studied contained between about 4 and 10 individuals, usually an adult male plus females and associated young.

Spider Monkeys. Spider monkeys are found throughout much of the Neotropics in rainforests and deciduous forests at a wide range of elevations. They are extremely arboreal, rarely descending to the ground. They stay mostly within a forest's upper canopy, moving quickly through trees using their fully prehensile tail as a fifth limb to climb, swing, and hang. Spider monkeys eat ripe fruit, young leaves, and flowers. Troops in Panama were observed to feed on 80% fruit and 20% leaves and other plant materials. During the day, troops, varying in size from 2 to 25 or more (groups of 100 or more have been reported), range over wide swaths of forest, but stay within a home range of 2.5 to 4.0 sq km (1 or 2 sq miles). Troops usually consist of an adult male and several females and their dependent offspring. Spider monkeys are commonly observed in small groups, often two animals, but frequently they are members of a larger troop; the troop breaks up daily into small foraging parties, then coalesces each evening at a mutual sleeping tree. They require large tracts of undisturbed forest (as opposed to the howlers, which can live in disturbed forests near human settlements), and therefore are usually spotted only in more remote, highly protected areas.

Breeding

Female monkeys in Central America usually produce a single young that is born furred and with its eyes open. In MANTLED and YUCATÁN BLACK HOWLERS, females reach sexual maturity at 3 to 5 years of age, and males at 6 to 8. Birth occurs following pregnancy of about 6 months. At 3 months, youngsters begin making brief trips away from their mothers, but until a year old, they continue to spend most of their time on their mother's back; they are nursed until they are 10 to 12 months old. Howlers have survived in the wild for up to 20 years. CENTRAL AMERICAN SPIDER MONKEYS appear to have no regular breeding seasons. Females reach sexual maturity at about 4 years old (males at about 5), then give birth every 2 to 4 years after pregnancies of about 7.5 months. Young, which weigh about 500 g (1 lb) at birth, are carried by the mother for up to 10 months and are nursed for up to a year. Upon reaching sexual maturity, young females leave their troops to find mates in other troops; males remain with their birth troop. Spider monkeys have lived for 33 years in captivity.

Ecological Interactions

A variety of predatory animals prey on the region's monkeys, including Boa Constrictors (Plate 13), birds of prey such as eagles, arboreal cats such as Jaguarundi and Margay (Plate 76), and people. Spider and howler monkeys attain a degree of safety because they are fairly large. Other causes of death are disease (for instance, many monkey populations in the Yucatán region crashed during the 1950s because of an outbreak of yellow fever) and parasite infestations, such as that by *botflies*. Botflies lay their eggs on mosquitos, monkeys being exposed when infected mosquitos land on them to feed. Botfly larvae burrow into a monkey's skin, move into the bloodstream, and often find their way to the neck region. Many howlers, for instance, are observed to have severe botfly infestations of their necks, seen as swollen lumps and the holes created when adult botflies emerge from the monkey's body. In one Panamanian study, members of a howler population were found each to have an average of two to five botfly parasites; several monkeys in the study died, apparently of high levels of botfly infestation.

Lore and Notes

People hunt monkeys for reasons other than food and the pet trade. Many monkey body parts – meat, blood, bones – in various regions of the world are believed to have effective medicinal or aphrodisiac value. Some monkeys are killed to be used as bait, for instance, in the Amazon Basin to lure large cats into traps. Monkeys are also killed for their skins and other body parts, which are made into ornaments. Last, some monkeys, including the New World's capuchins, are killed as crop pests.

Status

All New World Monkeys are listed in CITES Appendix I or II as endangered species (I) or species that, although they may not be currently threatened, need to be highly regulated in trade or they could soon become threatened (II). Main menaces to monkeys are deforestation – elimination of their natural habitats – and poaching for trade and meat. The larger monkeys especially – spiders, howlers – are often hunted for their meat, and therefore are usually rare near human settlements. (Those who eat monkeys claim that spider monkey is best.) The MANTLED HOWLER is listed by the USA ESA as endangered and the YUCATÁN BLACK HOWLER is listed as threatened. The CENTRAL AMERICAN SPIDER MONKEY is listed as endangered, but only in some countries – not Mexico. Large expanses of protected forest are the basic requirement for continued survival of southern Mexico's monkey populations.

Profiles
Yucatán Black Howler Monkey, *Alouatta pigra*, Plate 73a
Mantled Howler Monkey, *Alouatta palliata*, Plate 73b
Central American Spider Monkey, *Ateles geoffroyi*, Plate 73c

4. Anteaters and Armadillos

Anteaters and *armadillos*, together with *sloths*, comprise a group of very different and different-looking mammals that, somewhat surprisingly, are closely related. The group they belong to is the Order Edentata, meaning, literally, *without teeth*. Since all but the anteaters have some teeth, the name is a misnomer. The *edentates* are New World mammals specialized to eat ants and termites, or to eat leaves high in the forest canopy. Although the edentates might look and behave differently, they are grouped together because they share certain skeletal features and aspects of their circulatory and reproductive systems that indicate close relationships. Because anteaters and sloths are so diffferent and found only in the tropical and semi-tropical forests of Central and South America, they are perhaps the quintessential mammals of the region, the way that toucans and parrots are quintessential Neotropical birds.

Anteaters. The anteater family, Myrmecophagidae, has four species, all restricted to Neotropical forests. Two of the species occur in southern Mexico: the NORTHERN TAMANDUA and the very small SILKY ANTEATER (Plate 74). Because the latter species is a nocturnal tree-dweller, only the Northern Tamandua is likely to be seen. Fairly common inhabitants of rainforests, they have long cone-shaped snouts, large hooked claws on each front foot, noticeable ears, and a conspicuously dense coat of yellowish or brownish hair, with black on the belly and sides (a black *vest*, biologists say). Tamanduas weigh between 3 and 6 kg (6 to 13 lb). The anteater shape is unmistakable, and once you've studied a picture, you'll know one when you see one.

Armadillos. Armadillos are strange ground-dwelling beasts that, probably owing to the armor plating on their backs, are ecologically quite successful. The family, Dasypodidae, contains about 20 species that are distributed from the southern tip of South America to the central USA. Only one species occurs in the covered region, the NINE-BANDED ARMADILLO (Plate 74). They weigh from 3 to 4 kg (6 to 9 lb) and are grayish or yellowish with many crosswise plates of hard, horn-like material on their backs (bony plates underlie the outer horny covering). The plating produces a look that is unmistakable.

Sloths. What can one say? There is nothing else like a sloth. They vaguely resemble monkeys with longish, stiff hair, but their slow-motion lifestyle is the very antithesis of the primates' hyperkinetic life. There are two families of sloths, the two-toed and three-toed varieties, distinguished by the number of claws per foot. Sloths are active either nocturnally or both during the day and at night. They spend almost all their time in trees, feeding on leaves. None of the five Neotropical sloth species make it as far north as Mexico.

Natural History

Ecology and Behavior

Anteaters. Anteaters are mammals highly specialized to feed on ants and termites; some also dabble in bees. From an anteater's point of view, the main thing about these social insects is that they live in large colonies, so that finding one often means finding thousands. The anteaters' strong, sharp, front claws are put to use digging into ant colonies in or on the ground, and into termite nests in trees (the very abundant, dark, globular, often basket-ball sized *termitaries* attached to the trunks and branches of tropical trees); their long, thin snouts are used to get down into the excavation, and their extremely long tongues, coated with a special sticky saliva, are used to extract the little bugs. Anteaters have prehensile tails for hanging about and moving in trees, allowing them to get to hard-to-reach termite nests. Particular about their food, anteaters don't generally go after army ants or large, stinging ants that might do them harm. Tamanduas rest in hollow trees or other holes during midday, but are otherwise active, including nocturnally. They forage both on the ground and in trees, usually solitarily. Each individual's home range, the area in which it lives and seeks food, averages about 70 hectares (170 acres). Anteaters are fairly slow-moving animals and their metabolic rates low because, although ants and termites are plentiful and easy to find, they don't provide a high nutrition, high energy diet.

Armadillos. Some armadillos feed mainly on ants and termites, but the NINE-BANDED ARMADILLO is more omnivorous, eating many kinds of insects, small vertebrates, and also some plant parts. These armadillos, as is characteristic of their kind, have long claws for digging for food and for digging burrows. Usually they spend the day foraging alone, but several family members may share the same sleeping burrow. They are generally slow-moving creatures that, if not for their armor plating, would be easy prey for predators. When attacked, they curl up into a ball so that their armor faces the attacker, their soft abdomen protected at the center of the ball. Few natural predators can harm them. However, like opossums, they are frequently hit on roads by automobiles. Nine-banded armadillos can be found anywhere in the forest, but are more common in dryer areas.

Breeding

Female anteaters bear one offspring at a time, and lavish attention on it. At first the newborn is placed in a secure location, such as in a tree cavity, and the mother returns to it at intervals to nurse. Later, when it is old enough, the youngster rides on the mother's back. After several months, when the young is about half the mother's size, the two part ways. Breeding may be at any time of year. Female armadillos, after 70-day pregnancies, produce several young at a time, usually four. For some unknown reason, each litter of armadillo young arises from a single fertilized egg so that if a female has four young, they are always identical quadruplets.

Ecological Interactions

The theory of the ecological niche suggests that two or more species that are virtually identical in their lifestyles and resource use cannot coexist within the same habitat or, at least, not for long. Competition among the species for resources will eventually drive the poorer competitors to extinction. If true, how do several species of anteater all occur in the same tropical forest? After all, they all eat ants and termites. Ecologists believe that the SILKY and GIANT ANTEATERS (a very large, rare species that occurs in South and southern Central America), and the NORTHERN TAMANDUA, coexist in some of the same places because they are different sizes and, although they eat the same food, their activity patterns differ sufficiently so as to reduce competition. For instance, Silky Anteaters are strictly nocturnal, but Giant Anteaters and Tamandua are also active during the day. The Silky is arboreal, the Giant is ground-dwelling, and Tamandua are both.

Lore and Notes

People, it seems, have always had trouble deciding precisely what an armadillo is. Linnaeus, the Swedish botanist who originated our current method of scientifically naming organisms, came up with the armadillo genus name *Dasypus*, a term he derived from the Greek for *rabbit*. It seems Linnaeus was trying to incorporate into the scientific name the Aztec word for armadillo, which was *azotochtli*, and which translates as *turtle-rabbit*. Arguments went on for years in the Texas (USA) State Legislature over whether to name the armadillo as the state mammal; some legislators apparently balked at the idea because they felt that armadillos, instead of being pure mammals, might be crosses between mammals and reptiles.

Status

Overall, the edentate mammals are not doing badly, but all suffer population declines from habitat destruction. One problem in trying to determine the status of their populations is that many are nocturnal and some of the armadillos spend most of their time in burrows. The result is that nobody really knows the real health of some populations. The GIANT ANTEATER (CITES Appendix II listed) has been rare in southern Central America for over 100 years, and may be extinct now in most of the region. It is fairly common only on the savannas of Venezuela and Colombia. NORTHERN TAMANDUA are still fairly common animals. The SILKY ANTEATER is also thought to be fairly common, but because their populations naturally are sparse and also because they are so difficult to spot, good information on them is lacking. NINE-BANDED ARMADILLOS are very common over parts of their range, but because they are hunted for meat, their populations are often sparse around heavily settled areas; still, the species may be one of the most abundant mammals of Central American forests. In South America, the GIANT ARMADILLO, which weighs up to 30 kg (65 lb) and is killed for meat, is now endangered (CITES Appendix I and USA ESA listed) from overhunting.

Profiles

Silky Anteater, *Cyclopes didactylus*, Plate 74a
Northern Tamandua, *Tamandua mexicana*, Plate 74b
Nine-banded Armadillo, *Dasypus novemcinctus*, Plate 74c

5. Rodents

Ecotravellers discover among *rodents* an ecological paradox: although by far the most diverse and successful of the mammals, rodents are, with a few obvious exceptions in any region, relatively inconspicuous and rarely encountered. The number of living rodent species globally approaches 2000, fully 44% of the approximately 4500 known mammalian species. Probably in every region of the world save Antarctica, rodents – including the mice, rats, squirrels, chipmunks, marmots, gophers, beavers, and porcupines – are the most abundant land mammals. More individual rodents are estimated to be alive at any one time than individuals of all other types of mammals combined. Rodents' near-invisibility to people, particularly in the Neotropics, derives from the facts that most rodents are very small, most are secretive or nocturnal, and many live out their lives in subterranean burrows. That most rodents are rarely encountered, of course, many people do not consider much of a hardship.

Rodent ecological success is likely related to their efficient, specialized teeth and associated jaw muscles, and to their broad, nearly omnivorous diets. Rodents are characterized by having four large incisor teeth, one pair front-and-center in the upper jaw, one pair in the lower (other teeth, separated from the incisors, are located farther back in the mouth). With these strong, sharp, chisel-like front teeth, rodents "make their living": gnawing (*rodent* is from the Latin *rodere*, to gnaw), cutting, and slicing vegetation, fruit, and nuts, killing and eating small animals, digging burrows, and even, in the case of beaver, imitating lumberjacks.

Rodents are distributed throughout the world except for Antarctica and some Arctic islands. The Neotropics contain some of the largest and most interesting of the world's rodents. Rodents comprise fully 49% of Mexico's mammals – 220 of 450 species (about 100 of the 220 species are endemic to Mexico). Only a few of these are commonly spotted by visitors to southern Mexico and so warrant coverage here – several squirrels, a porcupine, and two larger rodent representatives, the PACA and CENTRAL AMERICAN AGOUTI (Plate 74). Squirrels are members of the family Sciuridae, a worldwide group of more than 350 species that occurs on all continents except Australia and Antarctica. The family includes ground, tree, and flying squirrels. The family Erethizontidae contains the 15 species of New World porcupines, which are distributed throughout the Americas except for the southern third of South America. A single species is native to the covered region. Last, two families that are restricted to tropical America contain a few large common rodents: Family Agoutidae contains the Paca, and Family Dasyproctidae includes the Agouti.

Most of the world's rodents are small mouse-like or rat-like mammals that weigh less than a kilogram (2.2 lb); they range, however, from tiny pygmy mice that weigh only a few grams to South America's pig-like CAPYBARA, behemoths at up to 50 kg (110 lb). The three common tree squirrels of southern Mexico are DEPPE'S, YUCATÁN, and RED-BELLIED SQUIRRELS (Plate 75), which are midsized squirrels with short ears and long, moderately bushy tails. FLYING SQUIRRELS (Plate 75), inhabitants of dryer forests, are small tree dwellers with soft fur. They have a flap of loose skin and muscle on each side that joins the arms and legs. They stretch the flap to form a flat surface that helps them aerodynamically as they glide from tree to tree – much like a parachute. Porcupines generally are fairly large, heavyset rodents, but the MEXICAN HAIRY PORCUPINE (Plate 75) is a relatively small, thin member of the group, weighing between 1.5 and 2.5 kg (3.3 to 5.5 lb). Their bodies are covered with long dark hair that covers most of their spines. They have short limbs, longish prehensile tails, small eyes, mostly hidden ears, and a hairless snout. AGOUTI and PACA are large, almost pig-like rodents, usually brownish, with long legs, short hair, and squirrel-like heads. Paca weigh up to 10 kg (22 lb), twice the weight of an Agouti. Males are slightly larger than females.

Natural History

Ecology and Behavior

DEPPE'S, RED-BELLIED, and YUCATÁN SQUIRRELS are day-active squirrels, generally seen in trees and, occasionally, foraging on the ground. Deppe's Squirrel, usually encountered moving about tree branches and vines in forests at various elevations, eats berries, acorns, and fungi. The larger Yucatán Squirrel, more restricted to lower elevation forests and semi-open areas with trees, eats flowers and fruit. Red-bellied Squirrels, which range over much of southern Mexico, including the pine–oak forests of Oaxaca and Chiapas, eat seeds and fruit. FLYING SQUIRRELS are active mostly at night. They eat, among other things, nuts, bark, fruits and berries. MEXICAN HAIRY PORCUPINES are solitary, nocturnal animals, almost always found in trees. They move slowly along branches, using their prehensile tail as a fifth limb, to feed on leaves, green tree shoots, and fruit. During the day they sleep in tree cavities or on branches hidden amid dense vegetation. They are found mostly in higher-elevation forests.

PACA are usually active only at night, foraging for fruit, nuts, seeds, and vegetation. Along with AGOUTI, they can sit up on their hind legs and eat, holding their food with their front paws, much like a squirrel or rat. They sleep away daylight hours in burrows. Several observers have noted that Paca, if startled or threatened, will freeze, and, if chased, will dive into nearby water, around which they are usually found. The smaller Agouti are naturally day-active, but have become increasingly nocturnal in their habits in areas where they are intensively hunted. Agouti mainly eat seeds and fruit, but also flowers, vegetation, and insects. Both Paca and Agouti appear to live in monogamous pairs on territories, although male and female tend to forage alone. When threatened or startled, Agouti usually run, giving warning calls or barks as they go, presumably to warn nearby relatives of danger. Agouti and Paca, large and tasty, are preyed on by a variety of mammals and reptiles, including large snakes and such carnivores as Jaguarundi (Plate 76).

Breeding

Relatively little is known of the breeding behavior of most Neotropical tree squirrels and of many other rodents. Tree squirrel nests consist of a bed of leaves placed in a tree cavity or a ball of leaves on a branch or in a tangle of vegetation. One to three young are born per litter, usually two. Young, born blind and naked about 40 days after mating occurs, stay in the nest for at least 6 weeks, and nursed for 8 to 10 weeks. Pregnancies in FLYING SQUIRRELS last about 40 days, after which two or three kids are born. Mexican Hairy Porcupines have one to three young per litter. Pregnancies are relatively long, and, as a result, the young are *precocial* – born eyes open and in an advanced state. They are therefore mobile and quickly able to follow the mother. Agouti and Paca also have precocial young, usually in litters of one or two. A day after their birth, a mother Agouti leads her young to a burrow where they hide, and to which she returns each day to feed them. Pregnancy durations for Agouti and Paca are 115 to 120 days.

Ecological Interactions

Rodents are important ecologically primarily because of their great abundance. They are so common that they make up a large proportion of the diets of many carnivores. For instance, in a recent study of Jaguars (Plate 76) in Costa Rica, it was discovered that rodents were the third most frequent prey of the large cats, after sloths and iguanas. In turn, rodents, owing to their ubiquitousness and numbers, are themselves important predators on seeds and fruit. That is, they eat seeds and seed-containing fruit, digesting or damaging the seeds, rendering them useless to the plants that produced them for reproduction. Of course, not every seed is damaged (some fall to the ground as rodents eat, others pass unscathed through their digestive tracts), and so rodents, at least occasionally, also act as seed dispersers. Burrowing is another aspect of rodent behavior that has significant ecological implications because of the sheer numbers of individuals that participate. When so many animals move soil around (rats and mice, especially), the effect is that over several years the entire topsoil of an area is turned, keeping soil loose and aerated, and therefore more suitable for plant growth.

Lore and Notes

Through the animals' constant gnawing, rodents' chisel-like incisors wear down rapidly. Fortunately for the rodents, their incisors, owing to some ingenious anatomy and physiology, continue to grow throughout their lives, unlike those of most other mammals.

Contrary to folk wisdom, porcupines cannot 'throw' their quills, or spines, at people or predators. Rather, the spines detach quite easily when touched, such that a predator snatching a porcupine in its mouth will be impaled with spines and hence rendered very unhappy. The spines have barbed ends, like fishhooks, which anchor them securely into the offending predator.

Paca meat is considered to be among the most superior of wild meats, because it is tasty, tender, and lacks much of an odor; as such, when it can be purchased, it is very expensive. Both Agouti and Paca are favorite game animals throughout their ranges.

Status

YUCATÁN, RED-BELLIED, and FLYING SQUIRRELS are common, widespread species, and although the larger squirrels are hunted for meat in some areas, they are not presently threatened. DEPPE'S SQUIRREL is now threatened in Costa Rica owing to deforestation (CITES Appendix III listed). The precise status of MEXICAN HAIRY PORCUPINE populations is unknown, but they are thought to be secure. The same is true for many other species of New World porcupines – little is known about them, but most appear not to be presently threatened. These nocturnal rodents are hunted for meat in some areas. One species, southeastern Brazil's BRISTLE-SPINED PORCUPINE, is highly endangered (USA ESA listed). PACA are hunted for meat throughout their broad geographic range, but, although scarce in heavily hunted areas, they are secretive enough and broadly enough distributed over Central and South America to still maintain many healthy populations. There are still good populations of AGOUTI in relatively undisturbed regions of southern Mexico. Several northern Mexican rodents are endangered, such as the MICHOACAN POCKET GOPHER and the MEXICAN PRAIRIE DOG (USA ESA listed).

Profiles

Paca, *Agouti paca*, Plate 74d
Central American Agouti, *Dasyprocta punctata*, Plate 74e
Mexican Hairy Porcupine, *Sphiggurus mexicanus*, Plate 75a
Deppe's Squirrel, *Sciurus deppei*, Plate 75b
Yucatán Squirrel, *Sciuris yucatanensis*, Plate 75c

Red-bellied Squirrel, *Sciurus aureogaster*, Plate 75d
Flying Squirrel, *Glaucomys volans*, Plate 75e

6. Carnivores

Carnivores are the ferocious mammals – the cat that sleeps on your pillow, the dog that takes table scraps from your hand – that are specialized to kill and eat other vertebrate animals. Four families within the Order Carnivora have representatives in southern Mexico: *felids* (cats), *canids* (dogs), *procyonids* (raccoons), and *mustelids* (weasel-like things). They have in common that they are primarily ground-dwelling animals and have teeth customized to grasp, rip, and tear flesh – witness their large, cone-shaped canines. Most are meat-eaters, but many are at least somewhat omnivorous, taking fruits and other plant materials. Only two wild members of the dog family, Canidae, occurs in the covered region, the COYOTE and GRAY FOX (Plate 77); because these are common mammals that range into North America, they are pictured but not detailed here. In total there are about 37 species of cat, Family Felidae, with representatives inhabiting all continents but Australia and Antarctica. Because all five species that occur in southern Mexico are fairly rare, most to the point of being endangered, and because of their mainly nocturnal habits, it is rare to see even a single wild cat on any brief trip. More than likely, all that will be observed of cats are traces; some tracks in the mud near a stream or scratch marks on a tree trunk or log.

All of the cats are easily recognized as such. They come in two varieties – spotted and not spotted. The three spotted species generally are yellowish, tan, or cinnamon on top and white below, with black spots and stripes on their heads, bodies, and legs. The smallest is the MARGAY (Plate 76), which is the size of a large house cat, weighing 3 to 5 kg (6 to 11 lb). OCELOTS (Plate 76) are the size of medium-sized dogs and weigh 7 to 14 kg (15 to 30 lb). Last, the JAGUAR (Plate 76) is the largest New World cat and the region's largest carnivore, sometimes nearly 2 m long (6.5 ft), and weighing between 60 and 120 kg (130 to 260 lb). The two unspotted cats are the mid-sized JAGUARUNDI (Plate 76), which is blackish, brown, gray, or reddish, and the PUMA (Plate 76), or MOUNTAIN LION, which is tan or grayish and almost as large as the Jaguar. Female cats often are smaller than males, up to a third smaller in the Jaguar.

The mustelid family comprises about 70 species of small and medium-sized, slender-bodied carnivores that are distributed globally except for Australia and Antarctica. Included in the family are the weasels, skunks, mink, otters, and badgers, animals that occupy diverse habitats, including, in the case of otters, the water. Mustelids generally have long, thin bodies, short legs, long tails, and soft, dense fur. Thirteen species occur in Mexico, but all are wary animals that, except for the skunks, are not commonly sighted. Occasionally seen, among a few others, are the TAYRA (TIE-rah) (Plate 78), a medium-sized (to 5 kg, or 11 lb) mink-like animal, long and slender and mostly black or brown; the NEOTROPICAL OTTER (Plate 78), short-legged, brownish, and aquatic (to 11 kg, or 25 lb); and the small LONG-TAILED WEASEL (Plate 78). Males are larger than females in many mustelids.

The raccoons, or Family Procyonidae, are a New World group of about 15 species (for several reasons, the Asian pandas were thought until recently to be procyonids). This is a very successful group of small and medium-sized mammals that occupy a range of habitats, usually where there are trees. Eight species occur in Mexico, about five of them in the covered region. The more visible ones are the NORTHERN RACCOON and WHITE-NOSED COATI (kah-WAH-tee) (Plate 79) on the mainland and the PYGMY RACCOON on Cozumel Island. In general, procyonids have long, pointed muzzles, short legs, and long tails that more often than not are noticeably ringed. Northern Raccoons (3 to 6 kg, 6 to 13 lb) are brownish or grayish with black-tipped outer hairs that produce a salt-and-pepper look. They have strongly banded tails and a distinguishing black mask that surrounds their eyes. Coatis (4 to 5 kg, 9 to 11 lb), brown or reddish with dark face masks, are more slender than raccoons and have very long, banded tails that, like a cat, they hold erect as they walk. As with the felids and mustelids, male procyonids are often a bit larger than females.

Natural History

Ecology and Behavior

Felids. The cats are finely adapted to be predators on vertebrate animals. Hunting methods are extremely similar among the various species. Cats do not run to chase prey for long distances. Rather, they slowly stalk their prey or wait in ambush, then capture the prey after pouncing on it or after a very brief, fast chase. Biologists are often impressed by the consistency in the manner that cats kill their prey. Almost always it is with a sharp bite to the neck or head, breaking the neck or crushing the skull. Retractile claws, in addition to their use in grabbing and holding prey, give cats good abilities to climb trees, and some of them are partially arboreal animals, foraging and even sleeping in trees. Aside from some highly social large cats of Africa, most cats are solitary animals, foraging alone, individuals coming together only to mate. Some species are territorial but in others individuals overlap in the areas in which they hunt. Cats, with their big eyes to gather light, are often nocturnal, especially those of rainforests, but some are also active by day. When inactive, they shelter in rock crevices or burrows dug by other animals. Cats are the most carnivorous of the carnivores; their diets are more centered on meat than any of the other families. MARGAYS are mostly arboreal forest cats; they forage in trees for rodents and birds. OCELOTS eat rodents, snakes, lizards, and birds. They are probably more common than Margays and, although quite secretive, they are the most frequently seen of the spotted cats. Active mainly at night, they often spend daylight hours asleep in trees. JAGUARUNDI are both day and night active, and are seen fairly frequently in forests. They eat small rodents, rabbits, and birds. JAGUARS can be active day or night. The big spotted cats inhabit low and middle elevation forests, hunting for large prey such as peccary and deer, but also monkeys, birds, lizards, even caiman. Studied extensively in Belize, male Jaguars were active at night and had overlapping home ranges of 28 to 40 sq km (11 to 15 sq miles). When hunting was good, a Jaguar would often stay in the same small area for up to 2 weeks. Prey taken was mostly armadillo, paca, and brocket deer. PUMA occupy various habitat types and prey on deer and other large mammals. They are rare in southern Mexico, the total population of the region probably no more than a few hundred.

Mustelids. Most of the mustelids are strongly carnivorous, although some, such as the ones detailed below, eat a number of other foods. These are powerful animals, sometimes capable of killing prey as large as, or even larger than, themselves. Like the cats, they kill with swift crushing bites to the head or neck. TAYRAS are tree climbers, active both day and night. Singly or sometimes in pairs or family groups, they search the ground and in trees for a variety of foods – fruit, bird eggs or nestlings, lizards, rodents, rabbits, and insects. Skunks occupy many kinds of habitats, although they avoid dense forest. They are active only at dusk and during the night, when they forage, usually solitarily. Skunks root about a good deal in the leaf litter and soil, looking for insects, snails, and small vertebrates such as rodents, lizards and perhaps snakes; occasionally they take fruit, and some species consume a good amount of vegetation. Skunks, what with their spray defenses (see below), usually move quite leisurely, apparently knowing that they are well protected from most predators. NEOTROPICAL OTTERS forage alone or in pairs. They are active both during the day and at night, hunting in streams, rivers, and ponds for fish and crustaceans such as crayfish. Although otters always remain in or near the water, they spend their inactive time in burrows on land. Adapted for moving swiftly and smoothly through water, they move on land awkwardly, with a duck-like waddle.

Procyonids. The distinctive ecological and behavioral traits of the procyonids are that (1) although classified as carnivores, they are omnivorous, (2) they are mostly nocturnal in their activities (except for the coati), and that (3) as a group, they have a great propensity to climb trees. NORTHERN RACCOONS forage either singly or in small groups composed of a mother and her young. They eat fruit and all sorts of small animals, both vertebrates and invertebrates, not to mention garbage around human settlements. Exhibiting a high degree of manual dexterity and sensitivity of their "fingers," raccoons actually search for food in ponds and streams by lowering only their front paws into the water and feeling for frogs, crabs, crayfish, etc. When not active, Raccoons seek shelter in burrows or in tree or rock crevices. WHITE-NOSED COATIS are known for their daylight activity and for the fact that, unlike others of the family, they are quite social. They usually group together in small bands, most commonly several adult females and their young. Occasionally coatis are seen in groups of 50 or more. Males tend to be solitary animals, joining a female band only for several weeks during the breeding season. Coatis are as comfortable foraging in trees as they are on the ground; they search for fruit, lizards, mice, insects and, in the great raccoon tradition, they are also denizens of trash heaps. KINKAJOUS (Plate 79) spend most of their time in trees, foraging for fruits and arboreal vertebrates. Alone among the procyonids, their tail is fully prehensile, permitting them to grasp branches with it and hang upside down. Kinkajous are nocturnal, and relatively little is known of their behavior in the wild. With a flashlight, they can often be spotted moving about tree branches at night, making squeaking sounds; several are often found feeding together in a single tree. CACOMISTLES (Plate 79) are nocturnal tree-climbers that feed mainly on fruit and insects. Adult procyonids probably have a relatively low rate of predation, but their enemies would include boas, raptors, cats, and Tayras.

Breeding
Felids. Male and female cats of the Neotropics come together only to mate; the female bears and raises her young alone. She gives birth in a den fashioned from a burrow, rock cave, or tree cavity. The young are sheltered in the den while the female forages; she returns periodically to nurse and bring the kittens prey to eat. Most of the cats have one or two young at a time, although PUMA and JAGUAR may have up to four. Pregnancy is about 75 days in the smaller cats, about 100 in the large ones. Juvenile Jaguars remain with their mother for up to 18 months, learning to be efficient hunters, before they go off on their own.

Mustelids. Female mustelids give birth in dens under rocks or in crevices, or in burrows under trees. Pregnancy for TAYRA, skunk, and otter usually lasts about 60 to 70 days. Tayra produce an average of two young per litter, skunk, three to five, and otter, two or three. As is true for many of the carnivores, mustelid young are born blind and helpless.

Procyonids. In all the southern Mexican procyonids, females raise young without help from males. Young are born in nests made in trees (NORTHERN RACCOONS in North America also give birth in rock crevices and in tree cavities). Duration of pregnancy varies from about 65 days in Raccoons, to 75 days in WHITE-NOSED COATIS, to about 115 days in KINKAJOUS. Raccoons have three to seven young per litter, coatis, one to five young, and Kinkajous, always only one.

Ecological Interactions

The five species of cats in southern Mexico are quite alike in form and behavior. Therefore, according to ecological theory, they should compete strongly for the same resources, competition that if unchecked, should drive some of the species to extinction. But are all the cats really so similar? One major difference is size. MARGAY are very small, JAGUARUNDI a bit bigger, and OCELOT larger than that. The two large cats, PUMA and JAGUAR, are somwhat similar in size, but Puma live in more diverse habitats than do Jaguar. Prey the animals take also varies. The smallest cats take small rodents and birds, the medium-sized cats take larger rodents and birds, and the large cats take larger prey such as large mammals. Biologists believe that these kinds of differences among similar species permit sufficient "ecological separation" to allow somewhat peaceful coexistence.

Lore and Notes

JAGUAR rarely attack people, who normally are given a wide berth; these cats tend to run away quickly when spotted. Recently there have been widely circulated reports of PUMA attacking people in the USA, but this seems more due to people moving to live in prime Puma habitat, which is increasingly limited, than to a desire on the cat's part for human prey. Large cats in the Neotropics are sometimes seen walking at night along forest trails or roads. General advice if you happen to stumble across one: do not run because that often stimulates a cat to chase. Face the cat, make yourself large by raising your arms, and make as much loud noise as you can.

Based on the excavated artworks of ancient Mayan and Aztec civilizations of Mexico and Central and South America, it is clear that felids in general, and especially the Jaguar, held important positions in the religious beliefs and cultures of these peoples. It is not hard to imagine why the Jaguar would occupy a central place in such cultures: the cat is a large, aggressive predator, a lone hunter, a night stalker with superior senses, one of the few competitors with people for larger prey animals such as deer. The Jaguar in ancient art was usually associated with kings, shamans, and warfare. The supreme Aztec god, Tezcatlipoca, in fact, appeared in one guise as a Jaguar.

Mustelids have a strong, characteristic odor, *musk*, that is produced by secretions from scent glands around their backsides. The secretions are used to communicate with other members of the species and to mark habitats, presumably also for signalling. In skunks, these glands produce particularly strong, foul-smelling fluids that with startling good aim can be violently squirted in a jet at potential predators. The fluids are not toxic, cannot cause blindness as is sometimes commonly believed, but they can cause temporary, severe irritation of eyes and nose. Predators, dogs, and small children that approach a skunk once rarely repeat the exercise.

Status

All of the Neotropical cats are now threatened or actually endangered. Their forest habitats are increasingly cleared for agricultural purposes, they were, and still are to a limited extent, hunted for their skins, and large cats are killed as potential predators on livestock and pets. The five species profiled here are CITES Appendix I and USA ESA listed, although the JAGUARUNDI and PUMA are listed for only parts of their ranges. Jaguar and Puma still roam more remote parts of southern Mexico. BOBCATS, which range from middle-Mexico to the USA and Canada, formerly had Oaxaca as the southernmost extent of their range, but they probably no longer occur there; all Mexican populations are CITES Appendix II listed.

Many mustelids in the past were trapped intensively for their fur, which is often soft, dense, and glossy, just the ticket, in fact, to create coats of otter or weasel, mink or marten, sable or fisher. River otters, although still widespread in the Americas, are sufficiently rare to be considered endangered (CITES I and USA ESA listed). Skunks are common, their populations healthy, and TAYRA are common animals that usually do well even where people disturb their natural habitats. Another Neotropical mustelid, the GRISON (Plate 78), a grayish and black weasel-like animal, is fairly uncommon throughout its range and CITES Appendix III listed by Costa Rica. Three island-bound procyonids in Mexico are considered endangered: the Cozumel Island Coati, the Cozumel Island (or Pygmy) Raccoon, and another raccoon that occurs only on the Tres Marias Islands off western Mexico.

Profiles

Jaguarundi, *Herpailurus yaguarondi*, Plate 76a
Ocelot, *Leopardus pardalis*, Plate 76b
Margay, *Leopardus wiedii*, Plate 76c
Jaguar, *Panthera onca*, Plate 76d
Puma, *Puma concolor*, Plate 76e
Coyote, *Canis latrans*, Plate 77a
Gray Fox, *Urocyon cinereoargenteus*, Plate 77b
Hooded Skunk, *Mephitis macroura*, Plate 77c
Hog-nosed Skunk, *Conepatus mesoleucus*, Plate 77d
Striped Hog-nosed Skunk, *Conepatus semistriatus*, Plate 77e
Tayra, *Eira barbara*, Plate 78a
Neotropical Otter, *Lontra longicaudis*, Plate 78b
Grison, *Galictis vittata*, Plate 78c
Long-tailed Weasel, *Mustela frenata*, Plate 78d
Northern Raccoon, *Procyon lotor*, Plate 79a
White-nosed Coati, *Nasua narica*, Plate 79b
Kinkajou, *Potos flavus*, Plate 79c
Cacomistle, *Bassariscus sumichrasti*, Plate 79d

7. Peccaries and Deer

Peccaries and *deer* are the two Neotropical representatives of the Artiodactyla, the globally distributed order of hoofed mammals (ungulates) that have an even number of toes on each foot. (The Perissodactyla are ungulates with odd numbers of toes; p. 232.) Other artiodactyls are pigs, hippos, giraffes, antelope, bison, buffalo, cattle, gazelles, goats, and sheep. In general, the group is specialized to feed on leaves, grass, and fallen fruit. The truth be told, peccaries look like mid-sized pigs; and deer are self-explanatory. Three peccary species comprise the Family Tayassuidae. They are confined in their distributions to the Neotropics, although one species pushes northwards into the southwestern USA; two species, the COLLARED and WHITE-LIPPED PECCARIES (Plate 80), occur in southern Mexico. Collared Peccaries are more abundant than the White-lipped, occur in more habitats, and are seen more frequently. Collareds are found in rainforests, deciduous forests, and areas of scattered trees and shrubs, including agricultural areas (where they raid crops). White-lippeds are denizens only of rainforests. The two deer species that occur in the region, WHITE-TAILED and RED BROCKET DEER (Plate 80) are members of the Family Cervidae, which is 36 species strong and distributed almost worldwide. PRONGHORN antelope, another member of the artiodactyl group, historically occupied dryer regions of Oaxaca, but no longer occur in Mexico so far south.

Peccaries are small to medium-sized hog-like animals covered with coarse longish hair, with slender legs, large heads, small ears, and short tails. They have enlarged, sharp and pointed, tusk-like canine teeth. The Collared Peccary is the smaller of the two species, adults typically weighing between 17 and 30 kg (35 to 75 lb). They come in black or gray as adults, with a band of lighter-colored hair at the neck that furnishes their name; youngsters are reddish-brown or buff-colored. The White-lipped Peccary, so named for the white patch of hair on its chin, weighs from 25 to 40 kg (55 to 85 lb). Deer are large mammals, reddish, brown, or gray. They have long, thin legs, short tails, and big ears. Males have antlers that they shed each year and regrow. White-tailed Deer have white markings around their eyes, on their muzzles, and, appropriately, on their tails. Southern Mexico's White-tailed Deer (30 to 50 kg, 65 to 110 lb) are, in general, slightly larger than Red Brocket Deer (24 to 48 kg, 50 to 100 lb), but usually smaller, by a third to a half, than members of their species in the USA and Canada. Very young deer are usually spotted with white.

Natural History

Ecology and Behavior

Peccaries are day-active, highly social animals, rarely encountered singly. COLLARED PECCARIES travel in small groups of three to 25 or so, most frequently six to nine; WHITE-LIPPED herds generally are larger, often 50 to 100 or more (smaller groups occur where they are heavily hunted). Peccaries travel single file along narrow forest paths, spreading out when good foraging sites are found. These animals are omnivores, but mainly they dig into the ground with their snouts, "rooting" for vegetation. They feed on roots, underground stems, and bulbs, but also leaves, fruit (especially the White-lipped), insects, and even small vertebrates that they stumble across. Because White-lipped Peccaries are larger than Collareds and travel in larger groups, they need to wander long distances each day to locate enough food. Like pigs, peccaries like to wallow in mud and shallow water, and there is usually a wallowing spot within their home ranges, the area within which a group lives and forages. During dry seasons, peccaries may gather in large numbers near lakes or streams. Because peccaries are hunted by people, they are usually quiet, wary, and therefore sometimes hard to notice or approach. Peccaries are preyed upon by large snakes such as Boa Constrictors (Plate 13), and probably by Puma and Jaguar (Plate 76).

Deer are *browsers* and *grazers*, that is, they eat leaves and twigs from trees and shrubs that they can reach from the ground (browsing), and grass (grazing). The RED BROCKET DEER, in particular, also eats fruit and flowers, chiefly those that have already fallen to the ground. WHITE-TAILED DEER inhabit open places and forest edge areas, rarely dense forest, whereas the Red Brocket Deer is a forest species that wanders through trailless terrain. The large, branched antlers of male White-tails make moving through dense forest a dubious business; male Red Brockets, on the other hand, have short, spike-like, rearwards-curving antlers – plainly better for maneuvering in their dense jungle habitats. Both White-tails and Red Brockets are active during daylight hours and also often at night; Red Brocket Deer are most commonly seen during early mornings and at dusk. White-tails travel either solitarily or in small groups, whereas Red Brockets are almost always solitary. Deer are *cud-chewers*. After foraging and filling a special chamber of their stomach, they find a sheltered area, rest, regurgitate the meal into their mouths and chew it well so that it can be digested. Predators on deer include the big cats – Puma and Jaguar; eagles may take young fawns.

Breeding
Female peccaries have either one or two young at a time, born 4 to 5 months after mating. The young are precocial, meaning that they can walk and follow their mother within a few days of birth. Deer, likewise, give birth to one or two young that, within a week or two, can follow the mother. Until that time, they stay in a sheltered spot while their mother forages, returning at intervals to nurse them.

Lore and Notes
Both species of peccary enjoy reputations for aggressiveness toward humans, but experts agree that the reputation is exaggerated. There are stories of herds panicking at the approach of people, stampeding, even chasing people. These are large enough beasts, with sufficiently large and sharp canine teeth, to do damage. If you spot peccaries, err on the side of caution; watch them from afar and leave them alone. Be quiet and they might take no notice of you; their vision apparently is poor. If you are charged, a rapid retreat into a tree could be a wise move.

When a WHITE-TAILED DEER spots a predator or person that has not yet spotted it, the deer slinks away with its head and tail down, the white patch under the tail concealed. But when the deer is alarmed – it spots a predator stalking it or hears a sudden noise – it bounds off with its tail raised, its white rump and white tail bottom exposed, almost like a white flag. Animal behaviorists believe that the white is a signal to the deer's party, relatives likely to be among them, that a potential predator has been spotted and that they should flee.

The Collared Peccary ranges northwards into the USA's Arizona, New Mexico, and Texas, where it is known locally simply as the PECCARY, or JAVELINA.

Status
Peccaries were hunted for food and hides long before the arrival of Europeans to the New World, and such hunting continues. COLLARED PECCARIES, listed in

CITES Appendix II, are still locally common in protected, wilderness, and more rural areas. WHITE-LIPPED PECCARIES, also CITES Appendix II listed, are less common, and less is known about their populations; they may soon be threatened. Deer are likewise hunted for meat, skins, and sport. Deer range widely in southern Mexico but, owing to hunting pressures, they are numerous only in protected areas. RED BROCKET DEER employ excellent anti-hunting tactics, being solitary animals that keep to dense forests. Both deer species profiled here are CITES Appendix III listed by Guatemala.

Profiles

Collared Peccary, *Tayassu tajacu*, Plate 80a
White-lipped Peccary, *Tayassu pecari*, Plate 80b
White-tailed Deer, *Odocoileus virginianus*, Plate 80c
Red Brocket Deer, *Mazama americana*, Plate 80d

8. Tapir

The *tapir* (TAE-peer) is a funny-looking relative of the horse and rhinoceros and the only member of that group to occur naturally in the New World. (Horses were brought from Asia by people.) Tapirs, horses, and rhinos belong to an order of mammals called the Perissodactyla, which refers to the fact that all of its members have an odd number of toes on each foot. Only four species of tapir comprise the family Tapiridae, three residing in the Neotropics and one in Asia. BAIRD'S TAPIR (Plate 80) is the only one to occur in southern Mexico, and it is now fairly rare; owing to its scarcity, spotting it is often a priority for visitors eager to see mammalian wildlife. This tapir used to be common in many habitats, including grassy swamps, rainforests, forested hillsides, and flooded grasslands. Now it is found only in small numbers, and then only where it is protected from hunting, mainly in biosphere reserves.

Baird's Tapir is the largest of the Neotropical tapirs, and a substantial animal, stocky and up to 2 m (6 ft) long, with short legs and a long snout somewhat reminiscent of a horse's. Weighing between 150 and 300 kg (330 to 660 lb), they have the distinction of being the region's largest native terrestrial mammal. Tapirs have short, sparse hair – from a distance they appear almost hairless – and short tails. The long snout, or *proboscis*, consists of an enlarged, elongated upper lip. Tapirs are blackish to dark- or reddish-brown in color. Youngsters have a characteristic lighter brown coloring with white or yellowish spots and stripes.

Natural History

Ecology and Behavior

Tapirs are mainly nocturnal, but are also seen foraging during daylight hours. They are herbivores, feasting on leaves, twigs, grass, fruit, and perhaps some seeds. As *browsers*, they walk along, stopping occasionally to munch on low plants. Apparently tapirs are very particular about the types of plants they consume, relying at least partially on a highly developed sense of smell to choose the right stuff; their vision is quite poor. In one study conducted in Belize, many more tapirs were sighted in open habitats such as partially logged forests than in dense, virgin, unlogged forests, apparently because many of the tapirs' preferred food plants grow more densely in open habitats. The distinctive tapir snout is used both to shovel food into the mouth and to reach food that the tongue and teeth cannot. Tapirs are not very social; they are usually encountered in ones or twos; for instance, a female and her young. These mammals have a strong affinity for water and are excellent swimmers; if they are disturbed, they often seek refuge in the water. Usually there is a bathing and wallowing site within their home range, the area within which they live and forage; sometimes tapirs sleep in the water.

Breeding
Relatively little is known about breeding in wild tapirs. A single offspring is born to a female BAIRD'S TAPIR 13 months after mating. At first, the youngster stays in a secluded spot while its mother forages elsewhere and periodically returns to nurse. After 10 days or so, the youngster can follow the mother, and stays with her for up to a year. When it is about two-thirds the mother's size, it goes its separate way. Mothers are reputed to attack people that threaten their dependent young.

Ecological Interactions
At least two researchers have noted that tapirs, which can become infested with bloodsucking, disease-carrying ticks, have let other mammals – a coati in one case, a tame peccary in the other – approach them and pick and eat the ticks on their bodies. If this occurs regularly, it is a mutualistic association: tapirs obviously benefit because they are freed temporarily from the harmful ticks, and the tick-eaters receive the nutritional value of the bloodsuckers.

Status
Because they are hunted for meat, and also due to deforestation, BAIRD'S TAPIR is now rare and considered endangered, listed by both CITES Appendix I and USA ESA. Tapirs are rarely seen because of their low population sizes, because they are active mostly at night, and because, owing to hunting, they are very shy and cautious animals.

Profile
Baird's Tapir, *Tapirus bairdii*, Plate 80e

9. Marine Mammals

Two kinds of large marine mammals, *manatees* and *dolphins*, occur in southern Mexico's coastal waters and both are seen fairly frequently. Manatees, or *sea cows*, are heavy-bodied, slow-moving mammals that inhabit Caribbean and Gulf coast areas, especially around river mouths, coastal lagoons, and some islands. The WEST INDIAN MANATEE (Plate 81) is one of four species placed within the Order Sirenia (along with African and South American manatee species, and an Australian member of the group, known as the DUGONG). The group is actually related more closely to elephants than to the whales and dolphins or to the seals and walruses. One subgroup of the West Indian Manatee occurs in the coastal waters of the southeastern USA, especially off Florida, and the other subgroup ranges throughout the Caribbean and along the eastern coast of Mexico and Central America, and the northern coast of South America. Manatees vaguely resemble walruses, but without the tusks. They are large, gray, cylindrical animals, tapered at the front and back. The hands are modified into flippers and the tail is a single, flattened fluke, or paddle. Manatees have thick, rough, mostly hairless skin, with some bristles near the mouth. Some grow to 3.5 m (11.5 ft) long and weigh up to 1000 kg (2200 lb).

Dolphins are smaller members of the Order Cetacea, which also includes whales and porpoises. The approximately 75 species of *cetaceans* occur throughout the world's oceans and some of the smaller dolphins also inhabit larger rivers and estuaries in Asia, Africa, and South America. Whether a given cetacean species is called a whale or a dolphin has to do with length: whales generally are at least 4.5 to 6 m (15 to 20 ft) long, while dolphins and porpoises are smaller. The differences between dolphins and porpoises? Dolphins have a beak-type nose and mouth, a backwards-curving dorsal fin, and sharp, pointed teeth; porpoises are more blunt-nosed with a triangular dorsal fin and blunt teeth. The three most commonly seen dolphins (family Delphinidae) off southern Mexico are the ATLANTIC SPOTTED and BOTTLE-NOSED DOLPHINS (Plate 81) (Atlantic coast) and COMMON DOLPHIN (Plate 81) (Pacific coast), all members of groups that are distributed throughout the world's tropical and warm-water seas. These dolphins are cigar-shaped, long and thin, and tapered at the ends. They have smooth, hairless skin and prominent beaks and dorsal fins. Their front limbs have been modified into paddles, they have no rear limbs, and the tail is flattened to form two rear paddles. Atlantic Spotted Dolphins are 1.2 to 3 m (4 to 10 ft) long, gray to black, with pale spots on their bodies; Bottle-nosed Dolphins are grayish or dark blue, 1.8 to 3.6 m (6 to 12 ft), with a distinctively rounded forehead; and Common Dolphins, mostly black and white with yellow, white and gray stripes along their sides, range up to 2.6 m (8.5 ft). (The largest member of the dolphin family is the KILLER WHALE, *Orcinus orca*; males grow to 9.5 m (31 ft) long, and weigh as much as 7000 kg (15,400 lb).)

Natural History

Ecology and Behavior

Manatees are considered semi-social: most often they are seen as solitary animals or female and calf pairs, moving slowly through the water, grazing on aquatic plants; at other times, groups congregate. The reasons for the groups are not always clear. In colder regions, manatees in winter sometimes associate in large groups in warm-water areas, such as near power-plant water outflows. Manatees, usually active both day and night, are the only aquatic mammals to be completely herbivorous. They feed on submerged, floating, and shoreline vegetation such as sea grasses, mangrove leaves, and water hyacinths. They remain underwater foraging for periods of up to 15 minutes before surfacing to breathe.

Because BOTTLE-NOSED DOLPHINS were the first dolphins to be kept in captivity for long periods (they are the species often seen in aquarium shows), and because they are often found close to shore and so are easily observed, more is known of their biology than of other species; spotted dolphins apparently lead lives that are very similar to those of Bottle-nosed Dolphins. Although sometimes found as solitary animals, these dolphins usually stay in groups, sometimes of up to 1000 or more. Large groups apparently consist of many smaller groups of about two to six individuals, which usually are quite stable in membership for several years. There are dominance hierarchies within groups, the largest male usually being top dolphin. COMMON DOLPHINS are considered to be among the most gregarious of mammals, routinely spotted in schools of 1000 or more, and seasonal groups of 300,000 have been known to occur. Large schools are believed to aid the dolphins in searching for and catching food, and to decrease the likelihood of the dolphins themselves becoming food. They eat primarily fish and squid, which they catch by making shallow dives into the water. They are fast swimmers, routinely jumping clear of the water when feeding or travelling. Dolphins use sounds as well as visual displays and touching to signal each other underwater; they also use sound for *echolocation*, like bats, for underwater navigation. Dolphins are considered highly intelligent and sometimes develop close affinities with people.

Breeding

Little is known about manatee courtship and mating in the wild. Adult males have been observed bumping and pushing each other, apparently in competition for females. Females give birth usually to a single calf after pregnancies of 12 to 13 months. Calves, at birth about 30 kg (66 lb) in weight and 1.2 m (4 ft) in length, stay with their mother for 1 to 2 years. Manatees apparently do not reach sexual maturity for at least 5 or 6 years, when they are about 2.7 m (9 ft) long. In Florida, breeding occurs year round, although newborns are more often spotted in spring and summer.

Dolphins usually produce a single young after pregnancies of about 12 months. When born, dolphins are about 1 m (3 ft) long. The mating systems of dolphins in the wild are not well known, for the obvious reason that it is difficult to observe underwater courtship and mating behaviors; also complicating observation is that male and female look much alike.

Lore and Notes

Although manatees, as endangered marine mammals, are not as well known today as their more celebrated cousins, the whales and dolphins, they have a long history of interactions with people. For instance, it is generally believed that legends of *mermaids* – beings half woman, half fish – arose with manatees, although the resemblance to human females is difficult to discern. The order of manatees, Sirenia, is named for these legendary female "sirens." Even Columbus referred to manatees in his ships' logs, supposedly complaining that the New World mermaids were not as attractive as their advance billing. Ancient Mayans hunted and ate manatees, as evidenced by the frequent renderings of the bulky sea creatures in their artworks. One species of manatee, the STELLER'S SEA COW, which reached 8 m (26 ft) in length and 6000 kg (13,000 lb) in weight, and that lived in the shallow, cold waters of the Bering Sea, was hunted to extinction in about 1770, less than 30 years after being first discovered.

Dolphins' intelligence and friendliness toward people have inspired artists and authors for thousands of years. Images of dolphins appear frequently on artworks and coins from at least 3500 years ago, and from both ancient Greece and Rome. Aristotle, 2300 years ago, noted that dolphins were mammals, not fish, and remarked on their intelligence and gentle personalities. Many other ancient writings tell stories of close relationships between people and dolphins. These animals are considered the only group, aside from humans, that regularly assists members of other species that are in distress. There have been many reports of dolphins supporting on the water's surface injured members of their own and other dolphin species, as well as helping people in the same way.

BOTTLE-NOSED DOLPHINS, among other claims to fame, were perhaps the first species, during the 1970s, to be studied using photographs of individuals that allowed biologists to track and study individual animals. In this case, close photos of dorsal fins permitted researchers to identify individuals and follow their activities for extended periods. This method of identifying individuals by photographs is now widely used to study the long-term behavior and movements of such other animals as whales, elephants, and lions.

Status

The WEST INDIAN MANATEE is considered endangered, listed by both CITES Appendix I and USA ESA; there is also a Florida law protecting manatees that declares the state a manatee sanctuary. In fact, all species of manatees and dugongs are threatened or endangered. The main threats to these animals are hunting – they are still taken in some parts of the world for meat, oil, and their skin – and collisions with boats and motorboat propellers. Unfortunately, the shallow, warm waters that manatees prefer are also usually the main sites people use for fishing, boating, tourism, and coastal development, so collisions are frequent. Even where manatees are strictly protected, conservation and population recovery is hampered owing to the lack of scientific information about their ecology and behavior in the wild, and by their slow breeding rate.

BOTTLE-NOSED, ATLANTIC SPOTTED, and COMMON DOLPHINS are CITES Appendix II listed as species not currently threatened but certainly vulnerable if protective measures are not taken. Spotted dolphins (genus *Stenella*) and Common Dolphins are among the dolphins most frequently caught accidentally in the nets of tuna fishermen, and hundreds of thousands have been killed in that way. Dolphins in some regions of the world are also sometimes killed by fishermen who consider them to be competitors for valuable fish, or to be used as bait – for instance, for crab fishing.

Profiles

West Indian Manatee, *Trichechus manatus*, Plate 81a
Atlantic Spotted Dolphin, *Stenella frontalis*, Plate 81b
Bottle-nosed Dolphin, *Tursiops truncatus*, Plate 81c
Common Dolphin, *Delphinus delphis*, Plate 81d

Environmental Close-Up 4
Endemism: Political System Advocating the End of the World, or Something More Interesting?

A few years ago I attended an international scientific conference on birds, held in New Zealand. The conference, a quadrennial event, moves from continent to continent, I suspect because this permits the participants – scientific researchers who usually double as birdwatchers – to see wild birds they have not seen before and cannot see back home. Indeed, overheard conversations at the conference centered on two topics: the awful cafeteria food at the host university and on seeing "endemics." People would ask each other "Which endemics have you seen so far?" and "Where would I go to see this or that endemic?". What they were referring to were New Zealand species of birds that occur nowhere else on Earth; seeing such unique species, for many a birder, is the paramount reason for visiting isolated spots of the world such as New Zealand. An organism is endemic to a place when it is found only in that place. But the size or type of place referred to is variable: a given species of frog, say, may be endemic to the Western Hemisphere, to a single continent such as South America, to a small mountainous region of Peru, or to a speck of an island off Peru's coast.

A species' history dictates its present distribution. When it's confined to a certain or small area, the reason is that (1) there are one or more barriers to further spread (an ocean, a mountain range, a thousand kilometers of tropical rainforest in the way), (2) the species evolved only recently and has not yet had time to spread, or (3) the species evolved long ago, spread long ago, and now has become extinct over much of its prior range. A history of isolation also matters: the longer animals and plants are isolated from their close relatives, the more time they have to evolve by themselves and to change into new, different, and unique groups. The best examples are on islands. Some islands once were attached to mainland areas but continental drift and/or changing sea levels led to their isolation in the middle of the ocean; other islands arose suddenly via volcanic activity beneath the seas. Take the island of Madagascar. Once attached to Africa and India, the organisms stranded on its shores when it became an island had probably 100 million years in isolation to develop into the highly endemic fauna and flora we see today. It's thought that about 80% of the island's plants and animals are endemic – half the bird species, about 800 butterflies, 8000 flowering plants, and essentially all the mammals and reptiles. Most of the species of lemurs of the world – small, primitive but cute primates – occur only on Madagascar, and an entire nature tourism industry has been built there around the idea of endemism: if you want to see lemurs in the wild, you must go there. Other examples of islands with high concentrations of endemic animals abound: Indonesia, where about 15% of the world's bird species occur, a quarter of them endemic; Papua New Guinea, where half the birds are endemic; the Philippines, where half the mammals are endemic.

Mexico, recent biological surveys reveal, supports a surprising number of endemic species. Mexico is, in fact, considered one of 10 or so "mega-diversity" countries, encompassing an area that harbors a very large number of plant and animal species; it's perhaps one of the top five countries in the world in terms of biodiversity (ahead of Congo – formerly Zaire, Madagascar, and Indonesia, but with less biodiversity than, say, Brazil). The main reason for Mexico's mega-diversity may be that the country straddles two major climate regions – the north temperate zone it shares with the USA and Canada, and the tropics it shares with Central America and northern South America – and so has plant and animal groups common to both. Moreover, due to its varied topography, multitude of habitat types, and some highly isolated habitats that act as "biological islands" (for example, inland highland areas surrounded by lower-lying regions – such as high-elevation forest habitats in Oaxaca's Sierra Madre del Sur, a southern mountain range), a high proportion of Mexico's wildlife is endemic to the country. Certain regions, such as the Sierra Madre del Sur area, support large numbers of endemics and so are considered "centers of endemism." Of the approximately 970 species of reptiles and amphibians that occur in Mexico, about 56% (174 amphibians and 375 reptiles) are endemic – many of them native to Oaxaca and Chiapas. Of the 182 amphibians and reptiles of the Yucatán Peninsula, about 25, or 14%, are endemic to the region. About 32% (142) of Mexico's 449 species of mammals are endemics (inlcuding 106 rodents, 14 bats, and eight rabbits). Many of these mammals, 23 species, are restricted to islands. About 770 species of birds breed in Mexico, about 100 of them, 13%, being endemics. These birds are concentrated in highland regions, particularly in the southern highlands of Oaxaca and Chiapas; 43 of the endemic birds occur in pine–oak forests and 30 species in cloud forests. The Yucatán Peninsula supports about 20 endemic birds, and 19 species are restricted to Mexican islands.

Why is a knowledge of endemism important for ecotravellers? Two main reasons, one practical, one environmental. First, like my friends the New Zealand birdwatchers, if there is a specific type of wildlife you'd like to see, you must first know where it occurs, then travel there: Madagascar for lemurs; Africa or Asia for elephants; Australia for koalas and wombats; the Neotropics for toucans; New Zealand for Yellow-eyed Penguins. If you wanted to visit a region where you might encounter large varieties of strange, exotic wildlife, a region with a high degree of endemism would be just the ticket – such as some of the "hot-spots" mentioned below. Second, species that are endemic to small areas often bear a special environmental vulnerability. Basically, when and if their numbers fall, these species or groups face a greater chance of extinction than others because they lack other places "to go," other populations in far-off places that might survive. Good examples are species that are endemic to islands. If a species of bird occurs only on a single island, all of its "eggs" are in one basket, so to speak: if there is a calamity there – a powerful hurricane, a volcanic eruption – the entire species could become extinct, because all individuals there die and there are no others elsewhere. This type of species extinction has apparently happened often on islands with birds over the past 400 years, as people colonized. People caused habitat destruction and brought animal predators that the native birds had no fear of or experience with. It's thought that about 108 bird species have become extinct in the last 400 years, 97 of them island endemics. (The problem persists: about 900 of the 9000 living bird species are island endemics, and so continually vulnerable.) Similarly, about 75% of mammals driven to extinction recently have been island dwellers.

Knowledge of the existence and distribution of endemic species is crucial for conservation of biodiversity. If we want to preserve biodiversity, then identifying areas with unique species (endemics) and areas with large numbers of unique species (centers of endemism), then targeting those areas for conservation attention is a potentially profitable strategy. In other words, we don't have to make much of an effort to conserve species that are distributed worldwide or hemisphere-wide: their broad ranges often provide protection against quick extinction; but endemics, with their restricted distributions, are inherently more vulnerable and so deserving of immediate attention. A recent concept in conservation biology has been the idea of "hot-spots" – relatively small areas of the world supporting very high numbers of endemic species, areas that should therefore receive priority conservation attention (meaning that time, effort, and funds allocated in these regions will result in greater conservation of biodiversity than efforts elsewhere). For instance, it's estimated that fully 20% of the Earth's endemic plants occur over just 0.5% of the world's land area: preserve that 0.5% and save 20% of endemic plants. Some reptile and amphibian hot-spots are Madagascar, certain regions of Colombia, and Atlantic coastal Brazil; chief mammal hot-spots are the Philippines, Madagascar, and northern Borneo.

Chapter 10

Coral Reef Wildlife
(by Richard Francis)

Coral reefs are the marine equivalents of tropical rainforests. As in rainforests, the diversity of life on a coral reef is truly mind-boggling and life's abundance palpable. In contrast to rainforests, however, where most of the action is beyond view and can be only vaguely appreciated, most of the life on a reef can be experienced directly. Equipped only with mask, snorkel and fins, you can explore environments so different from those in which we spend most of our lives that, by comparison, the Gobi desert, Antarctica, the Congo and New York City are just variations on a single theme.

Coral reefs can only develop in water that is warm and clear, but this alone is not sufficient. Corals only grow at depths to which significant amounts of light penetrate; as such they require a relatively shallow substrate, or bottom. The availability of suitable substrate largely determines the distribution of reefs in the tropics. Extensive reef systems, such as those found off the eastern shore of Mexico's Yucatán Peninsula and mainland Belize, only develop on continental shelves. *Barrier reefs*, which develop well offshore and parallel to the mainland, mark the boundaries of these shelves. The largest of these, the Great Barrier Reef, lies off northeastern Australia. The second largest barrier reef extends from the Yucatán Peninsula, just south of Cozumel, to the Bay Islands of Honduras. Inside the barrier, the water is relatively calm, even during the worst storms, and the conditions ripe for further reef development.

Most snorkeling is conducted in these calm waters, well inside the barrier, on *patch reefs* of various sizes. The patch reefs are separated by varying amounts of rubble and sand, which host a suite of animal inhabitants quite distinct from those you will find on the reefs themselves. The *coral rubble*, which often forms a transition habitat between the living reef and the sandy areas, also has a host of creatures unique to it. Another distinct community develops on the *grassbeds* which develop in the more protected shallow sandy areas. These grassbeds are important nurseries for many reef fishes, including *snappers*, *parrotfish* and some *wrasse*. The most important marine nurseries, however, are provided by the mosquito-infested mangroves (p. 23), which are also home to some of the favorite sportfish such as Tarpon and Bonefish (Plate 84). All of these habitats are worth exploring.

It is the coral reefs themselves, however, that are the main attraction for underwater explorers, so I will focus primarily on them. Aside from the *hard corals*, which form the foundation of these reefs, you will notice a tremendous variety of *gorgonians* (Plate 100) and other *soft corals*, including *sea fans* (Plate 100), *sea whips*, *sea rods* (Plate 101) and *sea plumes*. *Sponges* (Plate 104), including large tube and barrel forms, are another extremely important part of the reef community in this part of the Caribbean. The sponges, soft corals and hard corals provide the physical texture of the reef and the background colors, but it is the moving things that usually attract a snorkeler's attention first, and primary among these moving things, of course, are the fish. Their variety and number are dazzling. Loose schools of *surgeonfish* (Plate 83), sometimes a mixture of several species, and *damselfish* (Plate 87) of various sorts, will attract your attention. As you explore further, you will encounter the large Gray and French Angelfish (Plate 82), usually in pairs. These species are monogamous and pair for life, exhibiting a degree of fidelity that parsons envy. They are curious fish, and if you avoid rapid movements they will approach and inspect you from a distance, sometimes turning on their sides to get a better perspective. You will have to look more closely to find the much shyer Queen Angelfish (Plate 82), which never stray far from their nooks in the coral. These creatures are well worth the effort, as they are among the most magnificently beautiful fish you will find anywhere in the world.

Butterflyfish (Plate 82) are closely related to angelfish. These are among the most spectacular reef inhabitants. There are not nearly as many species of butterflyfish in the Caribbean as in the tropical Pacific, but they are still a prominent component of the reef community. Like the angelfish, some types of butterflyfish mate for life. Remarkably, they seem to pair up before they reach sexual maturity, a form of extended courtship. There is evidence that when they initially pair up, their sexes are not yet determined; only after pairing does one member of the pair "decide" to become a male, and the other a female. Some butterflyfish are among the few reef fish to eat live coral.

Although juvenile butterflyfish look like miniature versions of adults, juvenile angelfish look quite different from the adults, and, in the case of the Gray Angelfish, they are much more strikingly colored. The jet black body is decorated with circular white lines of increasing diameter, somewhat like a bulls-eye. These age-related color changes are common among reef fishes. Aside from angelfish, it is especially common among damselfish, wrasse (Plate 93) and parrotfish (Plate 92). So dramatic are the color differences between young and adult, that they were often classified as different species when first described scientifically. Among many wrasse and parrotfish the differences in color between juvenile and adult is further complicated by the dramatic sex differences in adult coloration. Again, males and females can look so different that, in some cases, they were once classified as different species. The females are typically fairly drab compared with the males, and hence harder to distinguish. Only experienced divers can discriminate between females of the various parrotfish; the males, however, are quite easy to identify.

To further complicate matters, some individuals don't appear to be either male or female, but rather like something intermediate between the two. And in fact they are. It turns out that many types of wrasse and parrotfish are *sex changers*. The Yellowhead Wrasse (Plate 93) is typical. In this species, all individuals mature first as females. Only when they are much older and achieve a certain (relative) size, do some individuals undergo sex change and become males. Sex changing species are often highly social and the sex change process is regulated by social interactions relating to dominance within the group. In general, only the most dominant individuals become males and enjoy the reproductive privileges that come from being one male among many females. Alas, in some species it gets even more complicated. The Bluehead Wrasse (Plate 93), for example, exhibits two distinct types of male. One type of male, referred to as "terminal phase," becomes male only after a sex change, as in the Yellowhead Wrasse. The second male type, or "initial phase," matures as a male without undergoing sex change. The initial phase males look just like females, which relates to their reproductive strategy. Whereas the large and colorful terminal phase males defend a territory and court females, the initial phase males like to sneak into the vicinity of the courting couple and spew their sperm all over them; they then hastily exit the arena before the terminal phase male can switch from a sexual to an aggressive, fighting mode. Not a terribly efficient way to fertilize eggs, but it works well enough.

In the Stoplight Parrotfish (Plate 92), one of the most common parrotfish species in this part of the Caribbean, there is yet another wrinkle to further complicate sexual matters. In the Bluehead Wrasse, initial phase males stay that way for life, but in the Stoplight Parrotfish, initial phase males can themselves become terminal phase males after they reach a certain size. So in this species there are two distinct routes to becoming a terminal phase male: either you mature as a female and later change sex, or you mature as an initial phase male and then become a terminal phase male. In either case there are not only dramatic color changes associated with these sexual changes, but internal changes – which include changes in the brain – as well. No wonder these fish gave headaches to early scientists who tried to classify them.

Aside from their complicated sex lives, parrotfish are notable for their unique feeding habits. The front teeth on each jaw are fused into a beak-like structure from whence they derive their name. With this extremely strong beak they can grind up the cement-like hard corals in order to extract the soft polyps and algae inside. The sounds they make in the process are quite audible underwater. After a hearty meal the parrotfish needs to rid itself of the non-nutritive coral matrix; it does so by excreting the ground-up stuff, now the consistency of fine sand, which forms the mysterious whispy clouds you will occasionally come across underwater. Parrotfish convert incredible amounts of coral into this white sand; and parrotfish excrement is a significant contributor to those beautiful white sand beaches that attract so many to the tropics.

Hamlets (Plates 89, 90) are small members of the seabass family which also exhibit noteworthy sex lives. They are true *hermaphrodites*, possessing both ovaries and testes. Mating, as you might expect, is somewhat unconventional, involving the alternate release of eggs and sperm by each member of a pair, during a process known as *egg trading*. The Yucatán and Belize reefs are particularly good places to find hamlets of several species, each of which is quite beautiful. Indigo Hamlets (Plate 89) and Blue Hamlets (Plate 90), for instance, are among my very favorite fish of the region.

Behaviorally, the most interesting group of fish are the damselfish, particularly during the nesting season. In most coral reef fish, the eggs, once fertilized, drift in the *plankton* (the oceans' tiny floating organisms) for varying periods before settling out on a suitable reef patch. Parental contribution to their lives ends with fertilization. Damselfish, however, lay eggs on the reef or sea floor in nests carefully prepared by the male. The male then protects the eggs from predators, especially from marauding wrasse and other damselfish, until they hatch. During this period the males attack anything and everything that comes near, including divers. What they lack in size they more than make up for in pugnacity, and if you come near a nest, they will come at you with a comically menacing demeanor. If you point your finger in their direction, they will try to nip it. It is also interesting to watch the males interact with each other in the form of constant border wars. Many damselfish, including Sergeant Majors (Plate 88), maintain territories in clusters, and because this means they have common borders, they are constantly stimulating each other to attack. At territorial boundaries you can observe a characteristic to and fro between males, in which one advances, and then retreats as it drifts into the neighbor's territory, at which point it instantly loses its confidence and the roles are reversed. The back and forth can go on for minutes, usually ending in a stalemate at the boundary, during which the two combatants just stare at each other across the divide; until that is, an invasion is detected from a different direction. Male damselfish use up tremendous energy in this way, and by the end of the breeding season they are understandably exhausted. The females take no part in protecting the territory or the eggs.

Gobies (Plates 94, 95), though they comprise the largest family of vertebrate animals, are often overlooked because of their small size and sedentary habits. Look for them in the rubble, cracks in coral heads, under coral heads, and inside the barrel sponges. Some species are quite beautiful, and several of the most attractive gobies act as *cleaners*, much like some Pacific wrasse. The Neon Goby (Plate 94), Yellownose Goby and the Cleaning Goby (Plate 94) are three of the more notable cleaners. They set up stations at which much larger fish, including *groupers* (Plate 90), snappers (Plate 86) and *grunts* (Plate 85) congregate, waiting patiently in line for the cleaners' attentions. This is not at all a frivolous matter; the cleaners remove external parasites and dead scales as they course over the body surface of their clients, as well as inside the mouths. To a first-time observer this looks like the fish equivalent of suicide, but after a while the cleaner will emerge from one of the gills, safe and sated. It has been shown that when cleaner fish are removed from a reef patch, the health of the large fish suffers significantly.

To fully appreciate the diversity of life on the reef, you will need to explore it at night as well as by day. When the sun goes down, a completely different cast of characters emerges, much as occurs on land. But the "shift changes" on land are much more gradual compared with similar transitions on the reef, where in as little as 15 minutes the daytime contingent disappears and the creatures of the night emerge. As the *anemones* (Plate 104) retract, the *basket stars* (Plate 102) unfurl; as the groupers and snappers retreat to their caves, the *moray eels* (Plates 97, 98) and *octopi* begin to prowl. Some wrasse bury themselves in the sand, while some parrotfish construct a giant mucous cocoon within which to sleep. Crevices that harbor *squirrelfish* (Plate 93) and *soldierfish* (Plate 93) by day are taken over by surgeonfish and butterflyfish as the light wanes. The myriad damselfish are seemingly absorbed by the coral.

No matter how many times you have dived or snorkeled these reefs, you can expect to see something new each time you enter the water. Start with the most obvious – the sponges, sea fans and angelfish; then, as you become more experienced, begin looking for the little, less obvious creatures, such as the *tunicates* (Plate 103), *brittle stars* (Plate 102), *shrimps* (Plate 101) and gobies. You will be amply rewarded.

References

Boag D. (1982) *The Kingfisher*. Blandford Press, Poole, UK.

Bowes A. L. (1964) *Birds of the Mayas*. West-of-the-Wind Publications, Big Moose, NY, USA.

Emmons L. H. (1997) *Neotropical Rainforest Mammals: A Field Guide*, 2nd ed. University of Chicago Press, Chicago, USA.

Greene H. W. (1997) *Snakes: The Evolution of Mystery in Nature*. University of California Press, Berkeley, USA.

Greene H. W. and R. L. Seib (1983) *Micrurus nigrocinctus* (Coral Snake). In: *Costa Rican Natural History*, D. H. Janzen (Ed) (pp. 406–408). University of Chicago Press, Chicago, USA.

Hairston N. G. (1994) *Vertebrate Zoology: An Experimental Field Approach*. Cambridge University Press, Cambridge, UK.

Halhead V. (1984) *The Forests of Mexico: The Resource and the Politics of Utilisation*. M.A. thesis, University of Edinburgh, UK.

Howell S. N. G. and S. Webb (1995) *A Guide to the Birds of Mexico and Northern Central America*. Oxford University Press, New York, USA.

Janzen D. H. (1983) *Costa Rican Natural History*. University of Chicago Press, Chicago, USA. (an edited work with many contributors)

Janzen D. H. and D. E. Wilson (1983) Mammals. In: *Costa Rican Natural History*, D. H. Janzen (Ed) (pp. 426–442). University of Chicago Press, Chicago, USA.

Kricher J. C. (1989) *A Neotropical Companion: An Introduction to the Animals, Plants, and Ecosystems of the New World Tropics*. Princeton University Press, Princeton, NJ, USA.

Lee J. C. (1996) *The Amphibians and Reptiles of the Yucatán Peninsula*. Cornell University Press, Ithaca, NY, USA.

Levey D. J., T. C. Moermond and J. S. Denslow (1994) Frugivory: An overview. In: *La Selva: Ecology and Natural History of a Neotropical Rainforest*, L. A. McDade, K. S. Bawa, H. A. Hespenheide and G. S. Hartshorne (Eds) (pp. 282–294). University of Chicago Press, Chicago, USA.

Perrins C. M. (1985) Trogons. In: *The Encyclopedia of Birds*, C. M. Perrins and A. L. A. Middleton (Eds) (pp. 264–265). Facts on File Publications, New York, USA.

Peterson C. R., A. R. Gibson and M. E. Dorcas (1993) Snake thermal ecology. In: *Snakes: Ecology and Behavior*, R. A. Seigel and J. T. Collins (Eds) (pp. 241–314). McGraw-Hill, New York, USA.

Primack R. B. (1993) *Essentials of Conservation Biology*. Sinauer Associates, Sunderland, MA, USA.

Ramamoorthy T. P., R. Bye, A. Lot and J. Fa (1993) *The Biological Diversity of Mexico: Origins and Distribution*. Oxford University Press, New York, USA.

Rowland B. (1978) *Birds with Human Souls: A Guide to Bird Symbolism*. University of Tennessee Press, Knoxville, TN, USA.

Skutch A. F. (1983) *Birds of Tropical America*. University of Texas Press, Austin, TX, USA.

Stiles F. G. and D. H. Janzen (1983) *Cathartes aura* (Turkey Vulture). In: *Costa Rican Natural History*, D. H. Janzen (Ed) (pp. 560–562). University of Chicago Press, Chicago, USA.

Stiles F. G. and A. F. Skutch (1989) *A Field Guide to the Birds of Costa Rica*. Cornell University Press, Ithaca, NY, USA.

Strauch J. G. (1983) *Calidris alba* (Sanderling). In: *Costa Rican Natural History*, D. H. Janzen (Ed) (pp. 556–557). University of Chicago Press, Chicago, USA.

Zug G. (1983) *Bufo marinus* (Marine Toad). In: *Costa Rican Natural History*, D. H. Janzen (Ed) (pp. 386–387). University of Chicago Press, Chicago, USA.

Habitat Photos

1   Isla Contoy Bird Sanctuary, Quintana Roo. The tower in the background serves admirably for island wildlife watching.

2   Sian Ka'an Biosphere Reserve, Quintana Roo. A huge area of now-protected wetlands; try not to miss it if you're in the area.

3 Punta Laguna Wildlife Sanctuary, Quintana Roo. A small-scale, village-run ecotourism operation not far from Cancún.

4 Cobá Archaeological Site, Quintana Roo. An extensive shady, forested site not far from Cancún.

5 The justifiably famous Sacred Cenote at Chichén Itzá Archaeological Site, Yucatán State.

6 Bird-of-paradise flower, Oaxaca.

7 Mangroves as seen from a tourboat, Ría Celestún Special Biosphere Reserve, Yucatán State.

8 The Sacred Cenote at Dzibilchaltún Archaeological Site, just north of the city of Mérida, Yucatán State, doubles on weekends as a popular swimming hole.

9 Uxmal Archaeological Site, Yucatán State. Fairly spectacular ruins set in a fairly spectacular area.

10   Beach scrub habitat along the Gulf of Mexico, Campeche.

11   Pine forest habitat in the highlands east of Oaxaca City, near the village of Ayutla, Oaxaca.

12 Cactus forest at Hierve el Agua mineral springs, southeast of Oaxaca City, Oaxaca.

13 Dry mineral falls at Hierve el Agua and views beyond, southeast of Oaxaca City, Oaxaca.

14 Cactus-fringed trail at Yagul Archaeological Site, southeast of Oaxaca City, Oaxaca.

15 Magnificent highlands view from the road between Tuxtla Gutiérrez and San Cristóbal de Las Casas, Chiapas.

16 Palenque National Park, Chiapas. Mayan ruins set amid dense lowland tropical forest; one of the more spectacular pre-Columbian sites in the New World.

17 Ruins at Bonampak Natural Monument, Chiapas.

18 Waterfalls at Agua Azul Cascades National Park, Chiapas. A nice sight, but the park has the feel of a roadside tourist attraction.

19 Partially buried ruins amid the forests of Yaxchilán Natural Monument, Chiapas. Entry to the main ruins is through a long dark tunnel, the entryway of which is visible in the photo's center.

20 Stunning gorge scenery along the Grijalva River, Sumidero Canyon National Park, Chiapas.

21 One of the 50 or so lakes of Lagunas de Montebello National Park, set amid wild pine forests of southeastern Chiapas.

Identification Plates

Plates 1–104

Abbreviations on the Identification Plates are as follows:

M; male
F; female
IM; immature

The species pictured on any one plate are not necessarily to scale.

Plate 1a

Mexican Caecilian
Dermophis mexicanus
Dos Cabezas = two heads
Culebra de Dos Cabezas = two-headed snake

ID: Wormlike, limbless, long (to 60 cm, 23 in) and wide (to 2.5 cm, 1 in); head triangular and a bit flattened; small eyes covered by skin; body encircled by ringed creases; gray, purplish, or blackish back; lighter underneath.

HABITAT: Low elevation wet forests and warm, moist places (under leaves, garbage piles) in other habitats; found on or under the ground.

LOCATIONS: OAX-LO, CHIAP-LO, TAB, CAM

Plate 1b

Mexican Salamander
(also called Mexican Mushroom-tongue Salamander)
Bolitoglossa mexicana
Salamandra = salamander
Salamanquesa

ID: Small reddish-brown or red-orange salamander with irregular darker spots and blotches; sides usually brownish; some with two stripes running along back; webbed feet; to 7.5 cm (3 in) plus tail.

HABITAT: Low, middle, and some higher elevation wet forests; found in trees, vegetation (often in bromeliad epiphytes), on ground, and under rocks and logs.

LOCATIONS: OAX-HI, OAX-LO, CHIAP-HI, CHIAP-LO, TAB, QRO

Plate 1c

Yucatán Salamander
Bolitoglossa yucatana
Salamandra de Yucatán = Yucatán salamander
Salamanquesa

ID: Small salamander with grayish or brownish back with tan or cream-colored mottling; brown or blackish sides and belly; webbed feet; to 6 cm (2.5 in) plus tail.

HABITAT: Low elevation dry forests; found beneath leaf litter or logs, and on the ground, especially after heavy rains.

LOCATIONS: CAM, YUC, QRO

Plate 1d

Rufescent Salamander
(also called Northern Banana Salamander)
Bolitoglossa rufescens
Salamandra del Oriente = eastern salamander
Salamandra

ID: Very small, slender salamander with large eyes and webbed feet; many are brown with darker sides and stomach; others are light tan with darker spots; to 3.5 cm (1.5 in) plus tail.

HABITAT: Low and middle elevation wet forests; found in trees, often in bromeliad epiphytes, banana leaves; nocturnal.

LOCATIONS: OAX-HI, OAX-LO, CHIAP-HI, CHIAP-LO, TAB

Plate I 257

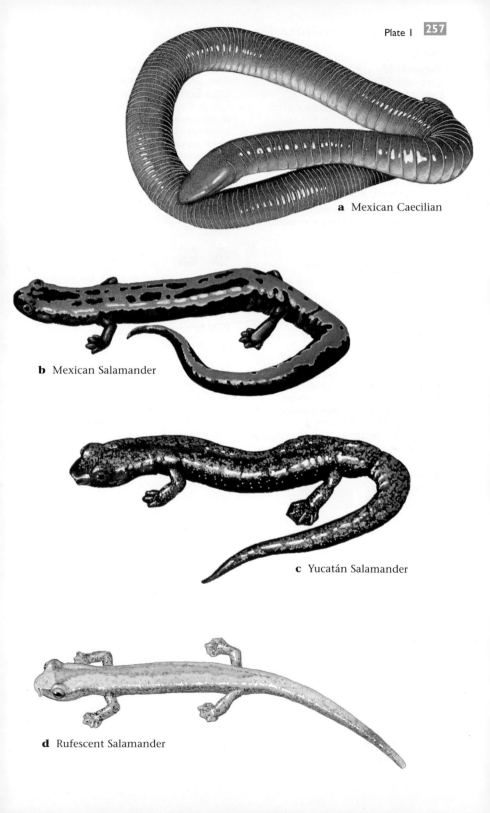

a Mexican Caecilian

b Mexican Salamander

c Yucatán Salamander

d Rufescent Salamander

Plate 2a

Marine Toad
(also called Cane Toad)
Bufo marinus
Sapo Grande, Sapo Gigante = giant toad
Sapo Cucarachero = cockroach toad

ID: A large, ugly, warted toad; large, triangular glands on each side of the head behind brown eyes; females are mottled, combinations of dusky brown, tan, and chocolate; males generally are uniformly brown; to 20 cm (8 in).

HABITAT: Low, middle, and some higher elevation forests; found in open and semi-open areas, often in and around buildings.

LOCATIONS: OAX-HI, OAX-LO, CHIAP-HI, CHIAP-LO, TAB, CAM, YUC, QRO

Plate 2b

Gulf Coast Toad
Bufo valliceps
Sapo Común = common toad
Sapo = toad

ID: Mid-sized gray, brown, yellowish, or reddish-orange toad with darker blotches; short legs; rough, warty skin; small ridges on head near eyes; to 10 cm (4 in).

HABITAT: Low, middle, and some higher elevation forests and more open sites such as grasslands, agricultural areas; found on ground and under rocks, logs.

LOCATIONS: OAX-HI, OAX-LO, CHIAP-HI, CHIAP-LO, TAB, CAM, YUC, QRO

Plate 2c

Marbled Toad
Bufo marmoreus
Sapo Jaspeado = mottled toad

ID: Small yellowish-brown, brown, or olive-green toad sometimes with faint or distinct stripes on back and sides; dark blotches often on back; rear legs dark with patches of dirty yellow; warts may or may not be conspicuous; to 6 cm (2.5 in).

HABITAT: Low and middle elevation wet and dry habitats, including deciduous forest, scrub forest, and savannah; found on the ground, on roads, and in streams, ponds, especially during and after rains.

LOCATIONS: OAX-HI, OAX-LO, CHIAP-HI, CHIAP-LO

Plate 2d

Mexican Burrowing Toad
Rhinophrynus dorsalis
Sapo Borracho = drunk toad
Ranita Boquita = little-mouth frog

ID: Medium-sized, odd-looking frog described as resembling a blob of jelly; smooth, moist skin of the back is gray, reddish-brown, dark brown, or purplish-black, with red, orange, or yellow lines and spots; often there is a single line down the center of the back; small eyes; fat legs; inflates like blowfish when scared; to 7 cm (2.8 in).

HABITAT: Lowland drier forests, scrub forests, wet pastures, fields; found on and under the ground.

LOCATIONS: OAX-LO, CHIAP-LO, TAB, CAM, YUC, QRO

Plate 2e

Tungara Frog
(also called Mudpuddle Frog, Foam Toad)
Physalaemus pustulosus
Sapito Tungara = little tungara toad
Rana = frog

ID: Small brownish frog with very rough (pustular) skin; resembles a toad, but lacks a toad's big glands on the head behind the eyes; to 3.5 cm (1.5 in).

HABITAT: Low and middle elevation forests, savannahs, grasslands, pastures; found on the ground, under leaf litter, or in or near small pools.

LOCATIONS: OAX-HI, OAX-LO, CHIAP-LO, TAB, CAM, QRO

Plate 2 259

a Marine Toad

e Tungara Frog

b Gulf Coast Toad

d Mexican
Burrowing Toad

c Marbled Toad

Plate 3a

Alfredo's Rainfrog
Eleutherodactylus alfredi
Rana de Lluvia = rainfrog

ID: Small tan or brownish frog with lighter spots on back; light brown legs often with darker spots; fingers and toes with large end disks; big protruding eyes; to about 4 cm (1.5 in).

HABITAT: Low and middle elevation wet forests; found on the ground or under leaf litter, sometimes near rocky areas.

LOCATIONS: OAX-HI, OAX-LO, CHIAP-HI, CHIAP-LO, TAB

Plate 3b

Polymorphic Robber Frog
(also called Lowland Rainfrog)
Eleutherodactylus rhodopis
Rana de Selva = jungle frog

ID: Small brown frog with darker markings on back and, usually, very dark eyestripes; legs brownish with dark bars; long, slender toes; to about 4 cm (1.5 in).

HABITAT: Low, middle, and some higher elevation wet forests; found on the ground, often near pools or streams.

LOCATIONS: OAX-HI, OAX-LO, CHIAP-HI, CHIAP-LO, TAB

Plate 3c

Black-backed Frog
(also called Sabinal Frog)
Leptodactylus melanonotus
Ranita Hojarasca = little leaf-litter frog
Rana = frog

ID: Small, stocky, dark brown or grayish frog with darker markings and large triangular patch on head between eyes; skin with small warts; to 4 cm (1.5 in).

HABITAT: Low and middle elevation wet and dry forests and more open areas such as savannah, scrub forests; found on ground near streams and ponds, and in marshes, wet pastures.

LOCATIONS: OAX-HI, OAX-LO, CHIAP-HI, CHIAP-LO, TAB, CAM, YUC, QRO

Plate 3d

White-lipped Frog
Leptodactylus labialis
Ranita Hojarasca = little leaf-litter frog
Ranita de Charco = little puddle frog

ID: Small tan, brown, or reddish-brown frog with darker, irregular blotches and markings on back and legs; white or cream-colored stripe along lip below eye; long, slender toes; to about 4 cm (1.5 in).

HABITAT: Low, middle, and some higher elevation wet and dry forests and more open areas such as savannah, scrub forests; found on the ground near marshes, streams and ponds, especially after rains.

LOCATIONS: OAX-HI, OAX-LO, CHIAP-HI, CHIAP-LO, TAB, CAM, YUC, QRO

Plate 3e

Red-eyed Leaf Frog
(also called Red-eyed Treefrog)
Agalychnis callidryas
Rana Verde = green frog
Rana Arbórea, Rana de Arbol = treefrog

ID: Largish treefrog, with large toes with considerable webbing; colors vary, but often has a pale or dark green back, sometimes with white or yellow spots; top of the thigh is green; blue-purple patches on rear of limbs; vertical bars on sides; often, hands and feet are orange; ruby red eyes; to 7 cm (3 in).

HABITAT: Low and some middle elevation wet forests; found in trees, small pools, swamps; nocturnal.

LOCATIONS: OAX-LO, CHIAP-HI, CHIAP-LO, TAB, CAM, YUC, QRO

Plate 3 261

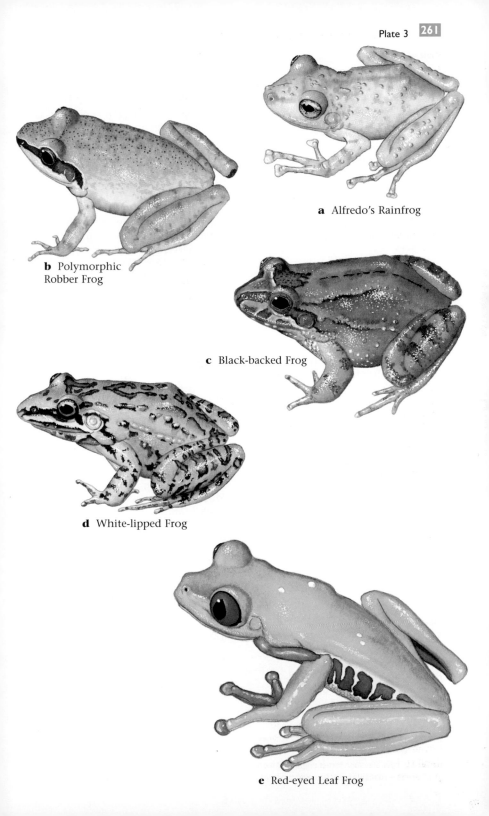

a Alfredo's Rainfrog

b Polymorphic Robber Frog

c Black-backed Frog

d White-lipped Frog

e Red-eyed Leaf Frog

Plate 4a

Variegated Treefrog
(also called Hourglass Treefrog)
Hyla ebraccata
Ranita Amarillenta = little yellowish frog
Rana = frog

ID: Small frog with a short, blunt snout, yellowish-tan or yellow back with or without large dark brown splotches (sometimes hourglass-shaped); yellowish thighs; often, dark bands on legs; in some, dark stripes start at the snout and run along both sides; webbed toes; to 3.5 cm (1.5 in).

HABITAT: Low elevation forests and forest edges; usually found on understory shrubs or in or near forest pools.

LOCATIONS: OAX-LO, CHIAP-LO, TAB, QRO

Plate 4b

Loquacious Treefrog
(also called Mahogany Treefrog, Red-footed Treefrog)
Hyla loquax
Rana Arborícola = treefrog
Rana = frog

ID: Mid-sized cream-colored, light brown, reddish brown, or grayish frog (more tan or yellowish at night) with brown bars or spots on limbs; some reddish webbing on toes and red on rear of thighs; to 5 cm (2 in).

HABITAT: Low, middle, and some higher elevation forests, including pine–oak forests, and more open areas such as forest edges; found in trees or in or near ponds.

LOCATIONS: OAX-HI, OAX-LO, CHIAP-HI, CHIAP-LO, TAB, CAM, YUC, QRO

Plate 4c

Yellow Treefrog
(also called Small-headed Treefrog)
Hyla microcephala
Rana Arborícola
Rana de Arbol = treefrog
Ranita Amarilla = little yellow frog

ID: Small yellowish, orangish, tan, or light brown frog with darker (often X-shaped) markings; yellowish throat; smooth skin; partially webbed toes; to about 3.5 cm (1.5 in).

HABITAT: Low elevation forest edges and more open areas – roadsides, ditches, wet pastures,

etc.; found in trees or in or near pools and puddles; nocturnal.

LOCATIONS: OAX-HI, OAX-LO, CHIAP-HI, CHIAP-LO, TAB, CAM, YUC, QRO

Plate 4d

Cricket Treefrog
(also called Painted Treefrog)
Hyla picta
Ranita Pintada = little painted frog
Rana Arborícola = treefrog

ID: Tiny yellowish, cream, or olive-brown frog with darker spots on back; a light (or light and dark) stripe runs through each eye and along sides; large eyes; smooth skin; partial webbing on toes; to 2.5 cm (1 in).

HABITAT: Low, middle, and higher elevation wet and dry forests, forest edges, and more open areas such as savannahs, roadside ditches, wet pastures; found on vegetation or in or near pools or ponds.

LOCATIONS: OAX-HI, OAX-LO, CHIAP-HI, CHIAP-LO, TAB, CAM, YUC, QRO

Plate 4e

Veined Treefrog
(also called Pepper Treefrog)
Phrynohyas venulosa
Rana Arbórea, Rana de Arbol = treefrog
Quech

ID: Largish, very rough-skinned frog; brown, cream, or tan with darker blotches; large eyes; partially webbed toes with large end disks; to 10 cm (4 in).

HABITAT: Low and middle elevation forests and more open areas such as forest edges, savannah, wet pastures and around human settlements; found in trees or on ground near water.

LOCATIONS: OAX-HI, OAX-LO, CHIAP-HI, CHIAP-LO, TAB, CAM, YUC, QRO

Note: This species produces noxious skin secretions.

Plate 4 263

a Variegated Treefrog

c Yellow Treefrog

b Loquacious Treefrog

d Cricket Treefrog

e Veined Treefrog

Plate 5a

Stauffer's Treefrog
Scinax staufferi
Rana de Arbol = treefrog
Ranita Arborícola = little treefrog

ID: Small gray, grayish-brown, or tan frog with darker bars or irregular stripes; many have a dark spot between eyes; longish, pointed snout; big eyes; toes but not fingers webbed; to about 3.5 cm (1.5 in).

HABITAT: Low and middle elevation open wooded areas such as forest edges and tree plantations, and open areas such as savannah, wet pastures, and around human settlements; found in trees, particularly in bromeliad epiphytes and banana leaves, or in or near pools and ponds.

LOCATIONS: OAX-HI, OAX-LO, CHIAP-HI, CHIAP-LO, TAB, CAM, YUC, QRO

Plate 5b

Mexican Treefrog
Smilisca baudinii
Ranita Arborícola Mexicana = little Mexican treefrog
Rana Arbórea, Rana de Arbol = treefrog

ID: Largish treefrog with short, bluntly rounded snout; pale green, tan, or brown above with darker blotches; limbs marked with dark bands; dark eye bar; often, vertical dark bars on upper lip; to 7 cm (2.75 in).

HABITAT: Low, middle, and higher elevation wet forests and dryer areas such as savannahs and pine–oak forests; found in trees, low vegetation, and near standing water.

LOCATIONS: OAX-HI, OAX-LO, CHIAP-HI, CHIAP-LO, TAB, CAM, YUC, QRO

Plate 5c

Blue-spotted Treefrog
(also called Blue-Spotted Mexican Treefrog)
Smilisca cyanosticta
Rana Arborícola de Manchas-azules = blue-spotted treefrog

ID: Largish tan or green frog, some with darker blotches; dark bars on limbs; light lip stripe and often, dark stripe through eye; light blue or green spots on rear of thighs; webbed toes; to 7 cm (2.75 in).

HABITAT: Low, middle, and some higher elevation wet forests; found in trees and in pools and ponds.

LOCATIONS: OAX-HI, OAX-LO, CHIAP-HI, CHIAP-LO

Plate 5d

Casque-headed Treefrog
(also called Yucatecan Shovel-headed Treefrog)
Triprion petasatus
Rana Arbórea = treefrog

ID: Largish olive, brown, or grayish frog with darker markings; head distinctively bony and triangular; large eyes; to about 7 cm (2.75 in).

HABITAT: Low and middle elevation drier forests and more open sites such as forest edges, scrub forest, and savannah; found in trees, other vegetation, and around pools, ponds.

LOCATIONS: CAM, YUC, QRO

Plate 5e

Fleischmann's Glass Frog
(also called Mexican Glass Frog)
Hyalinobatrachium fleischmanni
Ranita Verde = little green frog
Ranita de Vientre Transparente = little transparent-bellied frog

ID: Tiny translucent frog with a rounded snout; lime green with yellow spots and yellowish hands; organs/bones visible through abdominal skin; to 3 cm (1.2 in).

HABITAT: Low, middle and higher elevation wet forests; found in vegetation along rivers and streams.

LOCATIONS: OAX-HI, OAX-LO, CHIAP-HI, CHIAP-LO

Plate 5 265

a Stauffer's Treefrog

b Mexican Treefrog

c Blue-spotted Treefrog

e Fleischmann's Glass Frog

d Casque-headed Treefrog

Plate 6a

Vaillant's Frog
(also called Rainforest Frog)
Rana vaillanti
Rana de Vaillant = Vaillant's frog
Rana Verde = green frog

ID: A large frog with greenish head and shoulders; body and limbs shading to brown, bronze, or grayish with black flecks; sides sometimes gray or gray-brown with black spots; dark bars on rear legs; toes fully webbed; to 12 cm (4.75 in).

HABITAT: Low and middle elevation forests; found in quiet water of rivers and streams, and in ponds and lakes near shore.

LOCATIONS: OAX-HI, OAX-LO, CHIAP-HI, CHIAP-LO, TAB, CAM, QRO

Plate 6b

Río Grande Leopard Frog
Rana berlandieri
Rana Leopardo = leopard frog
Rana = frog

ID: Large tan, brown, or green/olive frog with dark circles and ovals on back; limbs with dark bands or spots; light stripe often runs through eye and down back; to about 11 cm (4.25 in).

HABITAT: Low, middle, and some higher elevation wet forests; found in and along rivers and streams, ponds, small pools, marshes.

LOCATIONS: OAX-HI, OAX-LO, CHIAP-HI, CHIAP-LO, TAB, CAM, YUC, QRO

Plate 6c

Forrer's Grass Frog
Rana forreri
Rana del Zacate = grass frog

ID: Large green, light brown, or dark brown frog with bold dark spots on back; light stripe above lip; long legs with dark bars or other markings; to about 11 cm (4.25 in).

HABITAT: Low elevation forests and more open sites, Pacific coastal regions; found in and along rivers and streams, ponds, roadside pools, irrigated fields, marshes.

LOCATIONS: OAX-LO, CHIAP-LO

Plate 6d

Highland Frog
Rana maculata
Rana Manchada = spotted frog
Rana = frog

ID: Large green, brown, or green-and-brown frog with spotted back, dark markings; light stripe along lip below eye; dark or dark and light eyestripes; legs with dark brown or black bars; webbed rear feet; to about 11 cm (4.25 in).

HABITAT: Mid and high elevation wet forests, pine–oak forests; found usually in and along streams.

LOCATIONS: OAX-HI, CHIAP-HI

Plate 6e

Sheep Frog
Hypopachus variolosus
Rana Ovejera = sheep frog
Rana Manglera = mangrove frog

ID: Small, squat, gray-brown or reddish-brown frog with darker markings on sides and limbs; a light thin line runs from snout along center of back; often, a light line runs from eye to throat; to 4.5 cm (1.75 in).

HABITAT: Low, middle, and some higher elevation forests and more open sites such as forest edges; found on or under ground, under leaf litter, beneath rocks, logs, and on roads after rains.

LOCATIONS: OAX-HI, OAX-LO, CHIAP-HI, CHIAP-LO, TAB, CAM, YUC, QRO

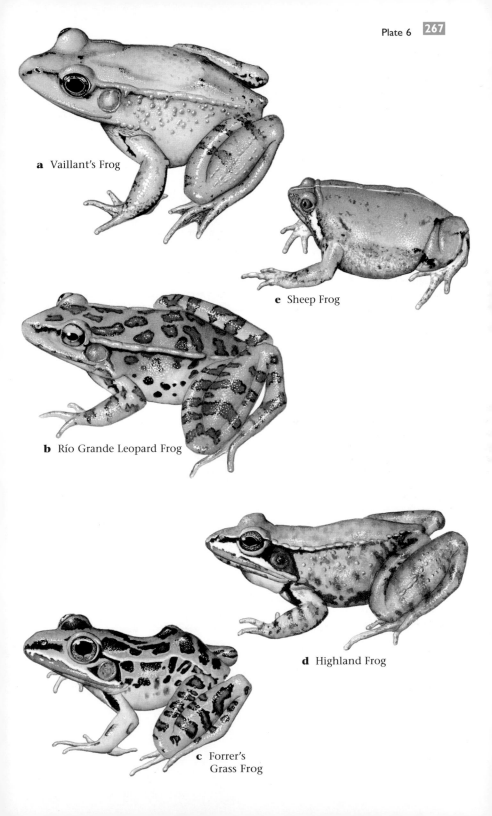

Plate 6 267

a Vaillant's Frog

e Sheep Frog

b Río Grande Leopard Frog

d Highland Frog

c Forrer's Grass Frog

Plate 7a

American Crocodile
Crocodylus acutus
Cocodrilo Amarillo = yellow crocodile
Lagarto = Alligator

ID: Large grayish, brown, or olive crocodile, to as long as 7 m (21 ft), but most are much smaller; individuals longer than 4 m (13 ft) are rare. Youngsters may show dark bands on back and tail. Distinguished from Morelet's Crocodile by (1) generally larger size; (2) narrower snout; and (3) typical habitat/locations.

HABITAT: Coastal lowlands and islands, in brackish or fresh water; found in or near swamps, mangrove swamps, estuaries, larger rivers.

LOCATIONS: OAX-LO, CHIAP-LO, TAB, CAM, YUC, QRO

Note: This species listed as endangered, CITES Appendix I and USA ESA.

Plate 7b

Morelet's Crocodile
Crocodylus moreleti
Cocodrilo Pardo = brown crocodile
Cocodrilo de Pantano = swamp crocodile
Lagarto = Alligator

ID: Dark brown or blackish crocodile with broad snout; to about 4 m (13 ft), but most are less than 2.5 m (8 ft). Youngsters are lighter-colored, olive or yellowish, with dark bands on body and tail. Distinguished from American Crocodile by (1) generally smaller size; (2) broader snout; and (3) typical habitat/locations.

HABITAT: Prefers inland lakes, ponds, lagoons, swamps, slow-moving rivers.

LOCATIONS: OAX-LO, CHIAP-LO, TAB, CAM, YUC, QRO

Note: This species listed as endangered, CITES Appendix I and USA ESA.

Plate 7c

Central American River Turtle
Dermatemys mawii
Tortuga Blanca = white turtle
Tortuga Aplanada = flat turtle

ID: Large brown or olive turtle with broad, smooth-looking top shell; several small plates on each side between top and bottom shells; small, flat head; webbed toes; to about 60 cm (2 ft).

HABITAT: Lowland rivers, larger streams, and freshwater lagoons.

LOCATIONS: OAX-LO, CHIAP-LO, TAB, CAM, QRO

Note: This species listed as endangered, USA ESA, and by CITES Appendix II.

Plate 7d

Snapping Turtle
Chelydra serpentina
Tortuga Lagarto, Tortuga Cocodrilo = alligator turtle
Chiquiguau

ID: A large, mean aquatic turtle; brown, tan, olive, or blackish back often caked with mud or aquatic vegetation; usually three ridges of sharp bumps along back; underneath yellow to tan; very large head with scales on its side and back; neck with pointed bumps; tail as long as shell; to 45 cm (18 in), but most are smaller.

HABITAT: Fresh and brackish water at low elevations; found in marshes, ponds, lakes, streams, rivers with abundant aquatic vegetation.

LOCATIONS: OAX-LO, CHIAP-LO, TAB, CAM

Plate 7e

Furrowed Wood Turtle
Rhinoclemmys areolata
Mojina

ID: Mid-sized tan, olive-brown, or blackish turtle, often with yellow markings; fairly high top shell; small head and long neck with red or yellow markings; yellow bottom shell; toes partially webbed; to about 20 cm (8 in).

HABITAT: Low elevation forests and more open areas such as savannahs and marshes; found on the ground.

LOCATIONS: OAX-LO, CHIAP-LO, TAB, CAM, YUC, QRO

Plate 7 269

a American Crocodile

b Morelet's Crocodile

c Central American River Turtle

d Snapping Turtle

e Furrowed Wood Turtle

Plate 8a

Common Slider
(also called Ornate Terrapin, Sun
Turtle)
Trachemys scripta
Tortuga de Agua = water turtle
Jicotea

ID: Large olive or brown turtle with yellowish and dark markings; limbs and head brown or green with yellowish or orange stripes; large head; webbed toes; to 60 cm (2 ft), but most are smaller.

HABITAT: Low and middle elevation forests and more open areas; found in or near lakes or ponds and in quieter waters of rivers, streams.

LOCATIONS: OAX-HI, OAX-LO, CHIAP-HI, CHIAP-LO, TAB, CAM, YUC, QRO

Plate 8b

Mexican Giant Musk Turtle
Staurotypus triporcatus
Tortuga de Tres Lomos = three-backed
turtle
Guao

ID: Large brown turtle with dark streaks and other markings; three conspicuous ridges (keels) run along back; large head with light spots; two pointed pieces of flesh (barbels) hang down from throat; gray legs; yellowish underneath; to 40 cm (16 in).

HABITAT: Low elevation marshes, lakes, quieter waters of rivers.

LOCATIONS: OAX-LO, CHIAP-LO, TAB, CAM, YUC, QRO

Plate 8c

Tabasco Mud Turtle
Kinosternon acutum
Pochitoque Jaquactero
Casquito = little cask

ID: Mid-sized brown or blackish turtle with darker seams; fairly high top shell; bottom shell yellowish with dark seams; head and limbs yellowish or grayish with dark markings and, often, red or yellow markings; male only with long tail and hooked "beak" on head; female with short tail; to 12 cm (5 in).

HABITAT: Low elevation forests and more open areas such as savannahs; found in or near lakes, streams, forest pools.

LOCATIONS: OAX-LO, CHIAP-LO, TAB, CAM, YUC, QRO

Plate 8d

White-lipped Mud Turtle
(also called White-faced Mud Turtle)
Kinosternon leucostomum
Tortuga de los Pantanos = swamp
turtle
Casquito = little cask
Pochitoque

ID: A mid-sized dark brown or blackish turtle; head brown with whitish jaws, sometimes with dark markings; a yellowish stripe on each side of the head from eye to neck; underneath yellow with darker seams; to 18 cm (7 in).

HABITAT: Low and some middle elevation forests and more open areas; found in quiet water with abundant vegetation in marshes, swamps, streams, rivers, ponds; also terrestrial.

LOCATIONS: OAX-LO, CHIAP-HI, CHIAP-LO, TAB, CAM, YUC, QRO

Plate 8e

Scorpion Mud Turtle
(also called Red-cheeked Mud Turtle)
Kinosternon scorpioides
Tortuga de los Pantanos = swamp
turtle
Casquito Pardo = little brown cask
Casquito Amarillo = little yellow cask
Pochitoque

ID: Mid-sized brown, tan, or yellowish turtle with fairly high top shell; brown head with reddish spots on sides; bottom shell yellowish with dark seams; male only with hooked upper jaw and long tail; female with short tail; to 18 cm (7 in).

HABITAT: Low and middle elevation lakes, ponds, and quieter waters of streams, rivers.

LOCATIONS: OAX-HI, OAX-LO, CHIAP-HI, CHIAP-LO, TAB, CAM, YUC, QRO

Plate 8 **271**

a Common Slider

b Mexican Giant Musk Turtle

c Tabasco Mud Turtle

d White-lipped Mud Turtle

e Scorpion Mud Turtle

Plate 9a

Green Sea Turtle
Chelonia mydas
Tortuga Verde = green turtle
Tortuga Blanca = white turtle
Parlama
Caguama

ID: A medium to large sea turtle with black, gray, greenish, or brown heart-shaped back, often with bold spots or streaks; yellowish white underneath; males' front legs each have one large, curved claw; name refers to greenish body fat; to 1.5 m (5 ft).

HABITAT: Ocean waters off mainland coast and islands; feeds in shallow water; lays eggs on beaches.

LOCATIONS: OAX-LO, CHIAP-LO, TAB, CAM, YUC, QRO

Note: This species is endangered, CITES Appendix I and USA ESA listed.

Plate 9b

Hawksbill Sea Turtle
Eretmochelys imbricata
Tortuga Carey = carey turtle
Carey

ID: A small to mid-sized sea turtle; shield-shaped back mainly dark greenish brown; yellow underneath; head scales brown or black; jaws yellowish with dark markings; chin and throat yellow; two claws on each front leg; narrow head and tapering hooked "beak" give the species its name; to 90 cm (35 in).

HABITAT: Feeds in clear, shallow ocean water near rocks and reefs, and also in shallow bays, estuaries, and lagoons; lays eggs on beaches.

LOCATIONS: OAX-LO, CHIAP-LO, TAB, CAM, YUC, QRO

Note: This species is endangered, CITES Appendix I and USA ESA listed.

Plate 9c

Loggerhead Sea Turtle
Caretta caretta
Caguama

ID: Large brown or reddish-brown sea turtle with heart-shaped top shell; tan or yellowish markings on plates at edge of top shell; head and limbs brownish with yellow or tan markings; bottom shell yellowish; to 2 m (6.5 ft).

HABITAT: Feeds often in shallow coastal areas and bays, occasionally wandering up larger rivers; lays eggs on beaches.

LOCATIONS: OAX-LO, CHIAP-LO, TAB, CAM, YUC, QRO

Note: This species listed as endangered, CITES Appendix I, and as threatened, USA ESA.

Plate 9d

Kemp's Ridley Sea Turtle
Lepidochelys kempii
Tortuga de Kemp = Kemp's turtle
Tortuga Lora = parrot turtle

ID: Relatively small greenish, gray, or yellowish sea turtle with heart-shaped or roundish top shell; central ridge runs along back; bottom shell white or yellowish; grayish or cream-colored head and legs, often with darker markings; to about 70 cm (28 in). (A closely related species, the Olive Ridley Sea Turtle, or Golfina, occurs off southern Mexico's Pacific Coast.)

HABITAT: Shallow coastal waters, bays, lagoons of the Gulf of Mexico; lays eggs on beaches.

LOCATIONS: TAB, CAM, YUC, QRO

Note: This species is critically endangered, CITES Appendix I and USA ESA listed.

Plate 9 273

a Green Sea Turtle

b Hawksbill Sea Turtle

c Loggerhead Sea Turtle

d Kemp's Ridley Sea Turtle

Plate 10a

Snail-eating Thirst Snake
(also called Short-faced Snail Sucker)
Dipsas brevifacies
Chupa Caracoles = snail sucker
Culebra de Sed = thirst snake

ID: Smallish, slender black snake with pink, orange, or white rings; large, protruding eyes; body slightly compressed from side to side; to about 56 cm (22 in).

HABITAT: Wet and dry forests of the Yucatán Peninsula; found on the ground; active at dusk and night.

LOCATIONS: CAM, YUC, QRO

Plate 10b

Black-striped Snake
Coniophanes imperialis
Culebra Rayada = striped snake

ID: Small, slender brown snake with darker sides; narrow light stripes run head to tail, as does a narrow dark line along the mid-back; brown head often with whitish eyestripe; rear belly often reddish; to 44 cm (17 in).

HABITAT: Low, middle, and high elevation forests and open areas such as forest edges, in and around human settlements; found on the ground; nocturnal.

LOCATIONS: OAX-HI, OAX-LO, CHIAP-HI, CHIAP-LO, TAB, CAM, YUC, QRO

Plate 10c

Roadguard
Conophis lineatus
Guarda Camino = roadguard
Sabanero = savannah snake

ID: Mid-sized olive or gray snake with dark eyestripes; pointed or cone-shaped head; smallish eyes; may have two narrow dark lines on back of neck; whitish or yellowish belly; in some regions with bold light and dark lengthwise stripes; to 1.2 m (4 ft).

HABITAT: Low and and some middle elevations; prefers open areas in forests, savannahs, pastures, roadsides, beaches; found on the ground; day-active.

LOCATIONS: OAX-HI, OAX-LO, CHIAP-HI, CHIAP-LO, TAB, CAM, YUC, QRO

Plate 10d

Brown Racer
(also called Lizard-eater)
Dryadophis melanolomus
Lagartijera Olivacea = olive lizard-eater

ID: Mid-sized tan or brown snake with black edging around scales; dark brown head; yellowish or cream lips and chin; to 1.2 m (4 ft).

HABITAT: Low and middle elevation forests and more open sites such as forest edges, savannahs; found on the ground; day-active.

LOCATIONS: OAX-HI, OAX-LO, CHIAP-HI, CHIAP-LO, TAB, CAM, YUC, QRO

Plate 10e

Indigo Snake
(also called Black-tailed Indigo Snake)
Drymarchon corais
Culebra Arroyera = stream snake

ID: A large, long snake, beige, brown, reddish-tan, or olive; last third of body darker; often has noticeable black lines radiating under eye and a short black bar just behind and to the side of the head; black eyes; to 4 m (13 ft).

HABITAT: Low and middle elevation forests and more open areas such as savannahs, agricultural sites; found on ground, along riverbeds, or in swamps, marshes; also climbs low plants; day-active.

LOCATIONS: OAX-HI, OAX-LO, CHIAP-HI, CHIAP-LO, TAB, CAM, YUC, QRO

Plate 10f

Speckled Racer
(also called Guinea Hen Snake)
Drymobius margaritiferus
Ranera = frog-eater

ID: A mid-sized black or green snake spotted all over with yellow, orange, or bluish dots; head black, often with yellow markings; eyes black; to about 1.2 m (4 ft).

HABITAT: Low, middle, and some higher elevation forests and more open areas; terrestrial, often found in thickets and near water; day-active.

LOCATIONS: OAX-HI, OAX-LO, CHIAP-HI, CHIAP-LO, TAB, CAM, YUC, QRO

Plate 10 275

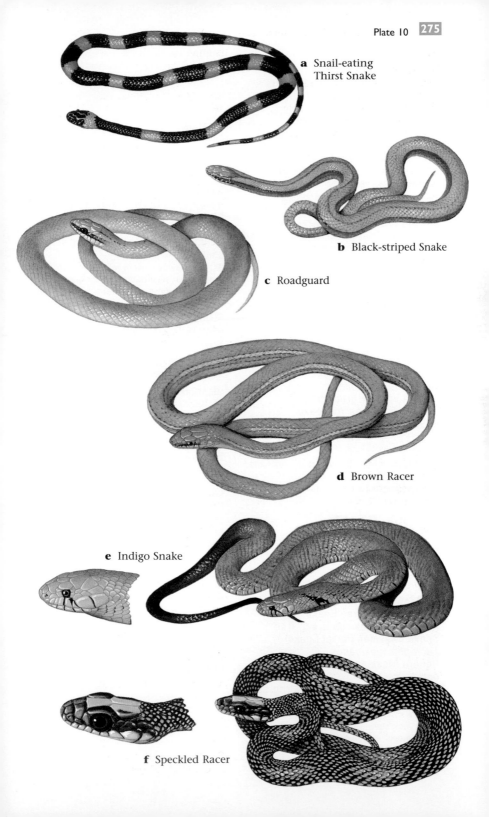

a Snail-eating Thirst Snake

b Black-striped Snake

c Roadguard

d Brown Racer

e Indigo Snake

f Speckled Racer

Plate 11a

Blunt-headed Tree Snake
(also called Chunk-headed Snake,
Blunt-headed Vine Snake)
Imantodes cenchoa
Bejuquilla = little vine snake
Cordelilla Manchada = little spotted
cord
Cordelilla = little cord

ID: Small, very slim snake, tan or light brown with wide dark brown bands; thin neck but noticeably wide, squarish head; very large, bulging eyes; body compressed side to side; to 1 m (3.3 ft).

HABITAT: Low and middle elevation forests; arboreal, usually found in small, outer branches of shrubs and trees; mostly nocturnal.

LOCATIONS: OAX-HI, OAX-LO, CHIAP-HI, CHIAP-LO, TAB, CAM, YUC, QRO

Plate 11b

Green Tree Snake
(also called Parrot Snake)
Leptophis ahaetulla
Ranera Verde = green frog-eater
Culebra Lora = parrot snake

ID: A slender snake, bright green on top; belly ranges from lighter green to whitish; large yellow and black eyes; to 2.2 m (7 ft).

HABITAT: Low and middle elevation forests and more open sites such as forest edges, clearings; found on ground and in small trees and shrubs, usually near water; day-active.

LOCATIONS: OAX-LO, CHIAP-HI, CHIAP-LO, TAB, CAM, YUC, QRO

Plate 11c

Green-headed Tree Snake
(also called Mexican Green Tree
Snake, Mexican Parrot Snake)
Leptophis mexicanus
Ranera Mexicana = Mexican frog-
eater
Ranera Bronceada = bronze frog-eater
Ranera = frog-eater

ID: Mid-sized slender snake with brown and green back; striking long green head with black stripe through eye; whitish lips, chin, and stomach; to 1.5 m (5 ft).

HABITAT: Low and middle elevation forest edges and savannahs; found in trees and shrubs; day-active.

LOCATIONS: OAX-LO, CHIAP-HI, CHIAP-LO, TAB, CAM, YUC, QRO

Plate 11d

Neotropical Vine Snake
(also called Brown Vine Snake,
Mexican Vine Snake)
Oxybelis aeneus
Bejuquilla Parda = little brown vine
snake

ID: A very slender snake with a slim, elongated head; brown, reddish-brown, or grayish, sometimes with small dark spots; sides of head and under head yellowish or whitish; black eyestripe; to 1.5 m (5 ft).

HABITAT: Low and middle elevation forests and forest edges; arboreal; day-active.

LOCATIONS: OAX-HI, OAX-LO, CHIAP-HI, CHIAP-LO, TAB, CAM, YUC, QRO

Plate 11e

Green Vine Snake
Oxybelis fulgidus
Bejuquilla Verde = little green vine
snake
Culebra Verde Arbórea = green tree
snake

ID: Largish slender green snake with lighter green chin; some with darker eyestripe; belly lighter green or yellowish, often with two light stripes running length of body; pointed head; to 2 m (6.5 ft).

HABITAT: Low and middle elevation forests and forest edges; arboreal; day-active.

LOCATIONS: OAX-LO, CHIAP-HI, CHIAP-LO, TAB, CAM, YUC, QRO

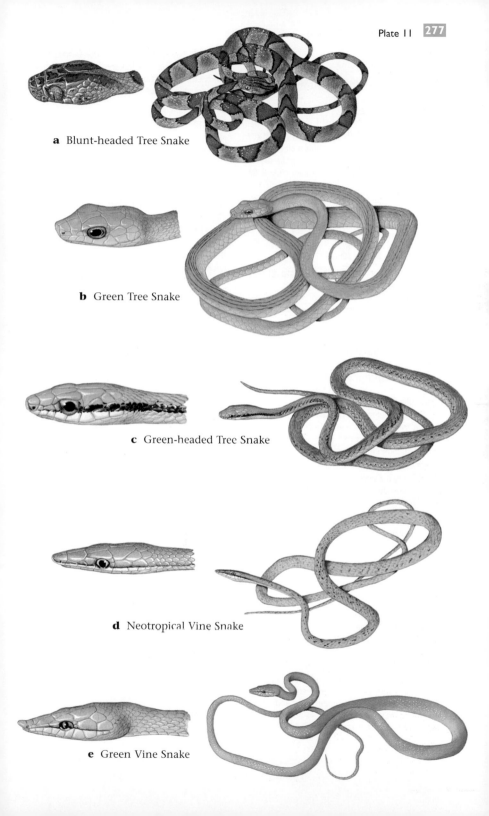

Plate 11 277

a Blunt-headed Tree Snake

b Green Tree Snake

c Green-headed Tree Snake

d Neotropical Vine Snake

e Green Vine Snake

Plate 12a

Cat-eyed Snake
(also called Rainforest Cat-eyed Snake)
Leptodeira frenata
Culebra Nocturna = night snake
Escombrera de Selva = jungle remover
ID: Small brown snake, sometimes a bit pinkish, with darker blotches; broad, flat, brown head with darker stripe behind eye; tan or cream chin; to 60 cm (2 ft).

HABITAT: Low and middle elevation forests and more open areas; found on the ground and in trees; nocturnal.

LOCATIONS: OAX-HI, OAX-LO, CHIAP-HI, CHIAP-LO, TAB, CAM, YUC, QRO

Plate 12b

Terrestrial Snail Sucker
(also called Sartorius's Snail Sucker)
Sibon sartorii
Culebra Negrinaranja = black-orange snake
Coralillo Falso = false coral snake
ID: Smallish but fairly broad black snake with red, orange, yellow, or white rings flecked with black; ring about the head is narrow on top, wider below, extending beneath the eye; to 60 cm (2 ft).

HABITAT: Low, middle, and some high elevation forests, forest edges, and more open areas such as roadsides; found on the ground; dusk- and night-active.

LOCATIONS: OAX-HI, OAX-LO, CHIAP-HI, CHIAP-LO, TAB, CAM, YUC, QRO

Plate 12c

Tropical Rat Snake
Spilotes pullatus
Culebra Voladora = flying snake
Chirrionera = hissing snake
Culebra Mica = monkey snake
ID: A mid-sized snake, variable in color, but often shiny black with yellow markings and bands; chin whitish and black; yellow under head and extending rearwards, changing to a black belly; large brown/black eyes; to 2.2 m (7.2 ft).

HABITAT: Low and middle elevation forests, forest edges, brushy woodland, savannahs, open areas around settlements; arboreal and terrestrial; often found near water.

LOCATIONS: OAX-HI, OAX-LO, CHIAP-HI, CHIAP-LO, TAB, CAM, YUC, QRO

Plate 12d

Scorpion-eating Snake
Stenorrhina freminvillei
Culebra Alacranera = scorpion-eating snake
ID: Mid-sized brownish, grayish, or orange snake; lengthwise stripes in some locations; light-colored belly; dark eyestripe; smallish eyes; to 80 cm (32 in).

HABITAT: Low, middle, and some high elevation forests and more open areas such as forest edges and savannas; found on and under the ground; nocturnal.

LOCATIONS: OAX-LO, CHIAP-HI, CHIAP-LO, CAM, YUC, QRO

Plate 12e

Ribbon Snake
Thamnophis proximus
Culebra de Agua = water snake
Culebra Palustre = marsh snake
ID: Smallish, slender olive-colored snake with yellowish lengthwise stripes; brown or olive head; white throat; yellowish belly; to 60 cm (2 ft).

HABITAT: Low, middle, and high elevation forests, including pine–oak forests, and more open sites such as savannahs; found in marshes, lakes, ponds or nearby in vegetation or on ground; active day or night.

LOCATIONS: OAX-HI, OAX-LO, CHIAP-HI, CHIAP-LO, TAB, CAM, YUC, QRO

Plate 12 279

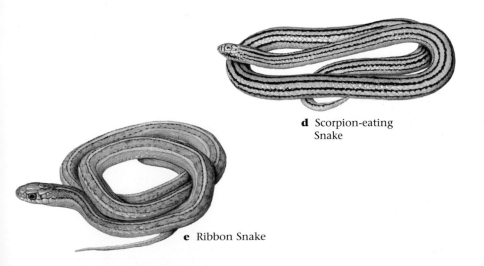

a Cat-eyed Snake

b Terrestrial
Snail Sucker

c Tropical Rat Snake

d Scorpion-eating
Snake

e Ribbon Snake

Plate 13a

Red Coffee Snake
(also called Red-back Coffee Snake)
Ninia sebae
Dormilona = sleepy snake
ID: Small red snake with black head; yellowish "collar" followed by wide black band; often, a few narrow black markings or bands on back; cream or grayish belly; a coral snake mimic (see p. 107); to 30 cm (1 ft).

HABITAT: Low and middle elevation forests and more open sites such as savannahs, agricultural areas; found on the ground; nocturnal.

LOCATIONS: OAX-HI, OAX-LO, CHIAP-HI, CHIAP-LO, TAB, CAM, YUC, QRO

Plate 13b

Tropical Kingsnake
(also called Tropical Milk Snake)
Lampropeltis triangulum
Coralillo Falso = false coral snake
ID: Mid-sized snake with repeated pattern of red, black, and yellow/orange rings, with red the widest and yellow/orange the narrowest; a coral snake mimic (see p. 107); to 1.7 m (5.5 ft).

HABITAT: Low, middle, and higher-elevation wet forests; moutainsides, hillsides, agricultural areas; found on the ground, often concealed in leaf litter.

LOCATIONS: OAX-HI, OAX-LO, CHIAP-HI, CHIAP-LO, TAB, CAM, YUC, QRO

Plate 13c

Boa Constrictor
(also called Imperial Boa)
Boa constrictor
Sierpe Cierva = deer serpent
Mazacuata
Boa
ID: A large, shiny, handsome snake with a long triangular head, often with a single stripe along its top, and dark eyestripes; tan, brown, or grayish body with dark blotches; the tail, often with browns, reds, and yellows, is usually brighter and more colorful than the body; to 3 m (10 ft).

HABITAT: Low and middle elevation forests and more open areas; found on the ground or in vegetation, trees, especially near human settlements.

LOCATIONS: OAX-LO, CHIAP-HI, CHIAP-LO, TAB, CAM, YUC, QRO

Note: This species regulated for conservation purposes, CITES Appendix II listed.

Plate 13d

Coral Snake
(also called Many-ringed Coral Snake, Variable Coral Snake)
Micrurus diastema
Coral
Coralillo = little coral
Coralillo Anillado = ringed coral snake
ID: Smallish snake with red, yellow, and black rings; red rings have black spots or markings; front part of head usually black; yellowish lips, chin; individuals in northern Yucatán and Quintana Roo may have fewer black and yellow rings, so wider red rings – to the point that sometimes most of the body is red; to 85 cm (34 in).

HABITAT: Low, middle, and some higher elevation forests and most other areas; found on the ground; active day or night.

LOCATIONS: OAX-LO, CHIAP-HI, CHIAP-LO, TAB, CAM, YUC, QRO

Plate 13e

Pelagic Sea Snake
Pelamis platurus
Serpiente de Mar = sea snake
Serpiente Marina = marine snake
Víbora de Mar = sea viper
ID: Typically yellow with a wide, dark stripe along its back, but some individuals are nearly all yellow and others, nearly all black; thin, smallish head; oar-shaped tail; rarely longer than 80 cm (32 in).

HABITAT: Pacific Ocean; usually lives fairly far from shore, but often found stranded or dead on beaches.

LOCATIONS: OAX-LO, CHIAP-LO

Plate 13   281

a Red Coffee Snake

b Tropical Kingsnake

c Boa Constrictor

d Coral Snake
(VENOMOUS)

e Pelagic Sea Snake
(VENOMOUS)

Plate 14a

Yucatán Cantil
Agkistrodon bilineatus
Víbora Freno = thicket viper

ID: Mid-sized, thick-bodied, brown, gray-brown, or reddish-brown snake with alternating lighter and darker crosswise bands outlined with whitish flecks; large, flattish, triangular brown head with whitish eyestripe and a second whitish stripe below eye; to 1.3 m (4 ft; to 1 m, or 3 ft, in the Yucatán).

HABITAT: Drier forests and more open areas such as savannahs; low elevations in northern Yucatán Peninsula and low and middle elevations in southern Oaxaca and southern and central Chiapas; found on the ground; nocturnal.

LOCATIONS: OAX-HI, OAX-LO, CHIAP-HI, CHIAP-LO, CAM, YUC, QRO

Plate 14b

Fer-de-lance (also called Terciopelo)
Bothrops asper
Barba Amarilla = yellow chin

ID: A large, fairly slender snake with a triangular head; brown, tan, or gray back covered with a series of beige and brown or black triangles on each side (when viewed from above the series of triangles can resemble hourglass figures or Xs); dark stripe on each side of head behind eye; yellowish under head; youngsters with yellowish tail tip; to 2.5 m (8 ft).

HABITAT: Low and middle elevation forests and open areas; somewhat arboreal as juveniles, terrestrial as adults.

LOCATIONS: OAX-LO, CHIAP-HI, CHIAP-LO, TAB, CAM, YUC, QRO

Plate 14c

Eyelash Viper
Bothriechis schlegelii
Nauyaca de Pestañas = eyelash pit-viper

ID: A slender snake with triangular head; individuals of different colors may occur in the same area, but most are green or grayish-green with brown, tan, or rust-colored markings on head and body; dark stripe behind eye; two or three horny, spine-like scales jut out over each eye (the "eyelashes"); small eyes; to 85 cm (33 in).

HABITAT: Low and middle elevation forests; arboreal, often hanging from vegetation with prehensile tail; nocturnal.

LOCATIONS: CHIAP-HI, CHIAP-LO, TAB

Plate 14d

Jumping Viper
Atropoides nummifer
Nauyaca Saltadora = jumping pit-viper

ID: A thickset, stocky pit-viper with a triangular brown head and dark eye stripe; tan, light brown, or grayish-brown body with dark brown or black blotches; the name refers to its rumored ability to launch itself and strike at long distance, but usually only over about half its body length; to 80 cm (32 in).

HABITAT: Low and middle elevation forests; found on the ground, but also low in trees.

LOCATIONS: OAX-LO, CHIAP-HI, CHIAP-LO, TAB, CAM, YUC, QRO

Plate 14e

Yucatán Hognosed Pit-viper
Porthidium yucatanicum
Víbora = viper

ID: Small, thick-bodied brown or grayish pit-viper with darker blotches; a narrow orange or yellowish line often runs down the mid-back; tip of snout of rounded trianglular head is strongly upturned, providing the basis for the name; tail end in youngsters is yellowish; to 60 cm (2 ft).

HABITAT: Drier forests of northern Yucatán Peninsula; found on the ground, especially after rains; nocturnal.

LOCATIONS: CAM, YUC, QRO

Plate 14f

Tropical Rattlesnake
Crotalus durissus
Cascabel Tropical = tropical rattler
Cascabel Diamante = diamond rattler

ID: A stout brown, gray, or yellowish rattlesnake with a triangular head; body with a pattern of dark triangles or diamonds; two stripes run from the top of the head along the neck; a noticeable ridge runs along the middle of the back; tail rattle; to 1.8 m (6 ft).

HABITAT: Low and middle elevation forests and more open sites such as forest clearings, savannahs, dry grasslands; found on the ground or in rock crevices.

LOCATIONS: OAX-LO, CHIAP-HI, CHIAP-LO, TAB, CAM, YUC, QRO

Plate 14 283

a Yucatán Cantil
(VENOMOUS)

b Fer-de-lance
(VENOMOUS)

c Eyelash Viper
(VENOMOUS)

d Jumping Viper
(VENOMOUS)

e Yucatán Hognosed
Pit-viper
(VENOMOUS)

f Tropical Rattlesnake
(VENOMOUS)

Plate 15a

Yucatán Banded Gecko
Coleonyx elegans
Geco Manchado = spotted gecko
Escorpión = scorpion
Perrito = little dog

ID: Large gecko with blackish, whitish, and brown bands and blotches; some have light and dark stripes down back; adults with blotched sides, youngsters with more solid brown sides; to 11 cm (4.5 in), plus tail.

HABITAT: Low and some mid-elevation forests; found on ground, in rotting logs, around archaeological ruins.

LOCATIONS: OAX-LO, CHIAP-LO, TAB, CAM, YUC, QRO

Plate 15b

House Gecko
Hemidactylus frenatus
Escorpión = scorpion

ID: Small brown, beige or grayish gecko with tiny spots or flecks; some have vague, wavy stripes running from eye to tail; 5 cm (2 in), plus tail.

HABITAT: Found in and around human settlements, in buildings, on ground, in trees; nocturnal.

LOCATIONS: CHIAP-LO, CAM, YUC, QRO

Note: This species is native to Asia but has now spread to many parts of the world through its association with people.

Plate 15c

Mediterranean Gecko
Hemidactylus turcicus
Gecko Pinto = speckled gecko
Escorpión = scorpion

ID: Small brown, beige, or grayish gecko with small spots; rows of small bumps, or "tubercles," run along back and tail; 5 cm (2 in), plus tail.

HABITAT: Found in and around human settlements, in buildings, on walls, ground; nocturnal.

LOCATIONS: TAB, CAM, YUC

Note: This species is native to the Mediterranean region but has been spread by people to southern Mexico and southeastern USA.

Plate 15d

Dwarf Gecko
(also called Least Gecko)
Sphaerodactylus glaucus
Gequillo Collarejo = collared gecko
Bota de la Cola = boot of the tail
Escorpión = scorpion
Gequillo = little gecko

ID: Tiny tan or grayish spotted house gecko; often with dark spot on neck and two spots at start of tail; to 3 cm (1.2 in), plus tail.

HABITAT: Low and middle elevation forests and more open areas; found in buildings, trees, on ground and under leaf litter.

LOCATIONS: OAX-HI, OAX-LO, CHIAP-HI, CHIAP-LO, TAB, CAM, YUC, QRO

Plate 15e

Central American Smooth Gecko
(also called Turnip-tail Gecko)
Thecadactylus rapicauda
Geco Patudo = bigfoot gecko
Escorpión = scorpion

ID: A large gecko with triangular head, distinct from neck; brown, tan, or gray with darker blotches or bands; black eyestripe running to shoulder; various kinds of scales on body; partly webbed hands and feet; newly grown tails (see p. 116) often greatly enlarged (the "turnip") and often lighter in color than body; to 11 cm (4.5 in), plus tail.

HABITAT: Low and middle elevation forests; found on rocks and trees near human settlements, around archaeological ruins, sometimes in buildings.

LOCATIONS: CHIAP-HI, CHIAP-LO, TAB, YUC, QRO

Plate 15 **285**

a Yucatán Banded Gecko

b House Gecko

c Mediterranean Gecko

d Dwarf Gecko

e Central American Smooth Gecko

Plate 16a

Green Iguana
(also called Common Iguana)
Iguana iguana
Iguana Verde = green iguana
Iguana Real = royal iguana
Iguana de Ribera = shore iguana
Guela

ID: Very large lizard with a tall crest (consisting of long, sickle-shaped scales) running from neck to tail; short head with prominent eye and large circular scale (larger than eye) at angle of the jaw; large dewlap (throat sac); greenish, brown, or gray with wavy black bands on body and tail; often some orange on head, body; to 1.8 m (6 ft), including very long tail.

HABITAT: Low and some middle elevation forests and forest edges; found on ground and in trees, often along streams, rivers, lakes; sometimes on sunny, leafy branches as high as 20 m (65 ft).

LOCATIONS: OAX-LO, CHIAP-LO, TAB, CAM, QRO – Cozumel only

Note: This species regulated for conservation purposes, CITES Appendix II listed.

Plate 16b

Striped Basilisk
(also called Brown Basilisk, Jesus Christ Lizard)
Basiliscus vittatus
Basilisco Rayado = striped basilisk
Lagartija Jesucristo = Jesus Christ lizard
Pasarrío = river-crosser
Basilisco

ID: Medium to large brown or olive lizard with dark cross bands along body, often a light stripe along lips and sides, and large, continuous crest along the head, back, and tail (much less prominent in females and young); small ones are good at running quickly across the surface of water (see p. 112), providing the irreverent name; to about 80 cm (32 in), including tail.

HABITAT: Low and middle elevations, region-wide; found especially along streams and bodies of water; terrestrial but also climbs trees, low vegetation.

LOCATIONS: OAX-HI, OAX-LO, CHIAP-HI, CHIAP-LO, TAB, CAM, YUC, QRO

Plate 16c

Spiny-tailed Iguana
(also called Ctenosaur, Black Iguana)
Ctenosaura similis
Iguana de Roca = rock iguana
Iguana Rayada = striped iguana
Iguana Negra = black iguana
Iguana Gris = gray iguana
Garrobo

ID: Large lizard with a tan, olive, olive-brown, or grayish body with dark crosswise bands; banded limbs; pale brown, weakly banded tail with circular rows of scales; back often with red/orange spots; old males have short crest of vertical scales along back; to 1.2 m (4 ft), including tail.

HABITAT: Low and middle elevation open areas, particularly around human settlements and beaches; found on the ground, on rocks, in trees, on buildings and around archaeological ruins; day-active.

LOCATIONS: OAX-HI, OAX-LO, CHIAP-HI, CHIAP-LO, TAB, CAM, YUC, QRO

Plate 16d

Helmeted Basilisk
(also called Helmeted Iguana)
Corytophanes hernandezii
Turipache de Montaña = mountain turipache
Turipache

ID: Mid-sized brown or greenish lizard with dark markings; wide dark eyestripe; whitish lips; body compressed from side to side; head with distinctive large "crest;" long, narrow toes; to about 40 cm (16 in), including long tail.

HABITAT: Low and middle elevation forests; found on trees, vines, shrubs.

LOCATIONS: OAX-HI, OAX-LO, CHIAP-HI, CHIAP-LO, TAB, QRO

Plate 16   287

a Green Iguana

b Striped Basilisk

c Spiny-tailed Iguana

d Helmeted Basilisk

Plate 17a

Ghost Anole
Anolis lemurinus
Lagartija Chipojo = chipojo lizard
Abaniquillo Lemurino = lemurino little fan

ID: Mid-sized brownish or grayish lizard, usually with dark blotches; light stripes on neck; dark band on head; some have I-shaped spots along mid-back; some have a dark stripe from eye to tail; males with reddish throat patch; to about 21 cm (8 in), including long tail.

HABITAT: Low and middle elevation forests; found in trees and on ground; day-active.

LOCATIONS: OAX-HI, OAX-LO, CHIAP-HI, CHIAP-LO, TAB, CAM, YUC, QRO

Plate 17b

Yucatán Smooth Anole
Anolis rodriguezii
Lagartija Chipojo = chipojo lizard
Chipojo Liso = smooth chipojo

ID: Small brown lizard with dark markings; some have I-shaped dark spots along mid-back; dark, pointed head; male with yellowish throat patch; to 14 cm (5.5 in), including tail.

HABITAT: Low and middle elevation forests and forest edges; found in vegetation, on the ground, on and around archaeological ruins; day-active.

LOCATIONS: OAX-HI, OAX-LO, CHIAP-HI, CHIAP-LO, TAB, CAM, YUC, QRO

Plate 17c

Brown Anole
Anolis sagrei
Lagartija Chipojo = chipojo lizard
Abaniquillo Pardo = brown little fan

ID: Mid-sized grayish or brown lizard with darker or lighter spots, bars, or V-shaped markings; some females with light stripe down back; male dewlap (throat sac) orangish; to 7 cm (2.75 in), plus tail.

HABITAT: Low elevations, usually around human settlements; found on fences, rock walls, buildings.

LOCATIONS: CHIAP-LO, TAB, CAM, YUC, QRO

Plate 17d

Silky Anole
Anolis sericeus
Abaniquillo Yanki = yankee little fan

ID: Small brown or gray lizard with darker markings; narrow head; some females with light stripe down back; male dewlap (throat sac) red or orangish with bluish spot; to about 5 cm (2 in), plus long tail.

HABITAT: Low and middle elevation open sites such as forest edges, savannahs, agricultural areas, roadsides; also pine–oak forests; found on low vegetation and ground; often align themselves on blade of grass when startled.

LOCATIONS: OAX-HI, OAX-LO, CHIAP-HI, CHIAP-LO, TAB, CAM, YUC, QRO

Plate 17e

Yellow-spotted Spiny Lizard
Sceloporus chrysostictus
Lagartija Escamosa = scaly lizard
Lagartija de Pintas Amarillas = yellow-spotted lizard

ID: Small gray, brown, or tan lizard with two rows of bars or V-shapes along back; males often with two light stripes down back and dark sides; to about 6 cm (2.5 in), plus tail.

HABITAT: Low elevation forest edges and other open areas; found on ground on rocks, logs, roadsides, etc.; day-active.

LOCATIONS: OAX-HI, OAX-LO, CHIAP-HI, CHIAP-LO, TAB, CAM, YUC, QRO

Plate 17 **289**

a Ghost Anole

b Yucatán Smooth Anole

c Brown Anole

e Yellow-spotted Spiny Lizard

d Silky Anole

Plate 18a

Cozumel Spiny Lizard
Sceloporus cozumelae
Lagartija Playera = beach lizard
ID: Small grayish or brownish lizard with indistinct light stripes running lengthwise along back, one at mid-back and one on each side; dark and light markings between stripes; back pattern less distinct in some males; reddish shoulders in some females; dark mark on side where front legs join body; white belly; to 6 cm (2.5 in), plus tail.

HABITAT: Beaches and other coastal sites of mainland and islands; found on the ground; day-active.

LOCATIONS: YUC, QRO

Plate 18b

Blue Spiny Lizard
Sceloporus serrifer
Lagartija Espinosa Azul = blue spiny lizard
Lagartija Escamoso Ocotero = ocote-pine scaly lizard
ID: Mid-sized light brown or grayish lizard with broad, dark crosswise bars with grayish-blue borders; dark band around neck; dark head with light spots; body scales with ridges and sharp projections produce spiny look; to 10 cm (4 in), plus tail.

HABITAT: Low, middle, and some higher elevation forests, including pine–oak forests, and forest edges; found on tree trunks, in rocky areas, and on and around archaeological ruins; day-active.

LOCATIONS: OAX-LO, CHIAP-HI, CHIAP-LO, TAB, CAM, YUC

Plate 18c

Teapen Rosebelly Lizard
Sceloporus teapensis
Lagartija Escamosos Variable = variable scaly lizard
Lagartija Teapeño = teapen lizard
ID: Small brown lizard with rows of dark markings running down back; dark sides with orange and blue spots; dark mark on side where front legs join body; adult males with two lengthwise light stripes along back; adult females with reddish head sides; body scales with ridges and projections produce somewhat spiny look; to 6 cm (2.5 in), plus tail.

HABITAT: Low elevation open areas such as open woodlands, forest edges, savannahs; found on the ground, on rocks, tree trunks; day-active.

LOCATIONS: OAX-LO, CHIAP-LO, TAB, CAM

Plate 18d

Ground Skink
(also called Brown Forest Skink, Litter Skink)
Sphenomorphus cherriei
Escíncela Parda = brown skink
Salamanquesa
ID: Small lizard with longish head, body, and tail; short legs; shiny bronze/brown back with dark spots; sides darker; dark eyestripes extend along neck to body; to 6.5 cm (2.5 in), plus tail.

HABITAT: Low and middle elevation wet forests and forest edges, tree groves, and other open areas; usually found on the ground, in leaf litter; day-active.

LOCATIONS: CHIAP-LO, TAB, CAM, YUC, QRO

Plate 18e

Shiny Skink
(also called Striped Skink, Central American Mabuya)
Mabuya brachypoda
(also called *Mabuya unimarginata*)
Lagartija Lisa = smooth lizard
Esquinco = skink
Sabandija de Rayas = small striped lizard
Salamanquesa
ID: Mid-sized shiny brown lizard; dark stripe along each side, with light stripe below and sometimes also above the dark stripe; narrow head; short legs; to about 8 cm (3 in), plus tail.

HABITAT: Low and middle elevations; prefers more open sites such as forest edges, savannahs, and near human settlements; found on ground and in trees.

LOCATIONS: OAX-HI, OAX-LO, CHIAP-HI, CHIAP-LO, TAB, CAM, YUC, QRO

Plate 18 291

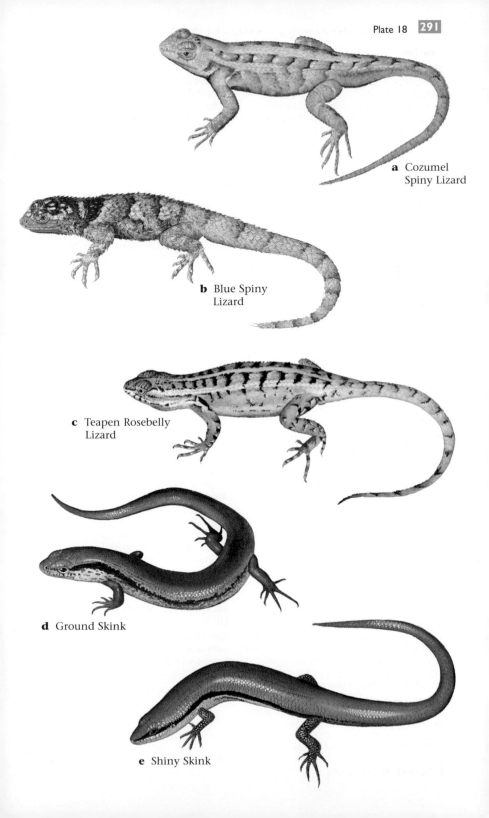

a Cozumel Spiny Lizard

b Blue Spiny Lizard

c Teapen Rosebelly Lizard

d Ground Skink

e Shiny Skink

Plate 19a
Barred Whiptail
(also called Rainbow Ameiva)
Ameiva undulata
Lagartija Metálica = metallic lizard
Ameiva Metálica = metallic ameiva
ID: Mid-sized brown or greenish-brown lizard with dark spots; sides brown sometimes with vertical green-blue bars; breeding males with bright yellowish, green, or orange face; to 10 cm (4 in), plus tail.

HABITAT: Low and middle elevation forest edges and other open sites; found on ground in leaf litter; also in thickets, agricultural areas, roadsides.

LOCATIONS: OAX-HI, OAX-LO, CHIAP-HI, CHIAP-LO, TAB, CAM, YUC, QRO

Plate 19b
Cozumel Whiptail
(also called Cozumel Racerunner)
Cnemidophorus cozumela
Huico de Cozumel = Cozumel huico
Huico
ID: Mid-sized slender tan, brownish, or gray lizard; several light stripes run along each side; to about 8 cm (3 in), plus tail.

HABITAT: Low elevation open areas such as savannahs, beaches, roadsides; found on ground; day-active.

LOCATIONS: CAM, QRO

Plate 19c
Yucatán Whiptail
(also called Narrow-headed Whiptail)
Cnemidophorus angusticeps
Huico Yucateco = Yucatán huico
Huico Rayado = striped huico
Lagartija Llanera = field lizard
ID: Mid-sized brown, reddish-brown, or blackish lizard with six yellowish stripes running from head to tail; some with many light spots that make stripes indistinct; usually with light spots or bars on sides; to 9 cm (3.5 in), plus very long tail.

HABITAT: Low elevation open areas such as open woodlands, forest edges, roadsides; found on the ground; day-active.

LOCATIONS: CAM, YUC, QRO

Plate 19d
Deppe's Whiptail
(also called Blackbelly Racerunner, Seven-lined Racerunner)
Cnemidophorus deppii
Lagartija Verdeazul = green-blue lizard
Lagartija Rayada de Panzanegra = striped blackbelly lizard
ID: Smallish brown, grayish, or blackish lizard with seven or eight narrow cream-colored stripes running from head to tail; dark legs often with lighter spots or bars; narrow pointed head; to 8 cm (3 in), plus long tail.

HABITAT: Low and middle elevation open areas such as forest edges (including some pine–oak forests), tree plantations, savannahs, beaches; found on the ground; day-active.

LOCATIONS: OAX-HI, OAX-LO, CHIAP-HI, CHIAP-LO, CAM

Plate 19e
Black-beaded Lizard
(also called Beaded Lizard)
Heloderma horridum
Escorpión = scorpion
Heloderma Negro = black heloderma
Lagarto Enchaquirado = beaded lizard
ID: Large, very thick, unmistakable brown or gray lizard; light spots and bars on body may or may not be conspicuous; youngsters dark with yellowish markings, and tail with alternating light and dark rings; to 70+ cm (28 in), including tail.

HABITAT: Low and middle elevation dry forests, including some pine–oak forests, and semi-open and open sites such as forest edges, scrub forests, savannahs, roadsides; found on or under the ground, and in trees; day and night active.

LOCATIONS: OAX-HI, OAX-LO, CHIAP-HI, CHIAP-LO

Note: This venomous species considered vulnerable to threat, with conspicuously declining populations in Mexico and Guatemala; CITES Appendix II listed.

Plate 19 293

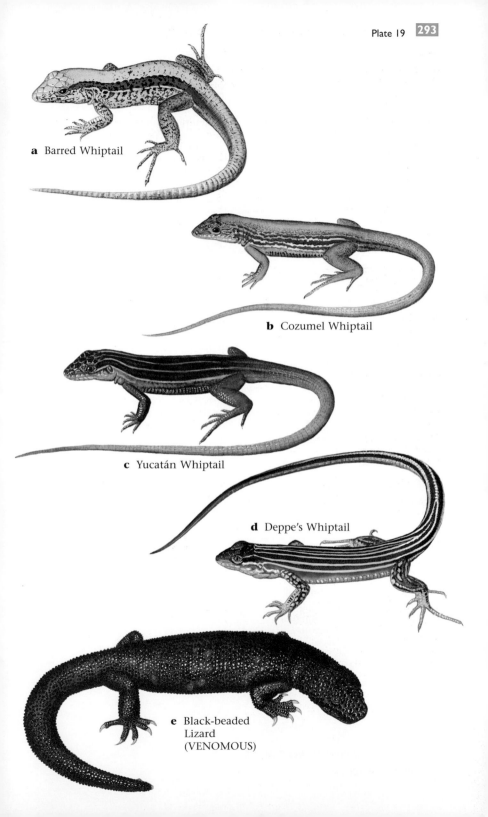

a Barred Whiptail

b Cozumel Whiptail

c Yucatán Whiptail

d Deppe's Whiptail

e Black-beaded
Lizard
(VENOMOUS)

Plate 20a

Brown Booby
Sula leucogaster
Bobo Café = brown stupid bird
Bobo Vientre-blanco = white-belly stupid bird

ID: A mid-sized seabird with brown back, brown neck, white belly, and yellowish, cone-shaped, sharply pointed bill; pointed wings; to 66 cm (26 in); wingspan to 1.4 m (4.7 ft).

HABITAT: Coastal; found around seashores and islands.

LOCATIONS: OAX-LO, CHIAP-LO, YUC, QRO

Plate 20b

Magnificent Frigatebird
Fregata magnificens
Fragata Magnífica = magnificent frigatebird
Rabihorcado

ID: Large black seabird with long, narrow, pointed wings; long, forked tail; long gray bill with down-curved tip; male with reddish throat pouch; female with white belly; immature bird has white head and belly; to 90 cm (3 ft); wingspan to 2 m (6.5 ft).

HABITAT: Coastal; found around seashores and islands; often seen soaring high over coastal areas.

LOCATIONS: OAX-LO, CHIAP-LO, TAB, CAM, YUC, QRO

Plate 20c

Laughing Gull
Larus atricilla
Gaviota Gritóna = shouting gull
Gaviota Risueña = smiling gull
Gaviota Reidora = laughing gull

ID: Smallish seabird, white with grayish back and tops of wings; white ring around eye; reddish-gray bill; adults during breeding have black heads and reddish bills; 40 cm (16 in); wingspan 1 m (3.3 ft).

HABITAT: Marine and fresh water; found along shores, beaches, islands; also seen inland on and near mudflats, river outlets, lake edges.

LOCATIONS: OAX-LO, CHIAP-LO, TAB, CAM, YUC, QRO

Plate 20d

Royal Tern
Sterna maxima
Golondrina Marina Real = royal sea swallow
Golondrina Real = royal swallow

ID: Largish white seabird with black head top; grayish wings; orange bill; black feet; forked tail; breeding adults have black head with noticeable crest; to 48 cm (19 in); wingspan to 1 m (3.3 ft).

HABITAT: Marine and fresh water; found along shores, beaches, islands; also seen inland near river outlets, lake edges.

LOCATIONS: OAX-LO, CHIAP-LO, TAB, CAM, YUC, QRO

Plate 20e

Brown Pelican
Pelecanus occidentalis
Pelícano Café = brown pelican
Pelícano Gris = gray pelican
Alcatraz Moreno = brown pelican

ID: Large brownish seabird with blackish (during breeding) or white (non-breeding) neck; very long bill with large throat pouch; head yellowish and bill reddish during breeding; immature bird is overall brown with lighter belly; to 1 m (3.3 ft); wingspan to 2.1 m (7 ft).

HABITAT: Coastal; found along shores, beaches, islands; occasionally inland.

LOCATIONS: OAX-LO, CHIAP-LO, TAB, CAM, YUC, QRO

Note: This species considered endangered over parts of its range, USA ESA listed.

Plate 20   295

b Magnificent Frigatebird

c Laughing Gull

a Brown Booby

e Brown Pelican

d Royal Tern

Plate 21a

Olivaceous Cormorant
(also called Neotropic Cormorant)
Phalacrocorax brasilianus
Cormorán Neotropical = neotropical
cormorant
Pato Cerdo = pig duck
ID: A blackish or brownish mid-sized waterbird
with brownish yellow facial skin; gray or black bill
with down-curved tip; long tail; 64 cm (2 ft);
wingspan to 1 m (3.3 ft).

HABITAT: Found in and around lakes, rivers,
lagoons, and near shore in the Caribbean; often
seen in rivers and lakes diving for fish.

LOCATIONS: OAX-HI, OAX-LO, CHIAP-HI, CHIAP-
LO, TAB, CAM, YUC, QRO

Plate 21b

Anhinga
Anhinga anhinga
Pato Aguja = needle duck
Anhinga Americana
ID: Large black waterbird with long, sharply
pointed bill; silvery-white streaks on wings; black
(males) or buffy-brown (females) neck and head;
to 90 cm (3 ft); wingspan to 1.1 m (3.8 ft).

HABITAT: Found in and around lakes, rivers,
lagoons, and marshes; often seen swimming with
only head and neck above water, or perched with
wings spread to dry.

LOCATIONS: OAX-LO, CHIAP-LO, TAB, CAM, YUC,
QRO

Plate 21c

White Ibis
Eudocimus albus
Ibis Blanco = white ibis
Ganchuda = hook bird
ID: Large white wading bird with thin, downward-
curved red bill and red legs; black wing tips
noticeable in flight; immature bird is brownish; to
62 cm (2 ft); wingspan to 97 cm (38 in).

HABITAT: Lowland aquatic sites; found along
shorelines, beaches, mangroves, streams and
rivers, ponds, lakes, and marshes.

LOCATIONS: OAX-LO, CHIAP-LO, TAB, CAM, YUC,
QRO

Plate 21d

Jabiru Stork
Jabiru mycteria
Jabirú
ID: A very large white wading bird with black
head and huge black bill; reddish area at bottom
of neck; to 1.4 m (4.5 ft) tall; wingspan to 2.3 m (7.5
ft).

HABITAT: Aquatic habitats in only a few lowland
areas; found around streams, rivers, ponds, lakes,
but most often in marshes.

LOCATIONS: TAB, CAM, QRO

Plate 21e

Wood Stork
Mycteria americana
Cigüeñón = big stork
Cigueña Americana = American stork
Galletán
ID: Large white wading bird with black head,
neck, and bill; black shows under wings in flight;
to 1.1 m (3.5 ft) tall; wingspan to 1.5 m (5 ft).

HABITAT: Aquatic habitats in some lowland
areas; found around streams, rivers, ponds, lakes,
but most often in marshes, including saltwater
marshes.

LOCATIONS: OAX-LO, CHIAP-LO, TAB, CAM, YUC,
QRO

Plate 21 297

e Wood Stork

b Anhinga

F

M

d Jabiru Stork

a Olivaceous Cormorant

IM

c White Ibis

Plate 22a

Roseate Spoonbill
Ajaia ajaja
Espátula Rosada = pink spatula
Chocolatera = chocolate bird
ID: A large pink or light red wading bird with white neck and large spoon-shaped bill; immature bird is whitish to slightly pink; to 80 cm (32 in); wingspan to 1.3 m (4.2 ft).

HABITAT: Lowland aquatic sites; found around ponds, lakes, marshes, including saltwater marshes.

LOCATIONS: OAX-LO, CHIAP-LO, TAB, CAM, YUC, QRO

Plate 22b

American Flamingo
Phoenicopterus ruber
Flamenco Americano
Flamenco
ID: Large pink wading bird with long neck and long pink legs; large, angled, black-tipped bill; youngsters are whitish or grayish with darker, streaked back; to 1.2 m (4 ft); wingspan to 1.5 m (5 ft).

HABITAT: Coastal lagoons and estuaries with salty water.

LOCATIONS: CAM, YUC, QRO

Note: All flamingo species are CITES Appendix II listed.

Plate 22c

Reddish Egret
Egretta rufescens
Garza Rojiza = reddish egret
Garza melenuda = long-haired egret
ID: Mid-sized bluish-gray heron with reddish-brown head and neck; pink bill with black tip; some are all white with pink bill with black tip; youngsters are grayish with gray bill; to 76 cm (30 in).

HABITAT: Coastal lagoons, estuaries, beaches.

LOCATIONS: OAX-LO, CHIAP-LO, TAB, CAM, YUC, QRO

Plate 22d

Little Blue Heron
Egretta caerulea
Garcita Azul = little blue heron
Garza Azul = blue heron
ID: Medium-sized blue-gray heron with purplish- or brownish-red head and neck and grayish bill; legs vary in color but often grayish or black; immature bird is white with dark-tipped grayish bill, lighter-colored legs; to 66 cm (26 in); wingspan to 1 m (3.3 ft).

HABITAT: Aquatic sites; found in coastal areas and in or near ponds, rivers, marshes, mangroves.

LOCATIONS: OAX-HI, OAX-LO, CHIAP-HI, CHIAP-LO, TAB, CAM, YUC, QRO

Plate 22e

Tricolored Heron
Egretta tricolor
Garza Tricolor = tricolored heron
Garcita Flaca = little skinny heron
ID: Medium-sized gray or bluish-gray heron with brownish purple neck and chest; white throat and belly; yellowish bill with darker tip; yellowish legs; 66 cm (26 in); wingspan to 91 cm (3 ft).

HABITAT: Aquatic sites; found in coastal areas and in or near ponds, rivers, marshes, mangroves.

LOCATIONS: OAX-LO, CHIAP-HI, CHIAP-LO, TAB, CAM, YUC, QRO

Plate 22 299

a Roseate Spoonbill

IM

b American Flamingo

IM

c Reddish Egret

IM

d Little Blue Heron

e Tricolored Heron

Plate 23a

Snowy Egret
Egretta thula
Garza Nívea = snowy egret
Garcita Blanca = little white egret
Garza Rizada = curly egret
ID: Medium-sized all-white heron with black bill and legs; immature bird is similar but back of legs yellowish-green and with some gray on bill; to 64 cm (25 in); wingspan to 1 m (3.3 ft).

HABITAT: Aquatic sites; found in coastal areas and in or near ponds, rivers, marshes.

LOCATIONS: OAX-HI, OAX-LO, CHIAP-HI, CHIAP-LO, TAB, CAM, YUC, QRO

Plate 23b

Cattle Egret
Bubulcus ibis
Garza Ganadera = cattle egret
Garcita de Ganado = cattle egret
Garza Vaquera = cowherd egret
ID: A smaller, white heron with thickish neck, yellow bill, and dark legs; during breeding, head, chest, and back with yellowish-buff color, and bill and legs reddish; immature bird is white with yellowish bill; 50 cm (20 in); wingspan to 91 cm (3 ft).

HABITAT: Low and middle elevation agricultural areas; found foraging in fields or following tractors; also marshes.

LOCATIONS: OAX-HI, OAX-LO, CHIAP-HI, CHIAP-LO, TAB, CAM, YUC, QRO

Plate 23c

Great Egret
Ardea alba
Garza Grande = big egret
Garza Blanca = white egret
ID: Large all-white heron with yellow bill and dark legs; 1 m (3.3 ft); wingspan to 1.3 m (4.3 ft).

HABITAT: Aquatic sites; found along coasts and inland around marshes, lakes, ponds, rivers.

LOCATIONS: OAX-HI, OAX-LO, CHIAP-HI, CHIAP-LO, TAB, CAM, YUC, QRO

Plate 23d

Green Heron
Butorides striatus
Garza Verde = green heron
Garcita Verde = little green heron
ID: Small heron with grayish-green back; maroon or reddish-brown neck; black on top of head; yellowish-orange legs; immature bird is darker and heavily streaked; to 45 cm (18 in); wingspan to 60 cm (2 ft).

HABITAT: Aquatic sites; found along coasts in mangroves and inland around marshes, lakes, ponds, rivers.

LOCATIONS: OAX-HI, OAX-LO, CHIAP-HI, CHIAP-LO, TAB, CAM, YUC, QRO

Plate 23e

Bare-throated Tiger Heron
Tigrisoma mexicanum
Garza Tigre = tiger heron
ID: Large brownish heron with fine black stripes on neck, back and sides; black and gray head; yellow throat; immature bird is more chestnut-brown, has brown and black mottled wings, and lacks gray and black on head; to 80 cm (32 in).

HABITAT: Aquatic sites; found along coasts and inland around marshes, lakes, ponds, rivers.

LOCATIONS: OAX-HI, OAX-LO, CHIAP-HI, CHIAP-LO, TAB, CAM, YUC, QRO

Plate 23 | 301

c Great Egret

IM

b Cattle Egret

a Snowy Egret

IM

d Green Heron

IM

e Bare-throated Tiger Heron

Plate 24a

Yellow-crowned Night-heron
Nycticorax violaceus
Garza-nocturna Coroniclara = light-crowned night-heron
Pedrete Enmascarado = masked pedrete
Pedrete Azul = blue pedrete

ID: Small gray heron with big black head; white forehead, head top, and behind eye; thick black bill; yellow legs; immature bird has brown streaks; to 60 cm (2 ft); wingspan to 1.1 m (3.5 ft).

HABITAT: Aquatic sites; found especially in and near mangroves, mudflats, riversides.

LOCATIONS: OAX-HI, OAX-LO, CHIAP-HI, CHIAP-LO, TAB, CAM, YUC, QRO

Plate 24b

Ruddy Crake
Laterallus ruber
Gallinita Rojiza = little reddish hen
Gallineta Enana = dwarf hen
Polluela Rojiza = small reddish hen

ID: Small reddish or reddish-brown chicken-like bird with gray head; black bill; olive-gray legs; to 16 cm (6.5 in).

HABITAT: Low and some middle elevation forests and savannahs; found in marshes, wet fields and pastures, roadside ditches.

LOCATIONS: OAX-LO, CHIAP-HI, CHIAP-LO, TAB, CAM, YUC, QRO

Plate 24c

Gray-necked Wood-rail
Aramides cajanea
Rascón Cuelligris = gray-necked rail
Tutupana

ID: Mid-sized marsh bird with gray head and neck; yellowish bill; olive/brownish back; reddish-brown chest; long, reddish legs; to 40 cm (16 in).

HABITAT: Low elevation forests; found in wet forest areas and around streams, rivers, marshes, mangroves.

LOCATIONS: OAX-LO, CHIAP-LO, TAB, CAM, YUC, QRO

Plate 24d

Purple Gallinule
Porphyrula martinica
Gallineta Morada = purple hen

ID: Striking mid-sized wading bird with bluish violet head, neck, and chest; green wings; red and yellow bill; light blue forehead; yellow legs; immature bird is light brown with greenish wings; to 40 cm (16 in).

HABITAT: Low and some middle elevation aquatic sites; found in marshes and along lake and pond shores.

LOCATIONS: OAX-LO, CHIAP-HI, CHIAP-LO, TAB, CAM, YUC, QRO

Plate 24e

Northern Jacana
Jacana spinosa
Jacana Mesoamericana = middle-american jacana
Cirujano = surgeon
Combatiente = fighter

ID: Smallish wading bird with black head, neck, and chest; bright brown wings, back, and belly; yellow bill and forehead; greenish legs and very long toes; yellow under wings seen in flight; immature bird is brown with white chest/belly, black eyestripe, and black on top of head; to 25 cm (10 in).

HABITAT: Low and middle elevations; found in rivers, ponds, marshes, wet fields.

LOCATIONS: OAX-HI, OAX-LO, CHIAP-HI, CHIAP-LO, TAB, CAM, YUC, QRO

Plate 24 **303**

IM

a Yellow-crowned
Night-Heron

d Purple Gallinule

c Gray-necked Wood-rail

IM

e Northern Jacana

b Ruddy Crake

Plate 25a

Black-bellied Whistling Duck
Dendrocygna autumnalis
Pijiji Aliblanco = white-winged pijiji
Pato Pijiji = pijiji duck
Pato Arbóreo = tree duck
ID: Slender, medium-sized duck, rust-colored with black belly; bill and feet red; immature bird is dull brown with grayish belly and dark bill; to 51 cm (20 in).

HABITAT: Aquatic sites; found in ponds, lakes, marshes, wet grassy areas.

LOCATIONS: OAX-LO, CHIAP-LO, TAB, CAM, YUC, QRO

Plate 25b

Muscovy Duck
Cairina moschata
Pato Real = royal duck
Pato Alas Blancas = white-winged duck
ID: Large, chunky duck, mostly black with greenish gloss, with white wing patches; male with feathered head crest and red "warts" on face; female lacks red warts, is a bit smaller and has a smaller crest; immature bird is brownish; to 85 cm (34 in).

HABITAT: Lowland forests; found in mangroves, wooded streams, rivers, and swamps.

LOCATIONS: OAX-LO, CHIAP-LO, TAB, CAM, YUC, QRO

Plate 25c

Blue-winged Teal
Anas discors
Cerceta Aliazul = blue-winged teal
Patillo Cariblanco = little white-faced duck
ID: Smallish brown duck with conspicuous blue wing patch; male with dark gray head, white crescent in front of eye, white patch near tail; female with dark eye-line; to 40 cm (16 in).

HABITAT: Fresh and brackish water, regionwide; lagoons, lakes, ponds, marshes.

LOCATIONS: OAX-HI, OAX-LO, CHIAP-HI, CHIAP-LO, TAB, CAM, YUC, QRO

Note: This species is a non-breeding seasonal migrant.

Plate 25d

Black-necked Stilt
Himantopus mexicanus
Candelero = candlestick
Zanquilargo
ID: Mid-sized, slender white and black marsh bird with long pink legs and long, thin, straight black bill; to 38 cm (15 in).

HABITAT: Low elevation aquatic sites; found in ponds, lagoons, estuaries, mudflats.

LOCATIONS: OAX-LO, CHIAP-LO, TAB, CAM, YUC, QRO

Plate 25e

Limpkin
Aramus guarauna
Carao
Correa = strap or belt
Totolaca
ID: Largish brown marsh bird with white streaking; long, straight, yellowish or pink/orange bill; gray legs; to 64 cm (25 in).

HABITAT: Lowland aquatic sites; found in rivers, ponds, marshes, swamps, mangroves.

LOCATIONS: OAX-LO, CHIAP-LO, TAB, CAM, YUC, QRO

Plate 25 305

b Muscovy Duck

c Blue-winged Teal

d Black-necked Stilt

a Black-bellied
Whistling Duck

e Limpkin

Plate 26a

Spotted Sandpiper
Actitis macularia
Playerito Alzacolita = beach tail-raiser
Alzacolita = tail-raiser
ID: Small brownish shorebird; white chest and belly, spotted with black during breeding; whitish eyestripe; straight brownish bill; yellowish legs; to 20 cm (8 in).

HABITAT: Aquatic habitats at low and middle elevations; found along ocean shores and around lakes, ponds, rivers; also mangroves, marshes.

LOCATIONS: OAX-HI, OAX-LO, CHIAP-HI, CHIAP-LO, TAB, CAM, YUC, QRO

Note: This species is a non-breeding seasonal migrant.

Plate 26b

Sanderling
Calidris alba
Playero Blanco = white beach-dweller
Playerito Correlón = little beach runner
Chichicuilote
ID: Small light gray shorebird with darker shoulder area and white head, chest and belly; straight black bill; black legs; to 20 cm (8 in).

HABITAT: Seashores; found on sandy shorelines and mudflats.

LOCATIONS: OAX-HI, OAX-LO, CHIAP-HI, CHIAP-LO, TAB, CAM, YUC, QRO

Note: This species is a non-breeding seasonal migrant.

Plate 26c

Black-bellied Plover
Pluvialis squatarola
Chorlito Gris = gray plover
Chorlo Axila Negra = black armpit plover
Avefría = plover
ID: Medium-sized brown, grayish, and white shorebird; grayish brown chest mixed with white; white belly; black bars on tail; short black bill; gray legs; black throat and chest during breeding; 30 cm (12 in).

HABITAT: Coastal areas; found on beaches, mudflats, mangroves; occasionally along rivers near coast.

LOCATIONS: OAX-LO, CHIAP-LO, TAB, CAM, YUC, QRO

Note: This species is a non-breeding seasonal migrant.

Plate 26d

Ruddy Turnstone
Arenaria interpres
Vuelvepiedras Rojizo = reddish turnstone
Vuelvepiedras Común = common turnstone
Chorlete
ID: Small brownish shorebird with white throat and belly, orange legs; dark U-shaped chest markings; smallish black bill; to 23 cm (9 in).

HABITAT: Coastal sites, sand and pebble beaches, mudflats around lagoons, estuaries.

LOCATIONS: OAX-LO, CHIAP-LO, TAB, CAM, YUC, QRO

Note: This species is a non-breeding seasonal migrant.

Plate 26e

Lesser Yellowlegs
Tringa flavipes
Patamarilla Menor = lesser yellowlegs
Tingüis
ID: Small gray or brownish-gray streaked shorebird with long yellow or yellow-orange legs; light eyestripe; whitish throat and belly; long, thin, straight, dark bill; to 25 cm (10 in).

HABITAT: Lakes, ponds, slow rivers, mudflats of lagoons, marshy areas and flooded fields.

LOCATIONS: OAX-HI, OAX-LO, CHIAP-HI, CHIAP-LO, TAB, CAM, YUC, QRO

Note: This species is a non-breeding seasonal migrant.

Plate 26 307

e Lesser Yellowlegs

a Spotted Sandpiper

c Black-bellied Plover

b Sanderling

d Ruddy Turnstone

Plate 27a
Plain Chachalaca
Ortalis vetula
Chachalaca Común = common chachalaca
Chachalaca Olivacea = olivaceous chachalaca

ID: Largish olive bird with small gray head and reddish throat; dark tail with light tip; blackish bill; to 53 cm (21 in).

HABITAT: Low and middle elevation forests; found in trees in more open forests, forest edges, and savannahs.

LOCATIONS: OAX-LO, CHIAP-HI, CHIAP-LO, TAB, CAM, YUC, QRO

Note: This species considered vulnerable in some areas; CITES Appendix III listed for neighboring Guatemala.

Plate 27b
Crested Guan
Penelope purpurascens
Pava Cojolita = little lame guan
Ajol

ID: Large, turkey-like brown bird with crest on head; red throat sac; white spots/streaks on chest; reddish legs; to 90 cm (3 ft).

HABITAT: Low and middle elevation forests; found in canopy or on ground.

LOCATIONS: OAX-HI, OAX-LO, CHIAP-HI, CHIAP-LO, TAB, CAM, YUC, QRO

Plate 27c
Highland Guan
(also called Black Penelopina)
Penelopina nigra
Chachalaca Negra = black chachalaca
Pajuil

ID: Male is large black bird with red bill, eyering, throat sac, legs, and dark brown belly; female is brown with fine barring pattern, lighter belly, gray bill, reddish legs; to 64 cm (25 in).

HABITAT: Higher-elevation forests, including pine–oak forests and cloud forests; found in trees and on ground.

LOCATIONS: OAX-HI, OAX-LO, CHIAP-HI, CHIAP-LO, TAB, CAM, YUC, QRO

Note: This species considered vulnerable; CITES Appendix III listed for neighboring Guatemala.

Plate 27d
Great Curassow
Crax rubra
Hocofaisán = pheasant
Faisán Real = royal pheasant

ID: Very large chicken-like bird with conspicuous, curly head crest and long tail; male is black with white belly and has yellow "knob" on bill; female is mostly brownish or reddish-brown with white and black barred head; to 91 cm (3 ft).

HABITAT: Low and middle elevation forests; found walking on ground in forest interior and at forest edges.

LOCATIONS: OAX-HI, OAX-LO, CHIAP-HI, CHIAP-LO, TAB, CAM, QRO

Note: This species considered vulnerable; CITES Appendix III listed for Guatemala, Honduras, and Costa Rica.

Plate 27e
Ocellated Turkey
Agriocharis ocellata
Guajolote Ocelado = ocellated turkey
Pavo Ocelado = ocellated turkey
Pavo de Monte = mountain turkey

ID: Large, metallic blue-green chicken-like bird with bare skin blue head with reddish warts; grayish tail with large blue and copper spots; female duller overall and without head warts; to 1 m (3.3 ft).

HABITAT: Low and middle elevation forests; found in more open sites such as forest edges, brushy and overgrown fields.

LOCATIONS: CAM, YUC, QRO

Note: This species considered vulnerable; CITES Appendix III listed for neighboring Guatemala.

Plate 27 **309**

a Plain Chachalaca

c Highland Guan

F

M

b Crested Guan

F

d Great Curassow

M

e Ocellated Turkey

Plate 28a

Black-throated Bobwhite
(also called Yucatán Bobwhite)
Colinus nigrogularis
Codorniz Yucateca = Yucatán quail
Codorniz = quail
Codorniz-cotui Yucateca
ID: Small, brownish chicken-like bird with striking black and white scalloped pattern on chest/belly; male with black and white head stripes, black throat; female with buffy/tan eyestripe and throat; to 20 cm (8 in).

HABITAT: Low and middle elevation open sites such as brushy fields and woodlands, clearings, savannahs.

LOCATIONS: CAM, YUC, QRO

Plate 28b

Slaty-breasted Tinamou
Crypturellus boucardi
Tinamú Jamuey = jamuey tinamou
Perdiz Jamuey = jamuey partridge
ID: Mid-sized thickset bird with dark gray head and chest, brownish back and wings; white throat; red legs; very small tail; slender bill; female with light brown/buffy wing bars; to 28 cm (11 in).

HABITAT: Low elevation wet forests; found on ground.

LOCATIONS: OAX-LO, CHIAP-LO, TAB, CAM, YUC, QRO

Plate 28c

Thicket Tinamou
Crypturellus cinnamomeus
Tinamú Canelo = cinnamon tinamou
Perdiz Canelo = cinnamon partridge
ID: Mid-sized thickset bird, brownish with conspicuous dark bars; chest and belly cinnamon; red legs; very small tail; slender bill; to 28 cm (11 in).

HABITAT: Low and middle elevation forests and forest edges; found on the ground.

LOCATIONS: OAX-LO, CHIAP-HI, CHIAP-LO, TAB, CAM, YUC, QRO

Plate 28d

Osprey
Pandion haliatus
Aguila Pescadora = fishing eagle
Gavilán Pescadora = fishing hawk
ID: Large brownish bird with white head; dark stripe through eye; gray legs; wing in flight has backward "bend;" underside of wing white with darker stripes and markings; to 60 cm (2 ft); wingspan to 1.8 m (6 ft).

HABITAT: Low, middle and some higher elevations; seen flying or perched in trees near water – ocean shores, mangroves, ponds, lakes.

LOCATIONS: OAX-HI, OAX-LO, CHIAP-HI, CHIAP-LO, TAB, CAM, YUC, QRO

Plate 28e

Crested Caracara
Polyborus plancus
Caracara Común = common caracara
Carroñero = carrion-eater
ID: A large black bird with barred neck; black, white, and reddish head; white throat; light-colored, hooked bill; yellow legs; to 60 cm (2 ft); wingspan to 1.3 m (4.3 ft).

HABITAT: Open areas – grasslands, savannahs, pastures, forest edges.

LOCATIONS: OAX-HI, OAX-LO, CHIAP-HI, CHIAP-LO, TAB, CAM, YUC, QRO

Plate 28 311

d Osprey

e Crested Caracara

c Thicket Tinamou

F

M

a Black-throated Bobwhite

M

F

b Slaty-breasted Tinamou

Plate 29a

Laughing Falcon
Herpetotheres cachinnans
Halcón Guaco = guaco falcon
Guaco

ID: Largish brown bird-of-prey with tawny or buffy head, chest, and belly; black mask around eyes; black tail with light bars; dark, hooked bill; to 55 cm (22 in); wingspan to 94 cm (37 in).

HABITAT: Low elevation open and semi-open areas; found in forest edge areas, grasslands, agricultural areas.

LOCATIONS: OAX-HI, OAX-LO, CHIAP-HI, CHIAP-LO, TAB, CAM, YUC, QRO

Plate 29b

Collared Forest Falcon
Micrastur semitorquatus
Halcón-selvático Collarejo = collared forest falcon
Halcón-selvático Grande = big forest falcon
Guaquillo Collarejo

ID: Largish bird-of-prey with dark, hooked bill, long, yellowish legs, and dark tail with light bars. Two forms: light form has black back and wings, white or tawny chest and belly, white face; dark form is mostly blackish, often with chest and belly barred with white or light brown; to 61 cm (24 in); wingspan to 94 cm (37 in).

HABITAT: Low and middle elevation wooded areas; found in forest interiors and edges.

LOCATIONS: OAX-HI, OAX-LO, CHIAP-HI, CHIAP-LO, TAB, CAM, YUC, QRO

Plate 29c

Bat Falcon
Falco rufigularis
Halcón Murcielaguero = bat falcon
Halcón Peqeño = little falcon

ID: Mid-sized blackish bird-of-prey with fine barring on belly and under wings; white throat, side of neck, chest; reddish-brown lower belly; yellow legs; to 28 cm (11 in); wingspan to 73 cm (29 in).

HABITAT: Low and middle elevations; found in semi-open sites such as forest edges, clearings, town parks.

LOCATIONS: OAX-HI, OAX-LO, CHIAP-HI, CHIAP-LO, TAB, CAM, YUC, QRO

Plate 29d

American Kestrel
Falco sparverius
Cernícalo Americano = American kestrel
Lic-lic

ID: Small reddish-brown falcon with black barring pattern; two vertical black stripes on each side of white face; black and white bands at tail end; small, dark, hooked bill; male with gray wings, black-spotted chest; female with white chest with brown streaks; to 28 cm (11 in); wingspan 61 cm (2 ft).

HABITAT: Low, middle and higher elevations, regionwide; open and semi-open areas, especially near roads and settlements; often on telephone lines, poles.

LOCATIONS: OAX-HI, OAX-LO, CHIAP-HI, CHIAP-LO, TAB, CAM, YUC, QRO

Plate 29e

Snail Kite
Rostrhamus sociabilis
Milano Caracolero = snail-eater kite
Gavilán Caracolero = snail-eater hawk

ID: Largish bird-of-prey with black and white barred tail; long, fine, down-curved hook on bill; male black with red skin at base of bill and reddish legs; female dark brown with light face, orange/yellow skin at base of bill, dark eyestripe, brown streaked chest/belly, orange legs; to 51 cm (20 in); wingspan to 1.2 m (3.8 ft).

HABITAT: Low elevation aquatic sites; found in and along waterways, lakes, marshes.

LOCATIONS: OAX-LO, CHIAP-LO, TAB, CAM, YUC, QRO

Plate 29 **313**

b Collared Forest Falcon

Dark

a Laughing Falcon

Light

d American Kestrel

F

M

M

F

c Bat Falcon

e Snail Kite

Plate 30a

Gray-headed Kite
Leptodon cayanensis
Milano Cabecigris = gray-headed kite
Gavilán Pantanero = marsh hawk
ID: Largish white and black bird-of-prey with gray head; black above, white below; underside of wing in flight is black in front, black with gray bars behind; tail with light and dark bars; dark, hooked bill; grayish legs; to 53 cm (21 in); wingspan to 1.1 m (3.6 ft).

HABITAT: Low elevation wooded areas; found in forest edge areas and along wooded waterways, mangroves

LOCATIONS: OAX-LO, CHIAP-LO, TAB, CAM, YUC, QRO

Plate 30b

Plumbeous Kite
Ictinia plumbea
Milano Plomizo = lead-colored kite
ID: Mid-sized blackish bird-of-prey with gray head, chest, and belly; reddish-brown patch under wing; bottom of tail with black and white bars; red eyes; reddish or yellowish legs; to 37 cm (14 in); wingspan to 94 cm (37 in).

HABITAT: Low and middle elevation forests and forest edges; often found near water.

LOCATIONS: OAX-HI, OAX-LO, CHIAP-HI, CHIAP-LO, TAB, CAM, QRO

Plate 30c

Roadside Hawk
Buteo magnirostris
Aguililla Caminera = little roadside eagle
Gavilán Lagartijero = lizard-eater hawk
ID: Mid-sized brown or brownish-gray bird-of-prey with grayish head and chest; gray and brown barred belly; tail with light and dark bars; yellow legs; to 40 cm (16 in); wingspan to 78 cm (31 in).

HABITAT: Low and middle elevations; found in open wooded areas, grasslands, forest edges, agricultural areas, roadsides.

LOCATIONS: OAX-HI, OAX-LO, CHIAP-HI, CHIAP-LO, TAB, CAM, YUC, QRO

Plate 30d

Gray Hawk
Buteo nitidus
Aguililla Gris = little gray eagle
ID: Largish gray hawk with lighter barred chest/belly; black tail with white bands; yellowish legs and base of bill; dark eyes; to 45 cm (18 in); wingspan to 90 cm (3 ft).

HABITAT: Low and middle elevation forested and semi-open sites, especially along forest edges, river edges, roadsides; often seen perched on trees, poles.

LOCATIONS: OAX-HI, OAX-LO, CHIAP-HI, CHIAP-LO, TAB, CAM, YUC, QRO

Plate 30e

Common Black Hawk
Buteogallus anthracinus
Aguililla Negra Menor = lesser black eagle
Gavilán Cangrejero = crab hawk
ID: Large, mostly black bird-of-prey with dark, hooked bill, yellow legs; shortish black tail with wide white bar; 56 cm (22 in); wingspan to 1.3 m (4.2 ft).

HABITAT: Low and middle elevation forests, forest edges; found often in trees near water – ocean shores, rivers, streams, marshes, mangroves.

LOCATIONS: OAX-HI, OAX-LO, CHIAP-HI, CHIAP-LO, TAB, CAM, YUC, QRO

Plate 30f

Short-tailed Hawk
Buteo brachyurus
Aguililla Colicorta = short-tailed little eagle
ID: Largish bird-of-prey with yellow at base of bill, yellow legs, whitish forehead. Two forms: light form is dark brown above with whitish chest/belly, dark tail with darker bands and white tip; dark form is all dark brown except for white seen under wings and tail in flight; to 45 cm (18 in); wingspan to 1 m (3.3 ft).

HABITAT: Low, middle, and some higher elevation forested and semi-open sites, regionwide; usually seen soaring, not perched.

LOCATIONS: OAX-HI, OAX-LO, CHIAP-HI, CHIAP-LO, TAB, CAM, YUC, QRO

Plate 30 **315**

a Gray-headed Kite

b Plumbeous Kite

c Roadside Hawk

d Gray Hawk

e Common Black Hawk

Dark

Light **f** Short-tailed Hawk

Plate 31a

Turkey Vulture
Cathartes aura
Aura Cabecirroja = red-headed vulture
Aura Común = common vulture

ID: Large black bird with featherless red head and neck, whitish bill, and yellowish, flesh, or light red legs; underside of wing in flight is black in front, grayish behind; wings held in shallow V during soaring flight; to 80 cm (32 in); wingspan to 1.8 m (5.8 ft). (Note: Lesser Yellow-headed Vulture is common over parts of Yucatán Peninsula, especially Campeche; it closely resembles Turkey Vulture but has yellow, orange, and blue head.)

HABITAT: Found region-wide; usually seen in the air, circling above open areas and many other habitats.

LOCATIONS: OAX-HI, OAX-LO, CHIAP-HI, CHIAP-LO, TAB, CAM, YUC, QRO

Plate 31b

Black Vulture
Coragyps atratus
Zopilote Negro = black vulture
Zope

ID: Large black bird with featherless black head and neck; whitish legs; underside of wing in flight shows a whitish area near wingtip; wings held flat out during soaring flight; to 66 cm (26 in); wingspan to 1.4 m (4.8 ft).

HABITAT: Found region-wide; usually spotted in the air, circling above villages, towns, garbage dumps, other open areas.

LOCATIONS: OAX-HI, OAX-LO, CHIAP-HI, CHIAP-LO, TAB, CAM, YUC, QRO

Plate 31c

King Vulture
Sarcoramphus papa
Zopilote Rey= king vulture

ID: Large white bird with featherless multi-colored head, black wings, gray neck, orange bill; underside of wing in flight is white in front, black behind; to 81 cm (32 in); wingspan to 2 m (6.5 ft).

HABITAT: Low, middle, and some higher elevation sites; usually seen in the air, circling above forested or partly wooded areas.

LOCATIONS: OAX-HI, OAX-LO, CHIAP-HI, CHIAP-LO, TAB, CAM, YUC, QRO

Plate 31 **317**

a Turkey Vulture

b Black Vulture

c King Vulture

Plate 32a

Band-tailed Pigeon
Columba fasciata
Paloma Encinera = oak pigeon
Paloma de Collar = collared pigeon
Paloma Ocotera = ocote-pine pigeon
ID: Large grayish or brownish pigeon with yellow bill with black tip and yellow legs; gray, pinkish-gray, or purplish-gray head; iridescent greenish patch with white bar at back of neck; to 35 cm (14 in).

HABITAT: Higher elevation forested areas, including oak and pine–oak forests; found high in trees, often in forest edge areas, tree plantations.

LOCATIONS: OAX-HI, CHIAP-HI

Plate 32b

Scaled Pigeon
Columba speciosa
Paloma Escamosa = scaly pigeon
ID: Large reddish-brown pigeon with bold dark (purple, green, or blackish) scaling on neck, back, and chest; belly light with light scaling; red bill with light tip; red eyering; female duller; to 33 cm (13 in).

HABITAT: Low and middle elevation forests and more open areas such as forest edges, clearings; found in trees.

LOCATIONS: OAX-HI, OAX-LO, CHIAP-HI, CHIAP-LO, TAB, CAM, QRO

Plate 32c

Red-billed Pigeon
Columba flavirostris
Paloma Piquirroja = red-billed pigeon
Paloma Morada = purple pigeon
ID: Large, dark reddish-purplish pigeon with brown back; gray belly and tail; whitish and red bill; reddish legs; to 33 cm (13 in).

HABITAT: Low and middle elevations; found in open wooded areas, forest edge, clearings, grassland, agricultural areas.

LOCATIONS: OAX-HI, OAX-LO, CHIAP-HI, CHIAP-LO, TAB, CAM, YUC, QRO

Plate 32d

Blue Ground-dove
Claravis pretiosa
Tórtola Azul = blue turtle dove
Tortolita Azul = little blue turtle dove
ID: Smallish dove with yellowish bill; male bluish-gray with lighter face, chest, and belly; dark spots and bars on wings; sides of tail black; female brownish with reddish-brown bars on wings; to 20 cm (8 in).

HABITAT: Low and middle elevation forest edges, open woodlands, pastures, fields; found usually on the ground.

LOCATIONS: OAX-HI, OAX-LO, CHIAP-HI, CHIAP-LO, TAB, CAM, YUC, QRO

Plate 32e

Ruddy Ground-dove
Columbina talpacoti
Tórtola Rojiza = reddish turtle dove
Tortolita Castaña = little chestnut turtle dove
ID: Smallish dove with yellow or brown bill; male reddish-brown with gray head; black spots on wings; female duller, less reddish; 17 cm (7 in).

HABITAT: Low and middle elevations; found usually on ground in open areas – pastures, fields, woodland clearings.

LOCATIONS: OAX-HI, OAX-LO, CHIAP-HI, CHIAP-LO, TAB, CAM, YUC, QRO

Plate 32 319

a Band-tailed Pigeon

b Scaled Pigeon

c Red-billed Pigeon

d Blue Ground-dove

M

F

e Ruddy Ground-dove

Plate 33a

Common Ground-dove
Columbina passerina
Tórtola Común = common turtle dove
Tortolita Común = little common turtle dove
Torobuey

ID: Small dove with reddish or yellowish bill with dark tip; male gray with black "scaling" on neck and chest, black spots on wings, reddish tinge on underparts; female brownish with less scaling; black spots on wings; 16 cm (6.5 in).

HABITAT: Open and semi-open sites such as parkland with shrubs, brushy and agricultural areas, near settlements; found on the ground.

LOCATIONS: OAX-HI, OAX-LO, CHIAP-HI, CHIAP-LO, CAM, YUC, QRO

Plate 33b

Inca Dove
Columbina inca
Tórtola Colilarga = long-tailed turtle dove
Tortolita Común= little common turtle dove

ID: Small pale gray dove with black linings on feathers that yields a "scaled" appearance; longish tail with white edges; reddish-brown wing patches seen in flight; to 20 cm (8 in).

HABITAT: Low, middle, and higher elevation open and semi-open sites; found usually on the ground in woodland clearings, forest edges, pastures, fields, lawns.

LOCATIONS: OAX-HI, OAX-LO, CHIAP-HI, CHIAP-LO, TAB

Plate 33c

White-tipped Dove
Leptotila verreauxi
Paloma Caminera = road pigeon
Paloma Arroyera = stream pigeon

ID: Mid-sized dove with gray or grayish-brown head; brown back and wings; lighter-colored face; pinkish or violet tinge to head and chest; tail dark with white edging at end; reddish skin around eye; black bill; male with iridescent sheen on neck; to 28 cm (11 in).

HABITAT: Low, middle and higher elevation forests, including pine–oak forests; found usually on ground in open areas such as agricultural sites, gardens, open woodlands, forest edges.

LOCATIONS: OAX-HI, OAX-LO, CHIAP-HI, CHIAP-LO, TAB, CAM, YUC, QRO

Plate 33d

White-winged Dove
Zenaida asiatica
Paloma Aliblanca = white-winged dove

ID: Mid-sized gray-brown dove with black spot on side of face below blue-ringed eye; grayish belly; white bar on wing, seen mostly in flight; sides of tail tipped with white; male with iridescent sheen on neck; to 28 cm (11 in).

HABITAT: Low, middle and some higher elevation forests and semi-open sites such as open woodlands, forest edges, brushy areas; found in trees, shrubs, and on the ground.

LOCATIONS: OAX-HI, OAX-LO, CHIAP-HI, CHIAP-LO, TAB, CAM, YUC, QRO

Plate 33e

White-faced Quail-dove
Geotrygon albifacies
Paloma-perdiz Cariblanca = white-faced quail-dove
Paloma-perdiz = quail-dove

ID: Mid-sized reddish-brown dove with whitish head; light- or grayish-brown neck and chest with dark lines; light brown belly; violet tinge on back; gray skin around eye; black bill; red legs; to 30 cm (12 in).

HABITAT: Cloud forests, especially in semi-open and agricultural areas; found on the ground or low in trees.

LOCATIONS: OAX-HI, CHIAP-HI

Plate 33 **321**

a Common Ground-dove

b Inca Dove

c White-tipped Dove

d White-winged Dove

e White-faced Quail-dove

Plate 34a
Scarlet Macaw
Ara macao
Guacamaya Roja = red macaw
ID: Very large, long-tailed red parrot; big patches of yellow on wings and blue on wings and tail; to 91 cm (3 ft).

HABITAT: Lowland wet forests; usually found high in tree canopy in wooded and forest edge areas, often near water; now occurs in Mexico only in eastern Chiapas, on Guatemalan border.

LOCATIONS: CHIAP-LO

Note: This species is endangered, CITES Appendix I listed.

Plate 34b
Mealy Parrot
Amazona farinosa
Loro Verde = green parrot
Loro Cabeza Azul = blue-headed parrot
Loro Verde de Cabeza Azul = green blue-headed parrot
ID: Large green parrot with blue tinge on top of head; touches of red and blue on wings, seen mostly in flight; end of tail lighter green or yellowish; whitish eyering; to 40 cm (16 in).

HABITAT: Low elevation forests; found in tree canopy in wooded areas, forest edges, tree plantations.

LOCATIONS: OAX-LO, CHIAP-LO, TAB, CAM, QRO

Plate 34c
Red-lored Parrot
(also called Yellow-cheeked Parrot)
Amazona autumnalis
Loro Frentirrojo = red-forehead parrot
Loro Cachete-amarillo, Loro Mejilla Amarilla = yellow-cheeked parrot
ID: Mid-sized green parrot with red forehead and blue tinge on head feathers; often some yellow on face; red and blue patches on wings, seen mostly in flight; to 34 cm (13 in).

HABITAT: Low and some middle elevation forests; found in tree canopy in wooded and semi-open areas, forest edges, tree plantations.

LOCATIONS: OAX-LO, CHIAP-LO, TAB, CAM, QRO

Plate 34d
Yucatán Parrot
(also called Yellow-lored Parrot)
Amazona xantholora
Loro Yucateco = Yucatán parrot
ID: Mid-sized green parrot; male with white forehead, small yellow patch behind bill, red eyering, red and blue patches on wings seen mostly in flight; female bluish on top of head, with small yellow patch behind bill, inconspicuous red and green eyering, only blue patches on wings; to 28 cm (11 in).

HABITAT: Low elevation forests and forest edges; seen in trees and flying low over forest.

LOCATIONS: CAM, YUC, QRO

Plate 34 323

b Mealy Parrot

a Scarlet Macaw

c Red-lored Parrot

M

F

d Yellow-lored Parrot

Plate 35a

White-fronted Parrot
Amazona albifrons
Loro Frentiblanco = white-fronted parrot
Perico Frentiblanco = white-fronted parakeet
Loro Manglero = mangrove parrot

ID: Mid-sized green parrot with red on face, including around eyes; white forehead; blue at top of head; red and blue patches on wings, seen mostly in flight; to 28 cm (11 in).

HABITAT: Low and middle elevation forests, forest edges, other semi-open sites, mangroves; seen in tree canopy and flying low over forest.

LOCATIONS: OAX-HI, OAX-LO, CHIAP-HI, CHIAP-LO, TAB, CAM, YUC, QRO

Plate 35b

Yellow-headed Parrot
Amazona oratrix
Loro Cabeciamarillo = yellow-headed parrot
Cotorra Cabeza Amarilla = small yellow-headed parrot
Loro Real = royal parrot

ID: Largish green parrot with yellow head; white skin around eye; red and blue patches on wings; to 38 cm (15 in).

HABITAT: Low elevation semi-open sites such forest edges, savannahs, river edges, and open areas with scattered trees.

LOCATIONS: OAX-LO, CHIAP-LO, TAB, CAM

Note: This species is considered endangered by some authorities.

Plate 35c

White-crowned Parrot
Pionus senilis
Loro Coroniblanco = white-crowned parrot
Perico Cabeza Blanca = white-headed parakeet

ID: A smallish dark green parrot with white forehead and white throat; blue or greenish-blue head and upper breast; brownish or bronze shoulder patches; blue under wings and red under tail, seen in flight; to 25 cm (10 in).

HABITAT: Low and middle elevation forests, including some pine–oak forests, forest edges, and other semi-open sites; seen in tree canopy and flying low over forests.

LOCATIONS: OAX-HI, OAX-LO, CHIAP-HI, CHIAP-LO, TAB, CAM, QRO

Plate 35d

Olive-throated Parakeet
(also called Aztec Parakeet)
Aratinga nana
Perico Pechisucio = dirty-chested parakeet
Cotorra Bosquera = small forest parrot

ID: Small green parrot with olive or brownish chest/belly; blue patches on wings; yellow underneath tail; white eyering; to 23 cm (9 in).

HABITAT: Low and middle elevation forests, forest edges, tree plantations, and other semi-open sites; seen in tree canopy and flying low over forests.

LOCATIONS: OAX-HI, OAX-LO, CHIAP-HI, CHIAP-LO, TAB, CAM, YUC, QRO

Plate 35e

Orange-fronted Parakeet
Aratinga canicularis
Perico Frentinaranja = orange-fronted parakeet
Periquillo Común = little common parakeet
Perico Atolero

ID: Smallish green parrot with orange forehead and blue on top of head; olive or brownish breast; light eyering; blue on wings, seen mostly in flight; long green tail yellowish underneath; to 24 cm (9.5 in).

HABITAT: Low elevation forests; found in tree canopy in wooded, semi-open, and forest edge areas.

LOCATIONS: OAX-LO, CHIAP-LO

Plate 35 325

b Yellow-headed Parrot

a White-fronted Parrot

c White-crowned Parrot

d Olive-throated Parakeet

e Orange-fronted Parakeet

Plate 36   327

c Yellow-billed Cuckoo

a Squirrel Cuckoo

b Groove-billed Ani

e Lesser Ground Cuckoo

d Lesser Roadrunner

Plate 37a

Ferruginous Pygmy-owl
Glaucidium brasilianum
Tecolotito Rayado = little striped owl
Tecolotito Común = little common owl
Vieja Común = common old woman
ID: A small grayish-brown or reddish-brown owl with white streaks on chest, belly; fine white streaks on head; brown or black tail with paler bars; black spot on side of neck; to 18 cm (7 in).

HABITAT: Low and middle elevations; found in wooded areas, and in semi-open sites such as forest edges, grassland areas with trees, and agricultural areas with trees; day-active.

LOCATIONS: OAX-HI, OAX-LO, CHIAP-HI, CHIAP-LO, TAB, CAM, YUC, QRO

Plate 37b

Mottled Owl
Ciccaba virgata
Búho Café (or Lechuza Café) = brown owl
Búho Tropical = tropical owl
ID: Mid-sized brownish or grayish owl with lighter bars/streaks; whitish eyebrows; whitish or buffy/yellowish throat and chest with darker mottling; streaked belly; to 38 cm (15 in).

HABITAT: Low, middle, and some higher elevation forests and semi-open sites such as forest edges, tree plantations; nocturnal.

LOCATIONS: OAX-HI, OAX-LO, CHIAP-HI, CHIAP-LO, TAB, CAM, YUC, QRO

Plate 37c

Vermiculated Screech Owl
Otus guatemalae
Tecolote Vermiculado = vermiculated screech owl
Tecolotito Maullador = little mewing screech owl
ID: Small dark brown or grayish streaked owl; lighter chest/belly with dark streaks; black border around face; white "eyelashes;" greenish bill; no feathers on feet; to 23 cm (9 in).

HABITAT: Low and middle elevation forests, forest edges, and open woodlands; nocturnal.

LOCATIONS: OAX-HI, OAX-LO, CHIAP-HI, CHIAP-LO, TAB, CAM, YUC, QRO

Plate 37d

Pauraque
Nyctidromus albicollis
Tapacaminos Picuyo = road-blocker picuyo
ID: Mid-sized brown bird with fine brown mottling and black spots or streaks; whitish band on throat; light brown chest/belly with fine black bars; longish tail with white stripes in male, white tips in female; white or light-colored band on wings, seen in flight; to 28 cm (11 in).

HABITAT: Low and middle elevation sites; found resting on the ground during the day in shady spots in grasslands, pastures, open woodlands, forest edges; found on the ground; nocturnal.

LOCATIONS: OAX-HI, OAX-LO, CHIAP-HI, CHIAP-LO, TAB, CAM, YUC, QRO

Plate 37e

Lesser Nighthawk
Chordeiles acutipennis
Chotacabras Menor = lesser nightjar
ID: Mid-sized gray-brown bird with light and dark brown and black markings; whitish band on throat; light brown chest/belly with fine black bars; white or light-colored band on wings, seen in flight; male has white band on tail; to 23 cm (9 in).

HABITAT: Low, middle and higher elevation forests and more open sites, often near water; forest edges, mangroves, beaches; found on the ground or on low tree branches; nocturnal.

LOCATIONS: OAX-HI, OAX-LO, CHIAP-HI, CHIAP-LO, TAB, CAM, YUC, QRO

Plate 37f

Mexican Whip-poor-will
Caprimulgus arizonae
Tapacaminos Cuerporruín = treacherous-body road-blocker
ID: Mid-sized brown or brownish-gray bird with black and reddish-brown markings; light chest band; light brown or buffy chest/belly with dark mottling; dark wings with reddish-brown bars; outer tail feathers with white tips; to 23 cm (9 in).

HABITAT: Middle and high elevation forests, including pine–oak forests; found in trees.

LOCATIONS: OAX-HI, CHIAP-HI

Plate 37 329

a Ferruginous Pygmy-owl

b Mottled Owl

c Vermiculated Screech Owl

e Lesser Nighthawk

f Mexican Whip-poor-will

d Pauraque

Plate 38a

White-collared Swift
Streptoprocne zonaris
Vencejo Cuelliblanco = white-collared swift
Vencejo Collarejo = collared swift

ID: Mid-sized black bird (for a swift, fairly large) with white ring encircling neck; squarish or slightly notched tail; long, slender, swept-back wings; to 21 cm (8.5 in).

HABITAT: Low, middle, and some higher elevations; seen flying above all types of habitats; flies continuously.

LOCATIONS: OAX-HI, OAX-LO, CHIAP-HI, CHIAP-LO, TAB

Plate 38b

Vaux's Swift
Chaetura vauxi
Vencejo de Vaux = Vaux's swift
Vencejo Alirrápido = fast-winged swift
Vencejito Común = little common swift

ID: Small, mostly blackish swift with slightly lighter chest and gray throat; very short tail; relatively short wings; to 11 cm (4.5 in).

HABITAT: Low, middle, and higher elevations; seen flying above all types of habitats; flies continuously.

LOCATIONS: OAX-HI, OAX-LO, CHIAP-HI, CHIAP-LO, TAB, CAM, YUC, QRO

Plate 38c

Chestnut-collared Swift
Cypseloides rutilus
Vencejo Cuellicastaño = chestnut-collared swift
Vencejillo Cuellicanelo = little cinnamon-collared swift

ID: Small blackish swift with conspicuous brown chest band; long, slender swept-back wings; squarish tail; to 14 cm (5.5. in).

HABITAT: Low, middle, and higher elevations; seen flying above all types of habitats, sometimes in large flocks; flies continuously.

LOCATIONS: OAX-HI, OAX-LO, CHIAP-HI, CHIAP-LO

Plate 38d

Gray-breasted Martin
Progne chalybea
Martín Pechigris = gray-breasted martin

ID: Small, glossy dark blue and blackish bird with brownish-gray throat, chest; whitish belly; forked tail; female duller with less blue; to 18 cm (7 in).

HABITAT: Low and middle elevation forests and more open areas such as forest edges, clearings, human settlements; seen flying or perched on wires, bridges, other structures.

LOCATIONS: OAX-HI, OAX-LO, CHIAP-HI, CHIAP-LO, TAB, CAM, YUC, QRO

Plate 38e

Barn Swallow
Hirundo rustica
Golondrina Ranchera = ranch swallow

ID: Small, slender, dark blue bird with narrow, swept-back wings and deeply forked tail; reddish-brown forehead, throat, and chest; light brown or tawny belly; thin dark band across chest; to 17 cm (6.5 in).

HABITAT: Low and middle elevations, particularly along coasts; often seen flying low over open areas – fields, pastures, lawns, waterways.

LOCATIONS: OAX-HI, OAX-LO, CHIAP-HI, CHIAP-LO, TAB, CAM, YUC, QRO

Note: This species is a non-breeding seasonal migrant.

Plate 38f

Northern Rough-winged Swallow
Stelgidopteryx serripennis
Golodrina Aliserrada = jagged-winged swallow

ID: Small brownish bird with darker wings and tail; light brown throat, chest; whitish belly; notched tail; to 13 cm (5 in). (Note: The rough-winged swallow in the Yucatán region may be a separate species, but appears much the same, if a bit darker.)

HABITAT: Low and middle elevation open areas, especially near waterways, roads.

LOCATIONS: OAX-HI, OAX-LO, CHIAP-HI, CHIAP-LO, TAB, CAM, YUC, QRO

Plate 38 331

a White-collared Swift

c Chestnut-collared Swift

b Vaux's Swift

d Gray-breasted Martin

e Barn Swallow

f Northern Rough-winged Swallow

Plate 39a

Mangrove Swallow
Tachycineta albilinea
Golondrina Manglera = mangrove swallow

ID: Small glossy dark green bird with dark wings and short, dark, notched tail; narrow white line over eye; white chest, belly, and rump; to 12 cm (4.5 in).

HABITAT: Low elevations, particularly coastal zones; seen flying over aquatic sites – rivers, large streams, ponds, lakes, lagoons, marshes, wet pastures.

LOCATIONS: OAX-LO, CHIAP-LO, TAB, CAM, YUC, QRO

Plate 39b

Cave Swallow
Hirundo fulva
Golondrina Pueblera = town swallow

ID: Small bluish-black bird with dark wings and tail; reddish-brown forehead and rump; light brown throat, chest, and side of face; whitish belly; square or slightly notched tail; to 14 cm (5.5 in).

HABITAT: Open areas, especially around human settlements, caves.

LOCATIONS: CHIAP-HI, CAM, YUC, QRO

Plate 39c

Little Hermit
Phaethornis longuemareus
Ermitaño Chico, Ermitaño Pequeño = little hermit
Chupaflor Ocrillo = ochre hummingbird

ID: Small brownish or greenish-brown hummingbird with down-curved bill; black and light facial stripes; cinnamon chest/belly; tail brownish with light tip; 9 cm (3.5 in).

HABITAT: Low and middle elevation forests; found near flowers in the forest interior and forest edge areas, particularly near watercourses; also in gardens.

LOCATIONS: OAX-HI, OAX-LO, CHIAP-HI, CHIAP-LO, TAB, CAM, YUC, QRO

Plate 39d

Fork-tailed Emerald
Chlorostilbon canivetti
Chupaflor Esmeralda = emerald hummingbird
Esmeralda de Canivet = Canivet's emerald
Esmeralda Tijereta = scissors emerald

ID: Small green hummingbird with straight bill; male with dark, deeply forked tail; female with white stripe behind eye, pale grayish chest/belly, dark, slightly forked tail with some white feather tips; to 9 cm (3.5 in). (Note: A nearly identical species, endemic to Cozumel Island, is the Cozumel Fork-tailed Emerald.)

HABITAT: Low and middle elevation semi-open sites such as open brushy woodlands, forest edges, clearings.

LOCATIONS: OAX-HI, OAX-LO, CHIAP-HI, CHIAP-LO, TAB, CAM, YUC, QRO

Plate 39e

Green-breasted Mango
Anthracothorax prevostii
Mango Pechiverde = green-chested mango
Chupaflor Pechiverde = green-chested hummingbird
Chupaflor Gargantinegra = black-throated hummingbird

ID: Mid-sized green hummingbird with down-curved bill; male with black throat, bluish center of chest; reddish/purplish tail; female with white throat/chest with dark/greenish central stripe, green and blackish tail with white tip; to 12 cm (4.5 in).

HABITAT: Low and middle elevations; found in semi-open sites such as forest edges, clearings, savannahs, plantations.

LOCATIONS: OAX-HI, OAX-LO, CHIAP-HI, CHIAP-LO, TAB, CAM, YUC, QRO

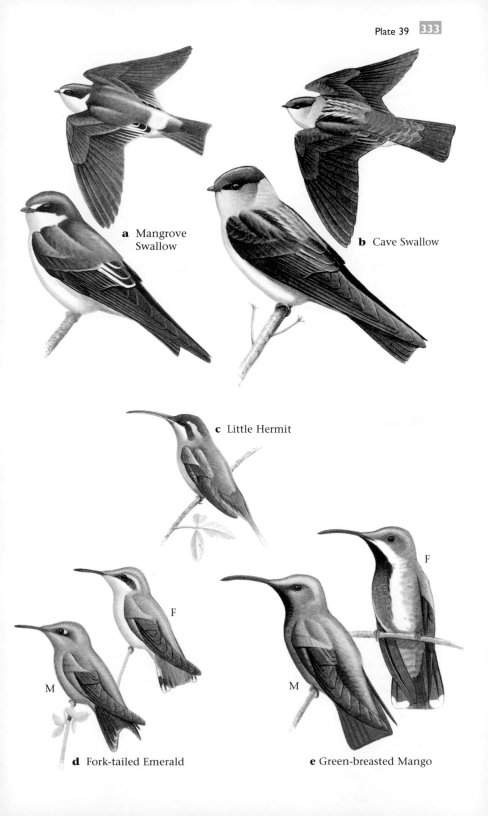

Plate 39   333

a Mangrove Swallow

b Cave Swallow

c Little Hermit

d Fork-tailed Emerald

M

F

e Green-breasted Mango

M

F

Plate 40a

White-bellied Emerald
Amazilia candida
Esmeralda Vientre-blanca = white-bellied emerald
Chupaflor Cándido = candid hummingbird
Chupaflor Esmeralda Petiblanco = white-breasted emerald hummingbird
ID: Small green hummingbird with straight reddish bill with dark tip; white chest/belly; greenish or brownish tail with some gray feather tips; to 9 cm (3.5 in).

HABITAT: Low and middle elevation forests and forest edges.

LOCATIONS: OAX-HI, OAX-LO, CHIAP-HI, CHIAP-LO, TAB, CAM, YUC, QRO

Plate 40b

Cinnamon Hummingbird
Amazilia rutila
Colibrí (or Chupaflor) Canelo = cinnamon hummingbird
Chupaflor Rojizo = reddish hummingbird
ID: Mid-sized green hummingbird with cinnamon chest/belly, straight bill, and reddish-brown tail; to 10 cm (4 in).

HABITAT: Low elevation drier coastal areas; found near flowers in mangrove forests and more open sites – forest edges, brushy areas.

LOCATIONS: OAX-LO, CHIAP-LO, YUC, QRO

Plate 40c

Rufous-tailed Hummingbird
Amazilia tzacatl
Colibrí (or Chupaflor) Colirrufo = rufous-tailed hummingbird
Chupaflor Pechigris = gray-breasted hummingbird
ID: Mid-sized green hummingbird with grayish belly and reddish-brown squared or slightly notched tail; straight red and black bill; to 10 cm (4 in).

HABITAT: Low and middle elevations; found near flowers in forest interior and more open sites – forest edges, scrub areas, gardens, plantations.

LOCATIONS: OAX-HI, OAX-LO, CHIAP-HI, CHIAP-LO, TAB, CAM, QRO

Plate 40d

Azure-crowned Hummingbird
Amazilia cyanocephala
Chupaflor Cabeciazul = blue-headed hummingbird
Colibrí Coroniazul = blue-crowned hummingbird
Chupaflor Serrano = mountain hummingbird
ID: Mid-sized green hummingbird with straight reddish bill with dark tip; blue/turquoise head top; white chest/belly; brownish-green lower back and tail; to 12 cm (4.5 in).

HABITAT: Low, middle, and higher elevation forests, including pine–oak forests, and semi-open areas such as forest edges, savannahs.

LOCATIONS: OAX-HI, OAX-LO, CHIAP-HI, CHIAP-LO

Plate 40e

Buff-bellied Hummingbird
Amazilia yucatanensis
Colibrí Vientre-canelo = cinnamon-bellied hummingbird
Chupaflor Yucateco = Yucatán hummingbird
Chupaflor Pechicastaño = chestnut-breasted hummingbird
ID: Mid-sized green hummingbird with buffy/cinnamon belly and reddish-brown, slightly notched tail; straight red bill; to 10 cm (4 in).

HABITAT: Low and middle elevation forests and semi-open sites such as forest edges, clearings.

LOCATIONS: OAX-HI, OAX-LO, CHIAP-HI, CHIAP-LO, TAB, CAM, YUC, QRO

Plate 40 335

a White-bellied Emerald

b Cinnamon Hummingbird

c Rufous-tailed Hummingbird

e Buff-bellied Hummingbird

d Azure-crowned Hummingbird

Plate 41a

Wedge-tailed Sabrewing
Campylopterus curvipennis
Fandangero Colicuña = wedge-tailed
sabrewing
Chupaflor Colicuña = wedge-tailed
hummingbird
Chupaflor Gritón = shouting
hummingbird

ID: Largish green hummingbird with long, slightly down-curved bill; violet forehead; grayish face, chest, belly; long greenish tail with dark edges; female with white tail tips; to 13 cm (5 in).

HABITAT: Low elevation forests and forest edges; found near flowers.

LOCATIONS: OAX-LO, CHIAP-LO, TAB, CAM, YUC, QRO

Plate 41b

Violet Sabrewing
Campylopterus hemileucurus
Chupaflor Morado = purple
hummingbird
Fandangero Morado = purple
sabrewing

ID: A large hummingbird with down-curved bill and white patches at end of tail; male with blue-violet head/chest/belly, dark green back, dark wings; female with blue violet throat, gray chest/belly; to 15 cm (6 in).

HABITAT: Low, middle, and some higher elevation wet forests; found near flowers in forest interior and more open sites at forest edges, tree plantations.

LOCATIONS: OAX-HI, OAX-LO, CHIAP-HI, CHIAP-LO, TAB

Plate 41c

Amethyst-throated Hummingbird
Lampornis amethystinus
Colibrí-serrano Gorjiamatisto =
amethyst-throated mountain-
hummingbird
Chupaflor Amatista = amethyst
hummingbird

ID: Mid-sized green hummer with pinkish throat; dark wings and tail; white line behind eye above dark patch; grayish belly; female throat is light brown to grayish; to 12 cm (5 in).

HABITAT: Middle and high elevation forests, including pine forest, and forest edges.

LOCATIONS: OAX-HI, CHIAP-HI

Plate 41d

Striped-tailed Hummingbird
Eupherusa eximia
Colibrí (or Chupaflor) Colirrayado =
striped-tailed hummingbird
Chupaflor Arroyero = stream
hummingbird

ID: Smallish green hummingbird with straight bill; reddish-brown on wings; dark tail with white lines; whitish belly; female with grayish-white throat/chest/belly and green spots on sides; 9 cm (3.5 in).

HABITAT: Middle and some higher elevation wet forests; found around flowers in forest interior but also semi-open sites such as forest edges, scrub areas, and tree plantations.

LOCATIONS: OAX-HI, OAX-LO, CHIAP-HI, CHIAP-LO, TAB, CAM, YUC, QRO

Plate 41e

Berylline Hummingbird
Amazilia beryllina
Colibrí (or Chupaflor) de Berilo =
berylline hummingbird
Chupaflor Cola Canela = cinnamon-
tailed hummingbird
Chupaflor Alicafé = brown-winged
hummingbird

ID: Mid-sized green hummingbird with reddish-brown on wings, reddish-brown, squarish tail, light brown belly; female with grayish belly; to 10 cm (4 in). (Note: In Chiapas, tail is purplish and male has green chest/belly, female greenish belly.)

HABITAT: Middle and higher elevation forests and more open sites such as forest edges, tree plantations, clearings.

LOCATIONS: OAX-HI, CHIAP-HI

Plate 41 337

b Violet Sabrewing

M

F

a Wedge-tailed
Sabrewing

e Berylline Hummingbird

F

M

d Striped-tailed
Hummingbird

F

M

c Amethyst-throated Hummingbird

Plate 42a

White-eared Hummingbird
Basilinna leucotis
Colibrí (or Chupaflor) Orejiblanco =
white-eared hummingbird

ID: Mid-sized green hummer with broad white
line behind eye, dark wings, straight red bill,
squarish tail; male with purple forehead and chin,
light green to whitish belly; female with whitish
throat with green spots; to 10 cm (4 in).

HABITAT: High elevation forests, including pine
and pine–oak forests, and forest edges, clearings.

LOCATIONS: OAX-HI, CHIAP-HI

Plate 42b

Blue-throated Hummingbird
Lampornis clemenciae
Colibrí Gorjiazul = blue-throated
hummingbird
Chupaflor Garganta Azul = blue-
throated hummingbird

ID: Largish green hummer with white lines above
and below eye; dark wings; blackish-blue tail with
white corners; male with blue throat, greenish-
gray belly; female with gray chest/belly; to 13 cm
(5 in).

HABITAT: High elevation forests, including oak
and pine–oak forests, and forest edges, clearings.

LOCATIONS: OAX-HI

Plate 42c

Magnificent Hummingbird
Eugenes fulgens
Colibrí (or Chupaflor) Magnífico =
magnificent hummingbird

ID: Largish green hummer with small white spot
behind eye, squarish green tail, straight dark bill;
male with purple forehead, bright green or bluish-
green throat, dark chest/belly; female with grayish
throat/chest/belly, outer tail feathers with white
tips; to 13 cm (5 in).

HABITAT: High elevation forests, including pine
and pine–oak forests, in more open sites such as
forest edges, clearings, rocky areas, grasslands.

LOCATIONS: OAX-HI, CHIAP-HI

Plate 42d

Garnet-throated Hummingbird
Lamprolaima rhami
Colibrí (or Chupaflor) Alicastaño =
chestnut-winged hummingbird
Chupaflor Real = royal hummingbird

ID: Largish green hummingbird with small white
spot behind eye, reddish-brown wings, straight
dark bill, notched dark tail; male with pinkish
throat, purple chest, dark belly; female with gray
throat/chest/belly, outer tail feathers with white
tips; to 12.5 cm (5 in).

HABITAT: High elevation forests including pine
and cloud forests, and semi-open areas such as
forest edges.

LOCATIONS: OAX-HI, CHIAP-HI

Plate 42e

Green Violet-ear
Colibri thalassinus
Orejavioleta Verde = green violet-ear
Chupaflor Orejavioleta, Colibrí de Oido
Violeta = violet-ear hummingbird
Chupaflor Pavito = little hen
hummingbird

ID: Mid-sized green hummingbird with purplish-
blue eyestripe; dark wings; bluish-green square
tail with dark band; dark, slightly down-curved bill;
some have small purplish-blue patch on chest; to
11.5 cm (4.5 in).

HABITAT: High elevation forests, including pine,
oak, and pine–oak forests, especially in semi-
open sites such as forests edges, clearings.

LOCATIONS: OAX-HI, CHIAP-HI

Plate 42 **339**

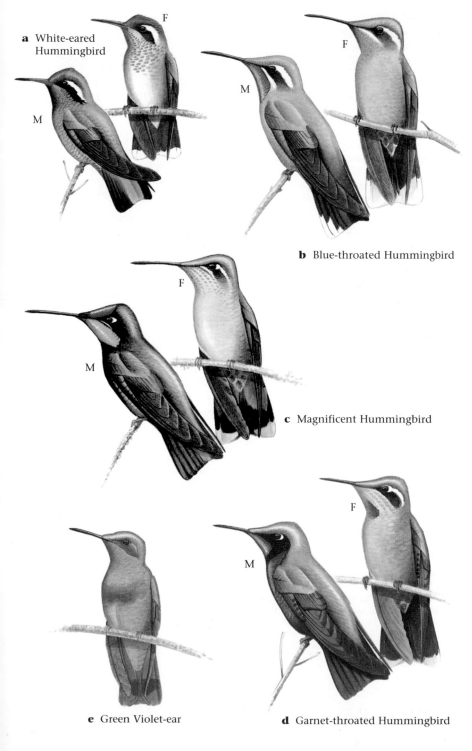

a White-eared Hummingbird

b Blue-throated Hummingbird

c Magnificent Hummingbird

e Green Violet-ear

d Garnet-throated Hummingbird

Plate 43a

Resplendent Quetzal
Pharomachrus mocinno
Quetzal Centroamericano = Central
American quetzal
Quetzal

ID: Male is unmistakable: a large bright green bird with red belly, green crest on head, white underneath tail, yellow bill, and a few extremely long tail feathers (streamers); female, lacking streamers, is duller green with brown-gray chest, red belly, white and black barred tail; to 40 cm (16 in), plus tail streamers.

HABITAT: High elevation wet forests; found in tree canopy in forest interior and edge areas.

LOCATIONS: OAX-HI, CHIAP-HI

Note: This species listed as endangered, CITES Appendix I and USA ESA.

Plate 43b

Slaty-tailed Trogon
Trogon massena
Trogón Colioscuro = dark-tailed trogon
Trogón Gigante = giant trogon

ID: Male is largish green bird with red belly, red/orange eyering and bill, and gray-black under tail; female is mostly gray with red belly and two-toned bill, black above, reddish below; to 35 cm (14 in).

HABITAT: Low elevation wet forests; found in tree canopy in forest interior and semi-open sites such as forest edges, tree plantations.

LOCATIONS: OAX-LO, CHIAP-LO, TAB, CAM, QRO

Plate 43c

Violaceous Trogon
Trogon violaceus
Trogón Violáceo = violaceous trogon

ID: Mid-sized bird with yellow belly, dark bill, black and white barred tail; male with black and dark blue/violet head and chest, greenish back, yellow eyering; female with gray head, back, and chest, white eyering; to 24 cm (9.5 in).

HABITAT: Low elevation forests; found in tree canopy in forest interior and in more open areas such as clearings, forest edges, mangroves, tree plantations, near streams and rivers.

LOCATIONS: OAX-HI, OAX-LO, CHIAP-HI, CHIAP-LO, TAB, CAM, YUC, QRO

Plate 43d

Black-headed Trogon
Trogon melanocephalus
Trogón Cabecinegro = black-headed trogon
Trogón Amarillo = yellow trogon

ID: Mid-sized bird with yellow belly, black/gray head and chest, pale bluish eyering, dark bill, and large white patches on black tail; male with greenish back and blue rump; female with grayish back; to 27 cm (10.5 in).

HABITAT: Low and middle elevation semi-open sites; found in tree canopy in forest edges, tree plantations, mangroves.

LOCATIONS: OAX-HI, OAX-LO, CHIAP-HI, CHIAP-LO, TAB, CAM, YUC, QRO

Plate 43e

Mountain Trogon
Trogon mexicanus
Trogón Mexicano = Mexican trogon
Trogón Ocotero = ocote-pine trogon
Coa Mexicana = Mexican coa

ID: Largish bird with red belly and white chest band; black-tipped tail; underneath tail black with white patches; male is green with black face, reddish eyering, yellow bill, gray wings; female is brown with white marks in front of and behind eye and two-tone bill, dark above, yellow below; to 31 cm (12 in).

HABITAT: High elevation forests, including cloud, pine, and pine–oak forests.

LOCATIONS: OAX-HI, CHIAP-HI

Plate 43 **341**

a Resplendent Quetzal

b Slaty-tailed Trogon

c Violaceous Trogon

d Black-headed Trogon

e Mountain Trogon

Plate 44a

Belted Kingfisher
Ceryle alcyon
Martín Pescador Norteño = northern martin fisher
Pescador Norteño = northern fisher
ID: Largish blue-gray bird with ragged head crest, white belly, white neck-band and throat, bluish bar across chest, and large, heavy bill; female has reddish-brown bar across belly; to 33 cm (13 in).

HABITAT: Low, middle, and high elevation aquatic sites; found along shoreline, streams, rivers, lakes, ponds, lagoons.

LOCATIONS: OAX-HI, OAX-LO, CHIAP-HI, CHIAP-LO, TAB, CAM, YUC, QRO

Note: This species is a non-breeding seasonal migrant.

Plate 44b

Ringed Kingfisher
Ceryle torquata
Martín Pescador Collarejo = necklace martin fisher
Martín Pescador Grande = big martin fisher
Pescador Gigante = giant fisher
ID: Large blue-gray bird with ragged head crest, brownish belly, white neck-band and throat, and large, heavy bill; female with bluish bar across chest; to 40 cm (16 in).

HABITAT: Low and middle elevation aquatic sites; found along streams, rivers, lakes, ponds, lagoons, estuaries, mangroves.

LOCATIONS: OAX-HI, OAX-LO, CHIAP-HI, CHIAP-LO, TAB, CAM, QRO

Plate 44c

Amazon Kingfisher
Chloroceryle amazona
Martín Pescador Amazona = Amazon martin fisher
Martín Pescador Verde = green martin fisher
Pescador Verde = green fisher
ID: Mid-sized green and white bird with long, heavy bill; conspicuous green head crest; white belly, white neck-band and throat; male has wide reddish-brown bar across chest; to 28 cm (11 in).

HABITAT: Low and middle elevation aquatic sites; found along rivers, larger streams, lakes, mangroves.

LOCATIONS: OAX-HI, OAX-LO, CHIAP-HI, CHIAP-LO, TAB, CAM, QRO

Plate 44d

Green Kingfisher
Chloroceryle americana
Martín Pescador Verde = green martin fisher
Martín Pescador Americano = American martin fisher
Pescadorcillo = little fisherman
ID: Smallish green and white bird with long, heavy bill; white belly, white neck-band and throat; male with wide reddish-brown bar across chest; female with greenish bands across chest; smaller than Amazon Kingfisher, with less conspicuous head crest; to 20 cm (8 in).

HABITAT: Low, middle, and high elevation aquatic sites; found along shoreline, rivers, larger streams, lakes, ponds, lagoons.

LOCATIONS: OAX-HI, OAX-LO, CHIAP-HI, CHIAP-LO, TAB, CAM, YUC, QRO

Plate 44e

American Pygmy Kingfisher
Chloroceryle aene
Martín Pescador Enano = dwarf martin fisher
Pescador Mosquito = mosquito fisher
Pescador Mínimo = least fisher
ID: Small green and reddish-brown bird with long, heavy bill; reddish-brown neck-band, throat, and chest; white lower belly; female with dark greenish bar across chest; to 13 cm (5 in).

HABITAT: Low elevation aquatic sites; found along small forest streams, ponds, pools, swamps.

LOCATIONS: OAX-LO, CHIAP-LO, TAB, CAM, YUC, QRO

Plate 44 343

b Ringed Kingfisher

a Belted Kingfisher

d Green Kingfisher

c Amazon Kingfisher

e American Pygmy Kingfisher

Plate 45a

Tody Motmot
Hylomanes momotula
Momoto Enano = dwarf motmot
Bobito = little stupid one
ID: Smallish green bird with darker green wings and tail; reddish-brown head top; small turquoise spot above eye; black stripe behind eye; whitish stripes on throat; light green chest; whitish belly; to 18 cm (7 in).

HABITAT: Low and middle elevation forests; found in tree canopy.

LOCATIONS: OAX-LO, CHIAP-HI, CHIAP-LO, TAB, CAM, QRO

Plate 45b

Blue-crowned Motmot
Momotus momota
Momoto Mayor = large motmot
Momoto Coroniazul = blue-crowned motmot
Péndulo = pendulum
ID: Large green bird with red eyes and long tail; black facial mask with blue edging; longish down-curved bill; blue on head; bluish-green throat; greenish to greenish-brown chest and belly; small dark spot on chest; tail usually with "tennis racquet" ends; to 41 cm (16 in).

HABITAT: Low, middle, and some higher elevation forests; found in forest interior and in more open areas such as forest edges, tree plantations, gardens, and along watercourses.

LOCATIONS: OAX-LO, CHIAP-HI, CHIAP-LO, TAB, CAM, YUC, QRO

Plate 45c

Turquoise-browed Motmot
Eumomota superciliosa
Momoto Cejiturquesa = turquoise-browed motmot
Pájaro Raqueta = racquet bird
Pájaro Reloj = clock bird
ID: Mid-sized green and brownish-green bird with long tail and longish down-curved bill; small black mask around eyes and black throat; turquoise or pale blue bar above eye; tail usually with "tennis racquet" ends; to 35 cm (14 in).

HABITAT: Low and middle elevation forests; found in forest interiors and more open areas such as forest edges, tree plantations, gardens, archaeological ruins.

LOCATIONS: CHIAP-LO, CAM, YUC, QRO

Plate 45d

Russet-crowned Motmot
Momotus mexicanus
Momoto Coronicafé = brown-crowned motmot
Pendulo Cabeza Naranja = orange-headed pendulum
Pájaro Reloj = clock bird
ID: Mid-sized bird with reddish-brown head, green back, and bluish-green wings; light green chest/belly with black chest spot; black eye-mask with purplish edges; tail usually with "tennis racquet" ends; to 35 cm (14 in).

HABITAT: Low, middle, and some higher elevation dryer woodlands and forest edges.

LOCATIONS: OAX-HI, OAX-LO, CHIAP-HI, CHIAP-LO

Plate 45 **345**

b Blue-crowned
Motmot

d Russet-crowned
Motmot

a Tody Motmot

c Turquoise-browed
Motmot

Plate 46a

Golden-fronted Woodpecker
Melanerpes aurifrons
Carpintero Frentidorado = golden-fronted woodpecker
Carpintero Común = common woodpecker

ID: Mid-sized black and white finely barred bird with grayish face, throat, chest; some red on belly; small red patch on forehead; male with red on head top and back of neck; female with red on back of neck; to 25 cm (10 in).

HABITAT: Low, middle, and higher elevation semi-open sites such as forest edges, tree plantations, shady gardens.

LOCATIONS: OAX-HI, OAX-LO, CHIAP-HI, CHIAP-LO, TAB, CAM, YUC, QRO

Plate 46b

Golden-olive Woodpecker
Piculus rubiginosus
Carpintero Verde = green woodpecker
Carpintero Oliváceo = olivaceous woodpecker
Picamadero Verde = green woodpecker

ID: Mid-sized olive-green bird with finely barred yellow and olive chest; gray head top; yellowish/buffy facial area; male with red back of neck and facial stripe; female lacks red facial stripe; to 22 cm (8.5 in).

HABITAT: Low, middle, and higher elevation forests and more open sites such as forest edges, tree plantations.

LOCATIONS: OAX-LO, CHIAP-HI, CHIAP-LO, TAB, CAM, YUC, QRO

Plate 46c

Yucatán Woodpecker (also called Red-vented Woodpecker)
Centurus pygmaeus
Carpintero Yucateco = Yucatán woodpecker

ID: Smallish black and white barred bird with gray head and small yellow patches above and below bill; grayish chest/belly; male with red on top of head and female with red on back of neck; to 20 cm (7.5 in).

HABITAT: Low elevation forests and semi-open sites such as forest edges, tree plantations, clearings.

LOCATIONS: CAM, YUC, QRO

Plate 46d

Ladder-backed Woodpecker
Picoides scalaris
Carpintero Listado = striped woodpecker
Carpintero Barrado = barred woodpecker

ID: Smallish black and white finely barred bird with black stripes on side of head; whitish, buffy, or light gray face, throat, chest; male red on top of head; female black on top of head; to 18 cm (7 in).

HABITAT: Low, middle, and high elevation drier semi-open sites such as forest edges, pine–oak forests, scrubby areas with scattered trees or cactus.

LOCATIONS: OAX-HI, OAX-LO, CHIAP-HI, CHIAP-LO, TAB, CAM, YUC, QRO

Plate 46 **347**

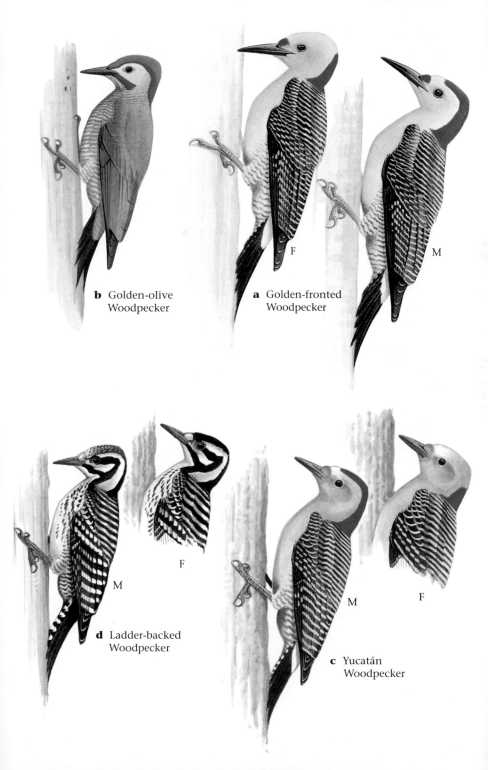

b Golden-olive
Woodpecker

a Golden-fronted
Woodpecker

F

M

d Ladder-backed
Woodpecker

M

F

c Yucatán
Woodpecker

M

F

Plate 47a

Chestnut-colored Woodpecker
Celeus castaneus
Carpintero Castaño = chestnut
woodpecker
Piocoa
ID: Mid-sized brown bird with black markings;
yellowish-brown head; yellowish bill; male with
red face mark under eye; to 23 cm (9 in).

HABITAT: Low and middle elevation forests.

LOCATIONS: OAX-HI, OAX-LO, CHIAP-HI, CHIAP-
LO, TAB, CAM, QRO

Plate 47b

Acorn Woodpecker
Melanerpes formicivorus
Carpintero Arlequín = harlequin
woodpecker
Picamadero Ocotero = ocote-pine
woodpecker
ID: Mid-sized black and white bird with longish
black bill; red on head top; white forehead; white
eyering on black facial area; black chin; whitish
throat; female with less red on top of head; to 23
cm (9 in).

HABITAT: Middle and high elevation oak and
pine–oak forests and woodlands.

LOCATIONS: OAX-HI, CHIAP-HI, TAB

Plate 47c

Lineated Woodpecker
Dryocopus lineatus
Carpintero Lineado = lineated
woodpecker
Carpintero Copeton = tufted
woodpecker
Carpintero Real = royal woodpecker
ID: Large black bird with red crest on head;
whitish stripe runs from edge of bill down along
neck; two whitish stripes on back; light-colored,
barred lower chest/belly; male has small red
stripe on side of throat; female has black
forehead; to 34 cm (13.5 in).

HABITAT: Low, middle, and some higher elevation
forests; found in tree canopy in open, wooded
sites such as forest edges and clearings.

LOCATIONS: OAX-HI, OAX-LO, CHIAP-HI, CHIAP-
LO, TAB, CAM, YUC, QRO

Plate 47d

Pale-billed Woodpecker
Campephilus guatemalensis
Carpintero Piquiclaro = light-billed
woodpecker
Carpintero Grande Cabecirrojo = large
red-headed woodpecker
Carpintero Real = royal woodpecker
ID: Large black bird with red head and crest; two
whitish stripes on back form V; light-colored,
barred lower chest/belly; female has black at front
of crest; to 37 cm (14.5 in).

HABITAT: Low, middle, and some higher
elevation forests; found in tree canopy in open,
wooded sites such as forest edges and clearings,
tree plantations, along rivers and streams.

LOCATIONS: OAX-HI, OAX-LO, CHIAP-HI, CHIAP-
LO, TAB, CAM, YUC, QRO

Plate 47e

Gray-breasted Woodpecker
Centurus hypopolius
Carpintero Pechigris = gray-breasted
woodpecker
Carpintero de Balsas = raft
woodpecker
ID: Mid-sized black and white finely barred bird
with grayish face, throat, chest; whitish forehead;
black eyering; reddish patch on side of head; male
with red on top of head; to 20 cm (8 in).

HABITAT: Higher elevation open and semi-open
sites such as scrubby woodlands, areas with
cactus or scattered trees.

LOCATIONS: OAX-HI

Plate 47 349

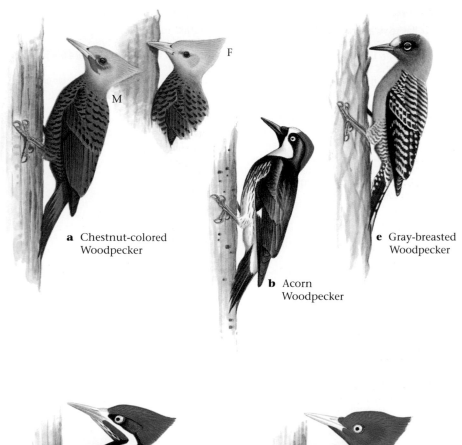

a Chestnut-colored Woodpecker

b Acorn Woodpecker

e Gray-breasted Woodpecker

c Lineated Woodpecker

d Pale-billed Woodpecker

Plate 48a

Collared Aracari
Pteroglossus torquatus
Tucancillo Collarejo = little necklace toucan
Tucancillo de Collar = little collared toucan
ID: Large black bird with huge bill light above, dark below; yellow lower chest/belly with red markings and central black spot; red skin around eye; to 41 cm (16 in).

HABITAT: Low elevation forests; found in tree canopy in more open habitats such as forest edges, tree plantations, clearings, along rivers and streams.

LOCATIONS: OAX-LO, CHIAP-LO, TAB, CAM, YUC, QRO

Plate 48b

Keel-billed Toucan
Ramphastos sulfuratus
Tucán Pico-multicolor = multi-colored-bill toucan
Tucán Cuello Amarillo = yellow-necked toucan
Tucán Real = royal toucan
ID: A large, mostly black bird with yellow face and chest, yellowish-green skin around eye, and that amazing, rainbow-colored (green-orange-blue) toucan's bill; to 56 cm (22 in).

HABITAT: Low and middle elevation forests; found in tree canopy in more open habitats such as forest edges, tree plantations, along rivers and streams.

LOCATIONS: OAX-HI, OAX-LO, CHIAP-HI, CHIAP-LO, TAB, CAM, YUC, QRO

Plate 48c

Emerald Toucanet
Aulacorhynchus prasinus
Tucancillo (or Tucaneta) Verde = little green toucan
ID: Large green bird with blue throat; tail green above, brown below; large bill yellow above, black below; to 35 cm (14 in).

HABITAT: Low, middle, and high elevation wet forests; found in tree canopy in forest interior, forest edges, clearings, tree plantations.

LOCATIONS: OAX-HI, CHIAP-HI, CHIAP-LO, TAB, CAM, QRO

Plate 48 351

a Collared Aracari

b Keel-billed Toucan

c Emerald Toucanet

Plate 49a

Tawny-winged Woodcreeper
Dendrocincla anabatina
Trepatroncos Alileonado = tawny-winged woodcreeper
Trepador Ala Bicolor = bicolored-winged woodcreeper
Trepador Franjeado = striped woodcreeper

ID: Small brown bird with buffy/tan throat; reddish-brown wings and tail; light stripe above eye; straight gray bill; to 19 cm (7.5 in).

HABITAT: Low and middle elevation forests; found climbing up tree trunks in forest interior.

LOCATIONS: OAX-HI, OAX-LO, CHIAP-HI, CHIAP-LO, TAB, CAM, QRO

Plate 49b

Olivaceous Woodcreeper
Sittasomus griseicapillus
Trepatroncos (or trepador) Oliváceo = olivaceous woodcreeper
Trepatroncos Gorjeador = warbling woodcreeper
Trepatroncos Cabeza Gris = gray-headed woodcreeper

ID: Small reddish-brown bird with gray head and chest; dark, slender, straight bill; to 15 cm (6 in).

HABITAT: Low, middle, and some higher elevation forests; found climbing up tree trunks in forest interior and in more open sites such as forest edges, tree plantations.

LOCATIONS: OAX-HI, OAX-LO, CHIAP-HI, CHIAP-LO, TAB, CAM, YUC, QRO

Plate 49c

Ivory-billed Woodcreeper
Xiphorhynchus flavigaster
Trepatroncos Piquiclaro = light-billed woodcreeper
Trepador Goteado = dripped woodcreeper
Trepador Dorsirrayado Mayor = larger lined-back woodcreeper

ID: Mid-sized brownish bird with bold buffy streaks on head, back, chest; reddish-brown wings, tail; tan/buffy throat; long pale bill; to 25 cm (10 in).

HABITAT: Low, middle, and some higher elevation forests; found climbing up tree trunks in forest interior and in more open sites such as forest edges, tree plantations, mangroves.

LOCATIONS: OAX-HI, OAX-LO, CHIAP-HI, CHIAP-LO, TAB, CAM, YUC, QRO

Plate 49d

Spot-crowned Woodcreeper
Lepidocolaptes affinis
Trepatroncos Corona-punteado = spot-crowned woodcreeper
Trepatroncos Montés = wild woodcreeper

ID: Smallish brown or grayish-brown bird with spotted head and buffy streaks on back, chest/belly; tan/buffy throat; reddish-brown wings; light gray bill; to 20 cm (8 in).

HABITAT: Higher elevation forests, including pine and pine–oak forests.

LOCATIONS: OAX-HI, CHIAP-HI

Plate 49   353

a Tawny-winged
Woodcreeper

b Olivaceous
Woodcreeper

c Ivory-billed
Woodcreeper

d Spot-crowned
Woodcreeper

Plate 50a

Barred Antshrike
Thamnophilus doliatus
Batará Barrada = barred batará
Pájaro Hormiguero = anteater bird

ID: Male is a small white-and-black barred bird with a mostly black head crest; female is reddish-brown with black streaks on face, buffy throat, and light brown chest/belly; 17 cm (6.5 in).

HABITAT: Low and middle elevation forests; found usually on the ground in open areas such as thickets but also in the forest interior.

LOCATIONS: OAX-HI, OAX-LO, CHIAP-HI, CHIAP-LO, TAB, CAM, YUC, QRO

Plate 50b

Black-faced Antthrush (also called Mexican Antthrush)
Formicarius analis
Hormiguero Gallito = little rooster anteater
Hormiguero Cara Negra = black-faced anteater
Colicorto = short-tail

ID: Mid-sized dark or olive-brown bird with black cheeks and throat; grayish chest and belly; small amount of pale bluish skin around eye; long legs, short tail; to 18 cm (7 in).

HABITAT: Low and middle elevation forests; found in forest interior or edge, on or near the ground.

LOCATIONS: OAX-HI, OAX-LO, CHIAP-HI, CHIAP-LO, TAB, CAM, QRO

Plate 50c

Dot-winged Antwren
Microrhopias quixensis
Hormiguerito Alipunteado = little dot-winged anteater
Hormiguero Marañero = thicket anteater

ID: Small black bird with white spots and bar on wing, white tail tips; male is almost all black; female more grayish with reddish-brown chest and belly; 11 cm (4.5 in).

HABITAT: Low elevation wet forests; found foraging on or near the ground in forest interior and more open sites such as forest edges, tree plantations, thickets.

LOCATIONS: OAX-LO, CHIAP-LO, TAB, CAM, QRO

Plate 50d

Dusky Antbird
Cercomacra tyrannina
Hormiguero Negruzco = blackish anteater
Hormiguero Tirano = tyrant anteater Matorralero

ID: Small bird with dark slender bill and longish tail; male is blackish/gray with white wing bars; female is brownish/olive with reddish-brown or tawny chest/belly and inconspicuous light eyering; to 14 cm (5.5 in).

HABITAT: Low and middle elevation forests and forest edges; found low in trees and in thickets.

LOCATIONS: OAX-HI, OAX-LO, CHIAP-HI, CHIAP-LO, TAB, CAM, QRO

Plate 50e

Scaled Antpitta
Grallaria guatimalensis
Hormiguero-cholino Escamoso
Cholina

ID: Small grayish-brown or grayish-olive bird with gray head and brown wings; light stripe below eye; cinnamon chest/belly with light band and black markings at upper chest; long legs; to 19 cm (7.5 in).

HABITAT: Low, middle and higher elevation forests, including pine–oak forests; found on the ground.

LOCATIONS: OAX-HI, OAX-LO, CHIAP-HI, CHIAP-LO, TAB

Plate 50 355

a Barred Antshrike

M

F

d Dusky Antbird

M

F

c Dot-winged Antwren

F

M

b Black-faced (Mexican) Antthrush

e Scaled Antpitta

Plate 51a

White-collared Manakin
Manacus candei
Saltarín Cuelliblanco = white-necked jumper
Turquito Cuello Blanco = little white-necked Turk
Quiebrapalito

ID: Female is a small olive-green bird with yellowish belly, orange legs, and small dark bill; male with black on wings and top of head; black band across back; white throat, chest, and upper back; olive-green rump; yellow belly; orange legs; small black bill; to 11 cm (4.5 in).

HABITAT: Low elevation wet forests; found in trees and thickets in more open wooded habitats – forest edges, tree plantations, along watercourses.

LOCATIONS: OAX-LO, CHIAP-LO, TAB, CAM, QRO

Plate 51b

Red-capped Manakin
Pipra mentalis
Saltarín Cabecirrojo = red-headed jumper
Turquito Sargento = little sergeant Turk
Turquito

ID: Female is small olive-green bird with lighter belly, brownish legs and small brownish bill; male is black with red head and yellow thighs; 10 cm (4 in).

HABITAT: Low elevation wet forests; found in trees in forest interior and in more open sites such as forest edges, clearings.

LOCATIONS: OAX-LO, CHIAP-LO, TAB, CAM, QRO

Plate 51c

Long-tailed Manakin
Chiroxiphia linearis
Saltarín Colilargo = long-tailed dancer
Pipra Colilargo = long-tailed pipra
Toledo

ID: Female is small olive-green bird with lighter belly, orange legs, small, black bill; male is black with blue back, red crest on head, and two very long, black tail feathers; to 12 cm (5 in), plus long tail feathers.

HABITAT: Low and middle elevation forests; found in trees in forest interior and more open areas such as forest edges, particularly along swamps and watercourses.

LOCATIONS: OAX-LO, CHIAP-HI, CHIAP-LO

Plate 51d

Rufous Piha
Lipaugus unirufus
Piha Rufa = rufous piha
Guardabosque = forest-ranger
Cotinga Silbadora = whistling cotinga

ID: Mid-sized reddish-brown bird with lighter throat and belly; dark, fairly thick bill, slightly paler throat, and grayish legs; to 25 cm (10 in).

HABITAT: Low elevation wet forests; found in tree canopy and lower, usually in more open sites such as forest edges and clearings.

LOCATIONS: OAX-LO, CHIAP-LO, TAB

Plate 51e

Lovely Cotinga
Cotinga amabilis
Cotinga Azuleja = bluish cotinga
Azulejo Real = royal bluebird

ID: Mid-sized bird with smallish dark bill; male is bright blue with purple throat and chest, blue band across chest, blackish wings and tail edges; female is gray-brown with scaled pattern; whitish chest/belly with brownish spots; brown wings and tail; to 19 cm (7.5 in).

HABITAT: Low elevation forests and semi-open sites such as forest edges, shady clearings; found in tree canopy.

LOCATIONS: OAX-LO, CHIAP-LO, TAB

Plate 51 357

a White-collared Manakin

M

F

b Red-capped Manakin

M

F

c Long-tailed Manakin

M

F

d Rufous Piha

e Lovely Cotinga

M

F

Plate 52a

Masked Tityra
Tityra semifasciata
Titira Enmascarada = masked tityra
Titira Puerquita = little pig tityra
ID: Mid-sized gray bird with black-tipped red bill, red skin around eye, broad black wingbar and tail; male with black on chin, forehead, and behind eye; white chest/belly; female is gray-brown with whitish throat and gray chest/belly; to 23 cm (9 in).

HABITAT: Low, middle, and some higher elevation forests; found in tree canopy, usually in more open sites such as forest edges, clearings, tree plantations, along waterways.

LOCATIONS: OAX-HI, OAX-LO, CHIAP-HI, CHIAP-LO, TAB, CAM, YUC, QRO

Plate 52b

Rose-throated Becard
Pachyramphus aglaiae
Cabezón Degollado = slashed-throat bighead
ID: Male is smallish gray or grayish-black bird with reddish throat patch; female is brown or cinnamon with lighter brown chest/belly and black head top; to 18 cm (7 in).

HABITAT: Low, middle, and high elevation forest edges, scrubby woodlands, clearings, and other open and semi-open areas.

LOCATIONS: OAX-HI, OAX-LO, CHIAP-HI, CHIAP-LO, TAB, CAM, YUC, QRO

Plate 52c

Vermilion Flycatcher
Pyrocephalus rubinus
Mosquero Cardenal = cardinal fly-eater
Cardenalito = little cardinal
ID: Male is small red bird with blackish brown back, wings, tail, and eye-mask; female is grayish to grayish-brown with light throat and chest with darker streaks; pinkish belly; pale eyestripe; to 14 cm (5.5 in).

HABITAT: Low, middle and some higher elevation semi-open and open sites such as agricultural areas, shady gardens, human settlements with scattered trees.

LOCATIONS: OAX-HI, OAX-LO, CHIAP-HI, CHIAP-LO, TAB, CAM, YUC

Plate 52d

Scissor-tailed Flycatcher
Tyrannus forficatus
Tirano-tijereta Rosada = liitle pink scissors tyrant
ID: A mid-sized silver-gray bird with very long, black, forked tail, black wings, reddish patch on side under wing, whitish chest/belly; to 20 cm (8 in), plus tail adds up to another 15 cm (6 in).

HABITAT: Low, middle and some higher elevations; found around very open sites such as fields, pastures, grasslands, marshes, human settlements.

LOCATIONS: OAX-HI, OAX-LO, CHIAP-HI, CHIAP-LO, TAB

Note: This species is a non-breeding seasonal migrant.

Plate 52e

Dusky Flycatcher
Empidonax oberholseri
Mosquero Oscuro = dusky fly-eater
ID: Small grayish-brown or grayish-olive bird with whitish throat, grayish-brown/dusky chest, and yellow belly; dark wings with light wingbars; white eyering; bill dark above, light below; fairly long dark tail; to 14 cm (5.5 in).

HABITAT: Middle and high elevation forest edges and other semi-open sites with scattered trees.

LOCATIONS: OAX-HI

Note: This species is a non-breeding seasonal migrant.

Plate 52f

Hammond's Flycatcher
Empidonax hammondii
Mosquero de Hammond = Hammond's fly-eater
ID: Small grayish or olive bird with light gray throat, brownish-gray/dusky chest, yellow belly; dark wings with light wingbars; whitish eyering; bill dark above, light below; shortish dark tail; to 14 cm (5.5 in).

HABITAT: High elevation forests and forest edges, including pine and pine–oak forests.

LOCATIONS: OAX-HI, CHIAP-HI

Note: This species is a non-breeding seasonal migrant.

Plate 52 359

b Rose-throated Becard

F

M

F

M

a Masked Tityra

e Dusky Flycatcher

f Hammond's Flycatcher

d Scissor-tailed Flycatcher

F

M

c Vermilion Flycatcher

Plate 53a

Tropical Kingbird
Tyrannus melancholicus
Tirano Tropical = tropical tyrant
ID: Mid-sized olive-colored bird with gray head; whitish throat; dark wings and dark notched tail; dark bar through eye; yellow belly; reddish patch on top of head usually concealed; to 23 cm (9 in).

HABITAT: Low, middle, and some higher elevation sites; found in trees in open wooded areas such as forest edges and tree plantations, and also in more open areas such as grasslands, pastures and other agricultural sites, beach areas, and near human settlements.

LOCATIONS: OAX-HI, OAX-LO, CHIAP-HI, CHIAP-LO, TAB, CAM, YUC, QRO

Plate 53b

Great Kiskadee
Pitangus sulphuratus
Luis Grande = big Luis
Luis Bienteveo = Luis good-to-see
ID: Mid-sized olive-brown bird with reddish-brown wings and bright yellow chest/belly; black and white head; yellow patch on top of head usually concealed; white throat; to 25 cm (10 in).

HABITAT: Low, middle, and some higher elevations; found in trees in more open sites such as grasslands, pastures, forest edges, gardens, and along waterways.

LOCATIONS: OAX-HI, OAX-LO, CHIAP-HI, CHIAP-LO, TAB, CAM, YUC, QRO

Plate 53c

Boat-billed Flycatcher
Megarhynchus pitangua
Luis Piquigrueso = thick-billed Luis
ID: Mid-sized olive or olive-brown bird with bright yellow chest/belly; black and white head; yellow or orange patch on top of head usually concealed; white throat; distinguished from Great Kiskadee by wing color and larger bill; to 24 cm (9.5 in).

HABITAT: Low and middle elevations; found in tree canopy in wooded areas and more open sites such as forest edges, tree plantations, gardens, and along waterways.

LOCATIONS: OAX-HI, OAX-LO, CHIAP-HI, CHIAP-LO, TAB, CAM, YUC, QRO

Plate 53d

Social Flycatcher
Myiozetetes similis
Luis Gregario = social Luis
ID: Smallish olive or olive-brown bird with dark gray/blackish and white head; dark eye-mask; reddish patch on top of head usually concealed; white throat; bright yellow chest/belly; very small black bill; to 18 cm (7 in).

HABITAT: Low, middle, and some higher elevation sites; found in trees in more open areas such as grasslands, pastures and agricultural sites, forest edges, along rivers and lakes, gardens.

LOCATIONS: OAX-HI, OAX-LO, CHIAP-HI, CHIAP-LO, TAB, CAM, YUC, QRO

Plate 53e

Sulphur-bellied Flycatcher
Myiodynastes luteiventris
Papamoscas Vientre-amarillo = yellow-bellied flycatcher
ID: Mid-sized grayish-brown or grayish-olive bird with darker mottled pattern; white stripes above and below eye; black stripe through eye; white throat and yellowish chest/belly all with dark markings; reddish-brown tail; thickish black bill; to 21 cm (8.5 in).

HABITAT: Low, middle, and some higher elevation forests and more open sites such as forest edges, clearings, tree plantations.

LOCATIONS: OAX-HI, OAX-LO, CHIAP-HI, CHIAP-LO, TAB, CAM, YUC, QRO

Plate 53f

Cassin's Kingbird
Tyrannus vociferans
Tirano de Cassin = Cassin's tyrant
ID: Mid-sized olive-colored bird with gray head and chest; whitish throat; dark wings and dark squarish tail; dark bar through eye; yellow belly; fairly short dark bill; reddish patch on top of head usually concealed; to 23 cm (9 in).

HABITAT: Low, middle, and high elevation open and semi-open sites with scattered trees, agricultural areas.

LOCATIONS: OAX-HI, OAX-LO, CHIAP-HI, CHIAP-LO

Plate 53 361

a Tropical Kingbird

b Great Kiskadee

c Boat-billed Flycatcher

d Social Flycatcher

f Cassin's Kingbird

e Sulphur-bellied Flycatcher

Plate 54a

Brown-crested Flycatcher
Myiarchus tyrannulus
Copetón Tirano = big-crested tyrant
Mosquero Copetón = big-crested fly-eater

ID: Mid-sized grayish-olive or grayish-brown bird with bushy head crest, gray throat and chest, yellowish belly; to 21 cm (8.5 in).

HABITAT: Low, middle, and some higher elevations; found in more open habitats such as forest edges, along waterways, open plantations and gardens.

LOCATIONS: OAX-HI, OAX-LO, CHIAP-HI, CHIAP-LO, TAB, CAM, YUC, QRO

Plate 54b

Dusky-capped Flycatcher
Myiarchus tuberculifer
Copetón Triste = sad big-crest
Papamoscas Copetón Común = common big-crested flycatcher

ID: Smallish olive or brownish-olive bird with very dark/blackish head, gray throat/chest, yellow belly; wings and tail brown or grayish-brown; to 18 cm (7 in).

HABITAT: Low, middle, and higher elevation forests; prefers more open parts of forests – edges, clearings, tree plantations, along waterways.

LOCATIONS: OAX-HI, OAX-LO, CHIAP-HI, CHIAP-LO, TAB, CAM, YUC, QRO

Plate 54c

Yucatán Flycatcher
Myiarchus yucatanensis
Copetón Yucateco = Yucatán big-crest

ID: Smallish grayish-olive bird with brownish-gray head with lighter area around eye; light gray throat/chest, yellow belly; dark wings with some white edging; dark tail; to 19 cm (7.5 in). (Note: The bird is overall darker on Cozumel Island.)

HABITAT: Low elevation forests, forest edges, scrubby woodlands and other semi-open sites with scattered trees.

LOCATIONS: CAM, YUC, QRO

Plate 54d

Tropical Pewee
Contopus cinereus
Pibí Tropical = tropical pewee
Tengofrío Tropical = tropical "I am cold"

ID: Small grayish-brown or brownish-olive bird with darker slight head crest; whitish throat; light chest; small bill dark above, orangish below; dark wings and tail; 14 cm (5.5 in).

HABITAT: Low and middle elevation semi-open areas such as forest edges, clearings.

LOCATIONS: OAX-HI, OAX-LO, CHIAP-HI, CHIAP-LO, TAB, CAM, YUC, QRO

Plate 54e

Tufted Flycatcher
Mitrephanes phaeocercus
Mosquero Penachudo = tufted fly-eater
Penachito = little tuft

ID: Small olive or brownish-olive bird with cinnamon face and chest/belly; conspicuous head crest; light eyering; dark wings with lighter wingbars; small bill dark above, orangish below; to 13 cm (5 in).

HABITAT: High elevation forests, including pine and pine–oak forests, and semi-open sites such as forest edges, clearings, tree plantations.

LOCATIONS: OAX-HI, CHIAP-HI

Plate 54f

Greater Pewee
Contopus pertinax
Pibí Mayor = greater pewee
Tengofrío Grande = big "I am cold"

ID: Smallish olive-gray or gray bird with conspicuous dark head crest; light throat; grayish chest; yellowish-gray belly; dark wings with light wingbars; bill dark above, orangish below; to 19 cm (7.5 in).

HABITAT: Middle and high elevation semi-open sites such as forest edges, clearings, including in pine–oak forests.

LOCATIONS: OAX-HI, OAX-LO, CHIAP-HI, CHIAP-LO, TAB, CAM, YUC, QRO

Plate 54 363

b Dusky-capped
Flycatcher

a Brown-crested
Flycatcher

c Yucatán Flycatcher

f Greater Pewee

e Tufted Flycatcher

d Tropical Pewee

Plate 55a

Spot-breasted Wren
Thryothorus maculipectus
Saltapared Pechimanchado = spot-breasted wren
ID: Small brown bird with whitish chest/belly with black spots; brown tail with black bars; light and dark striped face; 13 cm (5 in).

HABITAT: Low and middle elevation forests and semi-open areas such as forest edges, tree plantations.

LOCATIONS: OAX-HI, OAX-LO, CHIAP-HI, CHIAP-LO, TAB, CAM, YUC, QRO

Plate 55b

White-breasted Wood-wren
Henicorhina leucosticta
Saltapared-selvatico Pechiblanco = white-breasted wood-wren
ID: Small brown bird with streaked black, dark brown, and white head, white throat/chest, tawny/light brown belly, black-barred wings, and short, black-barred tail; 10 cm (4 in).

HABITAT: Low and some middle elevation forests; found low in trees in forest interior and more open areas such as forest edges, tree plantations. (Note: The Gray-breasted Wood-wren, which resembles the White-breasted but has gray throat/chest, occurs in higher elevation pine forests in Oaxaca and Chiapas.)

LOCATIONS: OAX-LO, CHIAP-LO, TAB, CAM, QRO

Plate 55c

Southern House Wren
Troglodytes aedon
Saltapared Común = common wren
Saltapared Cucarachero = cockroach wren
ID: A small brown bird with black-barred wings and tail; lighter brown chest/belly; pale stripe above eye; 11 cm (4.5 in). (Note: The Northern House Wren occurs over much of Oaxaca and the Cozumel Wren occurs on Cozumel – the sole wren on the island; both closely resemble the Southern House, but are somewhat lighter in color.)

HABITAT: Low, middle and higher elevations; found mostly around houses and other structures, and in open habitats such as pastures and low scrub.

LOCATIONS: OAX-LO, CHIAP-HI, CHIAP-LO, TAB, CAM, YUC, QRO

Plate 55d

Banded-backed Wren
Campylorhynchus zonatus
Matraca-barrado Tropical = tropical barred rattle
Carrasquita = little rattle
ID: Smallish black-and-whitish-barred bird (fairly large, however, for a wren) with white throat/chest with dark spots; light brown belly; light stripe above eye; thin bill grayish above, brownish below; to 19 cm (7.5 in).

HABITAT: Low, middle, and high elevation forests, including pine–oak forests; prefers semi-open sites such as forest edges, clearings, tree plantations, along waterways, vegetation around human settlements.

LOCATIONS: OAX-HI, OAX-LO, CHIAP-HI, CHIAP-LO, TAB

Plate 55e

White-bellied Wren
Uropsila leucogastra
Saltapared Vientre-blanco = white-bellied wren
Saltapared Cantarina = singing wren
ID: Small brown bird with whitish chest/belly; white stripe above eye and dark line behind eye; short dark-barred tail; to 10 cm (4 in).

HABITAT: Low elevation forests, scrubby woodlands, thickets, vegetation along waterways.

LOCATIONS: OAX-LO, CHIAP-LO, TAB, CAM, YUC, QRO

Plate 55f

Canyon Wren
Catherpes mexicanus
Saltapared Barranquero = canyon wren
Saltapared Risquero = cliff-wren
Saltarroca = rock-jumper
ID: Small reddish-brown bird with white spots on back; white throat/chest; chestnut belly with spots; reddish-brown tail with black bars; longish, dark, slightly down-curved bill; to 13 cm (5 in).

HABITAT: Middle and high elevation rocky sites such as cliffs, caves, canyons, and archaeological ruins, usually in more open areas.

LOCATIONS: OAX-HI, CHIAP-HI

Plate 55 **365**

a Spot-breasted Wren

b White-breasted Wood-wren

c Southern House Wren

d Banded-backed Wren

e White-bellied Wren

f Canyon Wren

Plate 56a

Wood Thrush
Hylocichla mustelina
Zorzal Maculado = spotted thrush
Zorzalito de Bosque = little forest
thrush
ID: Mid-sized brown bird with reddish-brown
head; white chest/belly with black spots; white
eyering; white and dark streaks on side of head;
bill dark above, light below; to 19 cm (7.5 in).

HABITAT: Low, middle, and some higher elevation
forests and forest edges; found low in trees or on
the ground.

LOCATIONS: OAX-LO, CHIAP-HI, CHIAP-LO, TAB,
CAM, YUC, QRO

Note: This species is a non-breeding seasonal
migrant.

Plate 56b

Clay-colored Robin
Turdus grayi
Zorzal Pardo = brown thrush
Mirlo Huertero = orchard blackbird
ID: Mid-sized brownish bird with lighter brown/
tawny chest and belly, dark-streaked throat,
yellowish bill, flesh-colored legs; to 25 cm (10 in).

HABITAT: Low, middle, and higher elevations;
inhabits open and semi-open areas such as forest
edges, tree plantations, pastures, agricultural
sites, gardens, lawns.

LOCATIONS: OAX-HI, OAX-LO, CHIAP-HI, CHIAP-
LO, TAB, CAM, YUC, QRO

Plate 56c

White-throated Thrush
Turdus assimilis
Zorzal Gorjiblanco = white-throated
thrush
Mirlo Bosquero = forest blackbird
ID: Mid-sized black or dark gray bird with light
gray chest/belly; white throat with dark streaks;
yellowish eyering, bill, and legs; to 25 cm (10 in).

HABITAT: Low, middle, and higher elevation
forests and semi-open sites such as forest edges
and along watercourses.

LOCATIONS: OAX-HI, OAX-LO, CHIAP-HI, CHIAP-
LO, TAB

Plate 56d

Tropical Mockingbird
Mimus gilvus
Cenzontle Tropical = tropical
mockingbird
Cenzontle Gris = gray mockingbird
ID: Mid-sized gray bird with whitish chest/belly;
black wings with white bars; long dark tail with
white edges; yellowish eye; to 25 cm (10 in).
(Note: The Northern Mockingbird, which closely
resembles Tropical Mockingbird, occurs in
Oaxaca.)

HABITAT: Low, middle, and some higher elevation
open and semi-open areas with scattered bushes
and trees, especially roadsides, gardens; found on
telephone wires, in bushes, trees, and on ground.

LOCATIONS: CHIAP-HI, CHIAP-LO, TAB, CAM,
YUC, QRO

Plate 56e

Black Catbird
Melanoptila glabrirostris
Pájaro-gato Negro = black catbird
ID: Mid-sized glossy black bird with dark reddish
eyes; slender black bill; black legs; to 20 cm (8 in).

HABITAT: Low elevation semi-open wooded sites
such as forest edges, wooded thickets,
woodlands adjacent to beaches.

LOCATIONS: CAM, YUC, QRO

Plate 56f

Blue Mockingbird
Melanotis caerulescens
Mulato Azul = blue mulato
Mulato Común = common mulato
ID: Mid-sized dull bluish bird with black eye-
mask; throat/chest slightly lighter blue with
streaks; red eyes; black legs; to 26 cm (10.5 in).
(Note: The Blue-and-White Mockingbird, which
occurs at high elevations in Chiapas, closely
resembles the Blue Mockingbird but has white
chest/belly – see color plate.)

HABITAT: Middle and higher elevation forests,
including pine forests and cloud forests, forest
edges, and scrubby woodlands.

LOCATIONS: OAX-HI, OAX-LO

Plate 56   367

c White-throated Thrush

a Wood Thrush

b Clay-colored Robin

e Black Catbird

d Tropical Mockingbird

Blue-and-white Mockingbird

f Blue Mockingbird

Plate 57a

Brown Jay
Cyanocorax morio
Urraca Papán= foolish magpie
Chara Papán = foolish jay
Chara Café = brown jay

ID: Large dark brown bird with whitish lower chest and belly; tail partially white-tipped; large black bill; to 40 cm (16 in).

HABITAT: Low and middle elevations; found in tree canopy in open wooded sites such as forest edges, tree plantations, near human settlements.

LOCATIONS: OAX-HI, OAX-LO, CHIAP-HI, CHIAP-LO, TAB, CAM, YUC, QRO

Plate 57b

Green Jay
Cyanocorax yncas
Chara Verde = green jay
Checla Verde
Queixque Verde

ID: Largish green bird with blue and black head, yellow chest/belly; blue-green tail with yellow edges; yellowish eyes; to 30 cm (12 in).

HABITAT: Low, middle, and higher elevation forests and semi-open sites such as forest edges, tree plantations.

LOCATIONS: OAX-HI, OAX-LO, CHIAP-HI, CHIAP-LO, TAB, CAM, YUC, QRO

Plate 57c

Yucatán Jay
Cyanocorax yucatanicus
Chara Yucateca = Yucatán jay
Cháchara

ID: Largish turquoise-blue bird with black head, chest, and belly; yellow legs; youngsters just out of nest are white with blue wings; to 33 cm (13 in).

HABITAT: Low elevation forests and semi-open sites such as forest edges, tree plantations.

LOCATIONS: CAM, YUC, QRO

Plate 57d

White-throated Magpie-jay
Calocitta formosa
Urraca-hermosa Cariblanca = beautiful white-faced magpie
Urraca Copetona = big-tufted magpie

ID: Large bird, blue above, white below; conspicuous crest on head; long blue tail; large black bill; to 51 cm (20 in).

HABITAT: Low and middle elevation drier regions; found in tree canopy in more open, wooded sites such as forest edges, trees along waterways.

LOCATIONS: OAX-HI, OAX-LO, CHIAP-HI, CHIAP-LO

Plate 57e

Steller's Jay
Cyanocitta stelleri
Chara de Steller = Steller's jay
Chara Copetona = big-tufted jay
Azulejo Ocotero = ocote-pine bluebird

ID: Mid-sized dark blue bird with darker head and conspicuous crest; small white marks around eye, on forehead, and below bill; black legs and bill; to 29 cm (11.5 in).

HABITAT: Higher elevation forests, including oak and pine–oak forests, and forest edges.

LOCATIONS: OAX-HI, CHIAP-HI

Plate 57f

Mexican Scrub Jay
Aphelocoma coerulescens
Chara Azuleja = bluish jay
Queixque de Ceja Blanca = white-browed queixque

ID: Mid-sized blue bird with light stripe above eye; white throat with darker streaks; grayish chest/belly; to 30 cm (12 in).

HABITAT: High elevation semi-open sites in drier regions; scrubby woodlands and areas with scattered trees.

LOCATIONS: OAX-HI

Plate 57 369

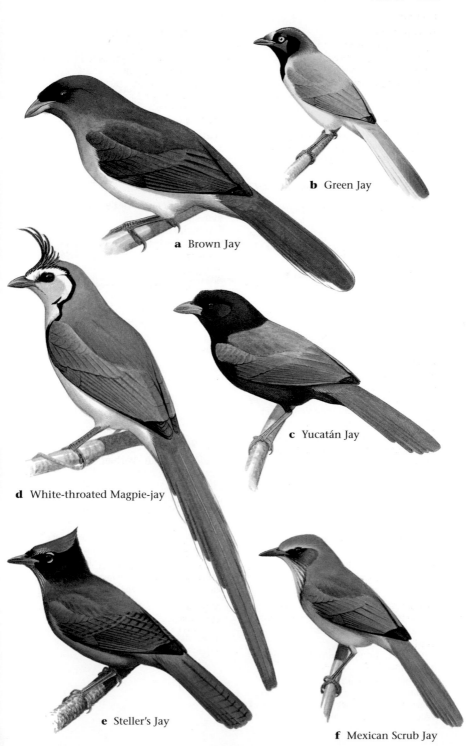

a Brown Jay

b Green Jay

c Yucatán Jay

d White-throated Magpie-jay

e Steller's Jay

f Mexican Scrub Jay

Plate 58a
Mangrove Warbler
Dendroica petechia
Chipe Manglera = mangrove chipe
Chipe Amarilla = yellow chipe
Gorjeador Manglero = mangrove warbler
Verdín Manglero

ID: Small greenish-olive bird with yellow chest/belly; light brown legs; male with reddish-brown chest streaks and head; female sometimes with reddish brown tinge on head top and a few chest streaks; to 13 cm (5 in).

HABITAT: Coastal; found in mangroves, brushy woodlands, scrubby areas.

LOCATIONS: OAX-LO, CHIAP-LO, TAB, CAM, YUC, QRO

Plate 58b
Magnolia Warbler
Dendroica magnolia
Chipe de Magnolia = magnolia chipe
Gorjeador Pechirayado = lined-breasted warbler
Verdín Pechirayado

ID: Small olive-green bird with black streaks on back; gray head; black tail; yellowish chest/belly with dark spots/streaks; white eyering; to 11.5 cm (4.5 in).

HABITAT: Low and middle elevation forests and semi-open sites such as forest edges, tree plantations; found in tree canopy.

LOCATIONS: OAX-HI, OAX-LO, CHIAP-HI, CHIAP-LO, TAB, CAM, YUC, QRO

Note: This species is a non-breeding seasonal migrant.

Plate 58c
Rufous-capped Warbler
Basileuterus rufifrons
Chipe Gorrirrufo = rufous-capped chipe
Larvitero Cabeciroja = red-headed larva-eater
Larvitero = larva-eater

ID: Small greenish-olive bird with reddish-brown head with white and black eyestripes; gray neck; yellow chest/belly; to 13 cm (5 in).

HABITAT: Low, middle, and higher elevations; prefers semi-open sites such as forest edges, scrubby areas with scattered trees, and thickets.

LOCATIONS: OAX-HI, OAX-LO, CHIAP-HI, CHIAP-LO, TAB, CAM, YUC, QRO

Plate 58d
Common Yellowthroat
Geothlypis trichas
Mascarita Común = little common mask
Tapaojito Común = little common eye-cover
Antifacito

ID: Small olive or brownish-olive bird with yellow throat/chest and flesh-colored legs; male with whitish-bordered black face mask and whitish to light yellow belly; female with light eyering and whitish to light brownish belly; to 13 cm (5 in).

HABITAT: Low, middle, and high elevation marshes, moist fields, and vegetation near water.

LOCATIONS: OAX-HI, OAX-LO, CHIAP-HI, CHIAP-LO, TAB, CAM, YUC, QRO

Note: This species is a non-breeding seasonal migrant.

Plate 58e
Golden-browed Warbler
Basileuterus belli
Chipe Cejidorado = golden-browed chipe
Larvitero Rayiamarilla = yellow-striped larva-eater

ID: Small olive bird with reddish-brown head top and eye-mask; black-bordered yellow stripe above eye; yellow chest/belly; to 13 cm (5 in).

HABITAT: High elevation forests, including pine and pine–oak forests, and forest edges.

LOCATIONS: OAX-HI, CHIAP-HI

Plate 58 371

c Rufous-capped Warbler

F

M

b Magnolia Warbler

F

M

a Mangrove Warbler

M

F

d Common Yellowthroat

e Golden-browed Warbler

Plate 59a
American Redstart
Setophaga ruticilla
Pavito Migratorio = little migratory turkey
Pavito Naranja = little orange turkey
Calandrita = little lark

ID: Male is small black bird with whitish belly; orange on sides of chest, on wingbars, and on sides of tail; female is olive-grayish with whitish throat and eyering; yellow on sides of chest, on wingbars, and on sides of tail; to 13 cm (5 in).

HABITAT: Low, middle, and some higher elevation forests and semi-open sites such as forest edges, light woodlands, scrubby areas with scattered trees.

LOCATIONS: OAX-HI, OAX-LO, CHIAP-HI, CHIAP-LO, TAB, CAM, YUC, QRO

Note: This species is a non-breeding seasonal migrant.

Plate 59b
Painted Redstart
Myioborus pictus
Pavito Aliblanco = little white-winged turkey
Pavito Ocotero = little ocote-pine turkey

ID: Small black bird with red belly and large white wing patches; tail black edged with white; small white mark below eye; black bill and legs; to 13 cm (5 in).

HABITAT: Middle and high elevation forests, including oak, pine, and pine–oak forests.

LOCATIONS: OAX-HI, CHIAP-HI

Plate 59c
Bananaquit
Coereba flaveola
Platanero = banana-guy
Reinita Amarilla = yellow little queen
Reinita = little queen

ID: Very small olive-grayish bird with white eyestripe, gray throat, yellow chest/belly, and short, pointed, down-curved black bill; 10 cm (4 in). (Note: Bananaquits on Cozumel Island have a darker back and white throat/chest.)

HABITAT: Low and middle elevation forests; found in tree canopy in more open wooded sites such as forest edges, gardens, plantations.

LOCATIONS: OAX-HI, OAX-LO, CHIAP-HI, CHIAP-LO, TAB, CAM, QRO

Plate 59d
Blue-gray Gnatcatcher
Polioptila caerulea
Perlita Grisilla = grayish little pearl
Perlita Común = common little pearl Grisilla

ID: Very small bluish-gray bird with whitish chest/belly; black tail with white edges; white eyering; male often with black stripe above eye; to 10 cm (4 in).

HABITAT: Low, middle, and high elevation forests and semi-open sites such as forest edges, open woodlands, scrubby areas with scattered trees.

LOCATIONS: OAX-HI, OAX-LO, CHIAP-HI, CHIAP-LO, TAB, CAM, YUC, QRO

Plate 59e
Slate-throated Redstart
Myioborus miniatus
Pavito Gorgigris = little gray-throated turkey
Pavito Selvático = little forest turkey

ID: Small, dark gray bird with black head, reddish chest/belly, and reddish patch on head top; tail blackish edged with white; black bill and legs; to 14 cm (5.5 in).

HABITAT: Middle and high elevation forests, including oak, pine, and pine–oak forests.

LOCATIONS: OAX-HI, CHIAP-HI

Plate 59 **373**

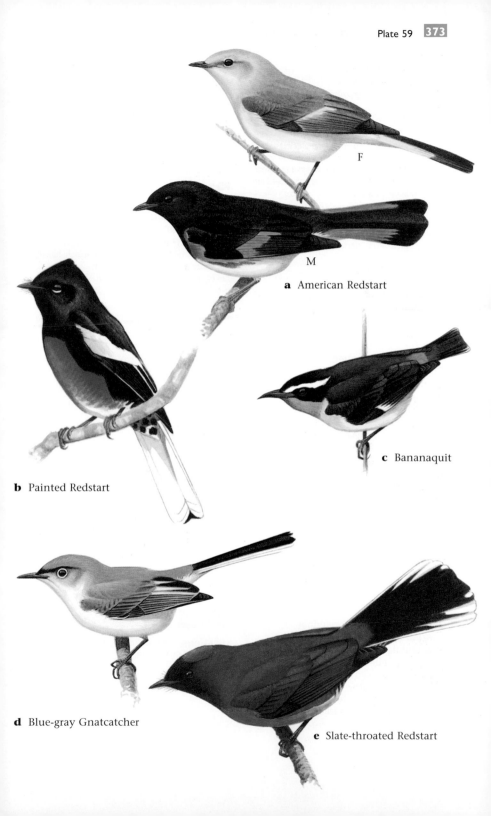

F

M

a American Redstart

b Painted Redstart

c Bananaquit

d Blue-gray Gnatcatcher

e Slate-throated Redstart

Plate 60a
Great-tailed Grackle
Quiscalus mexicanus
Clarinero = clarion-player
Zanate Mayor = large zanate
Zanate Mexicano
ID: Male is largish black bird with purple gloss, especially on head and back, long black tail that folds to V-shape, large black bill, yellowish eye; female is brownish with darker wings and lighter eyestripe, yellowish eye, black bill; male to 43 cm (17 in); female to 33 cm (13 in).

HABITAT: Low, middle, and higher elevations; found in almost all open and semi-open habitats such as open woodlands, tree plantations, gardens, grasslands, pastures, roadsides, towns.

LOCATIONS: OAX-HI, OAX-LO, CHIAP-HI, CHIAP-LO, TAB, CAM, YUC, QRO

Plate 60b
Montezuma's Oropendola
Psarocolius montezuma
Oropéndola de Moctezuma =
Montezuma's oropendola
Zacua Gigante = giant zacua
ID: Large brown bird with black head and chest; yellow-edged tail; large black bill with orange tip; blue patch under eye; to 50 cm (20 in); male larger than female.

HABITAT: Low and middle elevations; found in tree canopy in forest interior and in more open areas such as forest edges, clearings, tree plantations.

LOCATIONS: OAX-HI, OAX-LO, CHIAP-HI, CHIAP-LO, TAB, CAM, QRO

Plate 60c
Chestnut-headed Oropendola
Psarocolius wagleri
Oropéndola Cabecicastaña = chestnut-headed oropendola
Zacua Montañera = mountain zacua
ID: Large dark brown bird with black back and wings; yellow-edged tail; large whitish bill; to 35 cm (14 in); male larger than female.

HABITAT: Low and middle elevation forests; found in tree canopy in forest interior and in semi-open sites such as forest edges, clearings, tree plantations.

LOCATIONS: OAX-HI, OAX-LO, CHIAP-HI, CHIAP-LO, TAB

Plate 60d
Bronzed Cowbird
Molothrus aeneus
Vaquero Ojirrojo = red-eyed cowboy
Tordo Ojirrojo = red-eyed thrush
ID: Mid-sized blackish bird with greenish, bluish, or bronze gloss, black bill, red eyes; female is duller, browner; male especially has conspicuous area of raised feathers on back of neck; to 20 cm (8 in).

HABITAT: Low, middle, and higher elevation open and semi-open sites including open woodlands, agricultural areas, along roads, in town parks, near human settlements.

LOCATIONS: OAX-HI, OAX-LO, CHIAP-HI, CHIAP-LO, TAB, CAM, YUC, QRO

Plate 60e
Melodious Blackbird
Dives dives
Tordo Cantor = singing thrush
Clarinero = clarion-player
ID: Mid-sized all black bird with bluish gloss, pointed black bill, black eyes; female is a bit smaller and a bit duller; to 28 cm (11 in).

HABITAT: Low, middle, and some higher elevation open and semi-open sites with scattered trees such as forest edges, open woodlands, tree plantations.

LOCATIONS: OAX-HI, OAX-LO, CHIAP-HI, CHIAP-LO, TAB, CAM, YUC, QRO

Plate 60 375

a Great-tailed Grackle

F

M

e Melodious Blackbird

d Bronzed Cowbird

b Montezuma's Oropendola

c Chestnut-headed Oropendola

Plate 61a

Yellow-billed Cacique
Amblycercus holosericeus
Pico Blanco = white-bill
Cacique Piquiclaro = light-billed cacique
ID: Mid-sized all black bird with pointed yellow bill and yellow eyes; to 24 cm (9.5 in).

HABITAT: Low and middle elevations; found low in thickets with scattered trees and in open wooded areas such as forest edges and trees near water.

LOCATIONS: OAX-HI, OAX-LO, CHIAP-HI, CHIAP-LO, TAB, CAM, YUC, QRO

Plate 61b

Red-winged Blackbird
Agelaius phoeniceus
Sargento = sergeant
Tordo Sargento = sergeant thrush
Tordo de Charretera = epauletted thrush
Tunkil
ID: Male is mid-sized all black bird with red shoulder patches that can be covered; female is smaller, streaked-brown bird with whitsh/buffy eyestripe, whitish/buffy chest with dark streaks; to 24 cm (9.5 in).

HABITAT: Low and middle elevation marshes and other open sites such as fields, agricultural areas; often found near water.

LOCATIONS: OAX-HI, OAX-LO, CHIAP-HI, CHIAP-LO, TAB, CAM, YUC, QRO

Plate 61c

Black-cowled Oriole
Icterus dominicensis
Bolsero Cabecinegro = black-headed oriole
Bolsero Capucha-negra = black-hooded oriole
Calandria Capucha-negra = black-hooded lark
ID: Mid-sized yellow bird with black head, chest, wings, and tail; yellow shoulder patches can be almost covered; small, pointed, slightly down-curved blackish bill; female resembles male or has olive-yellow head top and back; to 20 cm (8 in).

HABITAT: Low elevation forests and more open areas such as forest edges, clearings, tree plantations.

LOCATIONS: OAX-HI, OAX-LO, CHIAP-HI, CHIAP-LO, TAB, CAM, YUC, QRO

Plate 61d

Yellow-backed Oriole
Icterus chrysater
Bolsero Dorsidorado = golden-backed oriole
Bolsero Espalda Amarilla = yellow-backed oriole
Calandria Espalda Amarilla = yellow-backed lark
Calandria Real = royal lark
ID: Mid-sized yellow bird with black throat, center of chest, and facial area around eyes; black wings and tail; smallish, straight, pointed, blackish bill; female a bit darker; to 23 cm (9 in).

HABITAT: Low, middle, and higher elevations; found in more open wooded areas such as forest edges, clearings, tree plantations.

LOCATIONS: CHIAP-HI, CAM, YUC, QRO

Plate 61e

Yellow-tailed Oriole
Icterus mesomelas
Bolsero Coliamarillo = yellow-tailed oriole
Calandria Coliamarilla = yellow-tailed lark
ID: Mid-sized yellow bird with black throat, center of chest, and facial area below eyes; black wings with yellow shoulder stripe; black tail with yellow edges; very slightly down-curved, pointed, blackish bill; to 23 cm (9 in).

HABITAT: Low elevation forests and forest edges; often found low in brushy areas and thickets.

LOCATIONS: OAX-LO, CHIAP-LO, TAB, CAM, YUC, QRO

Plate 61 377

a Yellow-billed Cacique

M

b Red-winged Blackbird

F

c Black-cowled Oriole

M

F

d Yellow-backed Oriole

e Yellow-tailed Oriole

Plate 62a

Black-vented Oriole
Icterus wagleri
Bolsero de Wagler = Wagler's oriole
ID: Mid-sized black bird with yellow belly, rump, and shoulder patch; orangish tinge between black chest and yellow belly; longish straight bill (sometimes slightly down-curved); to 22 cm (8.5 in).

HABITAT: Middle and high elevation semi-open areas with scattered trees, scrubby woodlands, agricultural sites.

LOCATIONS: OAX-HI, CHIAP-HI

Plate 62b

Audubon's Oriole
(also called Black-headed Oriole)
Icterus graduacauda
Bolsero de Audubon = Audubon's oriole
Bolsero Capucinegro = black-hooded oriole
ID: Mid-sized yellow bird with black head, chest, wings, and tail; shortish straight bill; female is duller overall with yellowish-olive back and, often, whitish edges on wing feathers; to 23 cm (9 in).

HABITAT: Middle and high elevation forests, including pine–oak forests, and semi-open sites such as forest edges, clearings, along waterways, in parks, gardens.

LOCATIONS: OAX-HI

Plate 62c

Hooded Oriole
Icterus cucullatus
Bolsero Cuculado = cuculado oriole
ID: Male is smallish orange bird with black back, throat, mid-chest patch, and eye-mask; black wings with white bars; black tail; slender, slightly down-curved blackish bill; female is yellowish-olive with darker wings and yellow chest/belly; to 19 cm (7.5 in). (Note: Hooded Orioles in Oaxaca, where males may be more yellowish, are seasonal migrants from the North.)

HABITAT: Low elevation open and semi-open sites with scattered trees and bushes such as forest edges, open woodlands, tree plantations.

LOCATIONS: OAX-HI, OAX-LO, CHIAP-LO, TAB, CAM, YUC, QRO

Plate 62d

Orange Oriole
Icterus auratus
Bolsero Yucateco = Yucatán oriole
Calandria Yucateca = Yucatán lark
ID: Mid-sized orangish bird with black in front of eye and center of chest; black wings with white bars and markings; black tail with white-tipped corners; smallish straight bill; female yellowish; to 19 cm (7.5 in).

HABITAT: Low elevation forest edges and semi-open sites with scattered trees, scrubby woodlands.

LOCATIONS: CAM, YUC, QRO

Plate 62e

Altamira Oriole
Icterus gularis
Bolsero Campero = country oriole
Bolsero de Altamira = Altamira oriole
ID: Mid-sized orange or yellowish-orange bird with black back, face, and center of chest; orangish shoulder patch; black wings with white bar and markings; black tail with white-tipped corners; stout straight bill; to 25 cm (10 in).

HABITAT: Low, middle, and some higher elevation open and semi-open sites with scattered trees and shrubs, scrubby woodlands, roadsides, agricultural areas.

LOCATIONS: OAX-HI, OAX-LO, CHIAP-HI, CHIAP-LO, TAB, CAM, YUC, QRO

Plate 62f

Streaked-backed Oriole
Icterus pustulatus
Bolsero Dorsirrayado = streaked-backed oriole
Calandria de Fuego = fire lark
ID: Mid-sized orange bird with black eye area and center of chest; black tail; black wings with white markings; black streaks on back; straight pointed bill; female a bit duller; to 20 cm (8 in). (Note: In some regions, male head and chest are reddish-orange.)

HABITAT: Low, middle, and higher elevation semi-open areas such as forest edges and sites with scattered trees and shrubs, scrubby woodlands, tree plantations

LOCATIONS: OAX-HI, OAX-LO, CHIAP-HI

Plate 62 379

b Audubon's Oriole

a Black-vented Oriole

F

M

c Hooded Oriole

F

M

d Orange Oriole

e Altamira Oriole

f Streaked-backed Oriole

Plate 63a

Red-legged Honeycreeper
Cyanerpes cyaneus
Mielero Patirrojo = red-legged honey-eater
Reinita Azul = little blue queen
ID: Male is small blue bird with black back, wings, tail, and eyestripe; turquoise head top; red legs; longish, thin, down-curved bill; female is green with light eyestripe; yellowish-green chest/belly; 12 cm (4.75 in).

HABITAT: Low, middle, and some higher elevation forests; found in tree canopy usually in more open sites such as forest edges, tree plantations, gardens, along waterways.

LOCATIONS: OAX-HI, OAX-LO, CHIAP-HI, CHIAP-LO, TAB, CAM, YUC, QRO

Plate 63b

Scrub Euphonia
Euphonia affinis
Eufonia Gorjinegra = black-throated euphonia
Tangarilla Gargantinegra = little black-throated tanager
Monjita Gargantinegra = little black-throated nun
ID: Male is very small bluish-black bird with black throat, yellow forehead, and yellow chest/belly; female is olive-greenish with lighter green-to-yellowish chest/belly, grayish on head top; 10 cm (4 in).

HABITAT: Low, middle, and some higher elevations; found in tree canopy in semi-open sites such as forest edges, trees along waterways, and in open habitats with scattered trees.

LOCATIONS: OAX-HI, OAX-LO, CHIAP-HI, CHIAP-LO, TAB, CAM, YUC, QRO

Plate 63c

Yellow-throated Euphonia
Euphonia hirundinacea
Eufonia Gorjiamarilla = yellow-throated euphonia
Tangarilla Gargantimarilla = little yellow-thoated tanager
ID: Male is very small bluish-black bird with yellow forehead and yellow throat, chest, and belly; female is greenish-olive with grayish chest/belly and yellowish sides; 11 cm (4.5 in).

HABITAT: Low, middle, and some higher elevations; found in tree canopy in forests and semi-open sites such as forest edges, clearings, tree plantations.

LOCATIONS: OAX-HI, OAX-LO, CHIAP-HI, CHIAP-LO, TAB, CAM, YUC, QRO

Plate 63d

Blue-crowned Chlorophonia
Clorophonia occipitalis
Chlorofonia Coroniazul = blue-crowned chlorophonia
Tángara Verde = green tanager
ID: Small green bird with bluish patch on head top; small dark bill; grayish legs; male with yellow chest/belly and narrow dark band across chest; female with yellow belly; to 13 cm (5 in).

HABITAT: High elevation forests, including cloud forests, and forest edges; found in tree canopy.

LOCATIONS: OAX-HI, CHIAP-HI

Plate 63 381

a Red-legged Honeycreeper

F

M

F

M

b Scrub Euphonia

F

M

c Yellow-throated Euphonia

F

M

d Blue-crowned Chlorophonia

Plate 64a

Common Bush-tanager
Chlorospingus ophthalmicus
Chinchinero Común
ID: Small greenish-olive bird with gray head, grayish-white throat, yellowish-olive chest, whitish belly; small white patch behind eye; small dark bill; gray legs; to 14 cm (5.5 in). (Note: Birds in Oaxaca have brownish head and buffy throat.)

HABITAT: High elevation forests, including cloud forests, and forest edges; found in trees.

LOCATIONS: OAX-HI, CHIAP-HI

Plate 64b

Gray-headed Tanager
Eucometis penicillata
Tángara Cabecigris = gray-headed tanager
ID: Smallish olive-green bird with gray, slightly ruffled or bushy head and yellow chest/belly; black bill; flesh-colored legs; to 18 cm (7 in).

HABITAT: Low elevation forests; found low in trees, often near ground.

LOCATIONS: OAX-LO, CHIAP-LO, TAB, CAM, QRO

Plate 64c

Yellow-winged Tanager
Thraupis abbas
Tángara Aliamarilla = yellow-winged tanager
Buscahigo = fig-searcher
ID: Smallish bird with bluish/lavender head, grayish-olive chest/belly; black wings with yellow patches; black tail; small black bill; gray legs; to 18 cm (7 in).

HABITAT: Low, middle and some higher elevation open wooded sites such as forest edges, tree plantations, gardens.

LOCATIONS: OAX-HI, OAX-LO, CHIAP-HI, CHIAP-LO, TAB, CAM, YUC, QRO

Plate 64d

Blue-gray Tanager
Thraupis episcopus
Tángara Azuligris = blue-gray tanager
Tángara Azulejo = bluish tanager
Obispillo = little bishop
ID: Small pale blue-gray bird, darker on back, with darker blue wings and tail; small dark bill; female duller; to 16 cm (6.5 in).

HABITAT: Low and middle elevation open and semi-open sites; found in forest edges, tree plantations, and in trees and shrubs in human settlements, town parks.

LOCATIONS: OAX-HI, OAX-LO, CHIAP-HI, CHIAP-LO, TAB, CAM, QRO

Plate 64   383

S. Chiapas form

Oaxaca form

a Common Bush-Tanager

b Gray-headed Tanager

c Yellow-winged Tanager

d Blue-gray Tanager

Plate 65a

Red-crowned Ant-tanager
Habia rubica
Tángara-hormiguera Coronirroja = red-crowned anteater tanager
Tángara Matorralero = thicket tanager
Tángara Hormiguera = anteater tanager
ID: Male is mid-sized dark red bird with reddish-gray belly; bright red patch with black borders on head top usually concealed; blackish bill; brownish legs; female is olive-brownish with lighter chest/belly and usually concealed yellowish/tawny patch on head top; to 20 cm (8 in).

HABITAT: Low and middle elevation forests; found in trees, shrubs, and thickets.

LOCATIONS: OAX-HI, OAX-LO, CHIAP-HI, CHIAP-LO, TAB, CAM, QRO

Plate 65b

Red-throated Ant-tanager
Habia fuscicauda
Tángara-hormiguera Gorjirroja = red-throated ant-eater tanager
Tángara Hormiguera = ant-eater tanager
ID: Male is mid-sized dark red bird with bright red throat, paler red belly; bright red patch on head top usually concealed; very dark between eye and black bill; brownish legs; female is dark reddish-brown with paler chest/belly and yellowish throat; to 21 cm (8.5 in).

HABITAT: Low and middle elevation forests and more open sites such as forest edges; often found in thickets and low in trees.

LOCATIONS: OAX-HI, OAX-LO, CHIAP-HI, CHIAP-LO, TAB, CAM, YUC, QRO

Plate 65c

Crimson-collared Tanager
Ramphocelus sanguinolentus
Tángara Cuellirroja = red-collared tanager
Tángara Huelguista = striker tanager
ID: Smallish black bird with red neck, chest, and head top, and red at base of tail; pale bluish white or silvery white bill; to 19 cm (7.5 in).

HABITAT: Low and middle elevation forests and forest edges; found in tree canopy.

LOCATIONS: OAX-HI, OAX-LO, CHIAP-HI, CHIAP-LO

Plate 65d

Scarlet-rumped Tanager
Ramphocelus passerinii
Tángara Terciopelo = velvet tanager
Rabadilla Escarlata = scarlet-rump
ID: Male is smallish black bird with red rump and light gray or pale blue bill with dark tip; female is brownish or olive with grayish head and orangish/tawny rump, chest, and belly; to 18 cm (7 in).

HABITAT: Low elevation thickets and open wooded areas such as forest edges, tree plantations; also found in trees and shrubs in gardens.

LOCATIONS: OAX-LO, CHIAP-LO, TAB

Plate 65 **385**

a Red-crowned Ant-tanager

b Red-throated Ant-tanager

d Scarlet-rumped Tanager

c Crimson-collared Tanager

Plate 66a

Hepatic Tanager
Piranga flava
Tángara Encinera = oak tanager

ID: Male is mid-sized reddish-gray bird with reddish head top and chest/belly; dark gray bill, legs; small light patch behind eye; female is olive or olive-grayish with olive-yellowish chest belly; to 20 cm (8 in).

HABITAT: Higher elevation forests, including pine and pine–oak forests, and savannahs.

LOCATIONS: OAX-HI, CHIAP-HI

Plate 66b

Summer Tanager
Piranga rubra
Tángara Roja = red tanager
Cardenal Avispero = wasp-eater cardinal

ID: Male is mid-sized bright red bird with slightly darker back and wings; yellowish-beige bill; gray legs; female is yellowish-olive with yellow chest/belly; to 20 cm (8 in).

HABITAT: Low, middle, and higher elevation forests, including pine–oak forests, and more open woodlands, tree plantations, forest edges.

LOCATIONS: OAX-HI, OAX-LO, CHIAP-HI, CHIAP-LO, TAB, CAM, YUC, QRO

Note: This species is a non-breeding seasonal migrant.

Plate 66c

Flame-colored Tanager
Piranga bidentata
Tángara Dorsirrayada = striped-back tanager
Tángara Rayada = striped tanager

ID: Smallish red or red-orange bird with dark patch behind eye; dark streaks on back; dark wings wth whitish bars and markings; dark tail with white-tipped corners; gray bill; female is yellowish with olive back with dark streaks; to 18 cm (7 in).

HABITAT: Middle and high elevation forests, including oak and pine–oak forests, and semi-open sites such as forest edges, tree plantations.

LOCATIONS: OAX-HI, CHIAP-HI

Plate 66d

Rose-throated Tanager
Piranga roseogularis
Tángara Yucateca = Yucatán tanager
Aguacatero = avocado-eater

ID: Male is small gray bird with reddish head top, wings, and tail; pinkish-red throat; white eyering; gray legs and bill; female is olive-yellowish instead of reddish; to 16 cm (6 in).

HABITAT: Low elevation forests and forest edges; found in tree canopy.

LOCATIONS: CAM, QRO

Plate 66 387

M

M

F

d Rose-throated Tanager

M

F

a Hepatic Tanager

M

F

b Summer Tanager

M

F

c Flame-colored Tanager

Plate 67a

Grayish Saltator
Saltator coerulescens
Saltador Grisáceo = gray saltator
ID: Mid-sized olive-gray or bluish-gray bird with grayish-brown to light brown chest/belly; whitish eyestripe; whitish throat with black borders; brownish tail; short, thick, black bill; to 23 cm (9 in).

HABITAT: Low, middle, and some higher elevations; found in semi-open and open sites such as forest edges, tree plantations, parks, gardens.

LOCATIONS: OAX-HI, OAX-LO, CHIAP-HI, CHIAP-LO, TAB, CAM, YUC, QRO

Plate 67b

Black-headed Saltator
Saltator atriceps
Saltador Cabecinegro = black-headed saltator
Saltador de Pechera = bibbed saltator
ID: Mid-sized olive-green bird with blackish head; white eyestripe; white throat with black borders; gray lower chest and belly; short, thick, black bill; to 27 cm (10.5 in).

HABITAT: Low, middle, and some higher elevation forest edges and more open sites such as thickets, brushy pastures.

LOCATIONS: OAX-HI, OAX-LO, CHIAP-HI, CHIAP-LO, TAB, CAM, YUC, QRO

Plate 67c

Northern Cardinal
Cardinalis cardinalis
Cardenal Norteño = northern cardinal
ID: Male is mid-sized red bird with black facial area, red head crest, and reddish bill; female is grayish with darker facial area, reddish head crest and bill, reddish tinge on wings and tail, pale brown/buffy chest/belly; to 22 cm (8.5 in).

HABITAT: Low, middle, and some higher elevation semi-open areas such as forest edges, river edges, scrubby woodlands, thickets, clearings.

LOCATIONS: OAX-HI, OAX-LO, CHIAP-HI, CHIAP-LO, TAB, CAM, YUC, QRO

Plate 67d

Yellow-eyed Junco
Junco phaeonotus
Junco Ojilumbre = fire-eyed junco
Ojilumbre Mexicano = Mexican fire-eye
ID: Small bird with gray head, reddish-brown back, light gray chest, whitish belly; black patch in front of yellow eye; dark wings and tail; white patches on tail seen in flight; bill dark above, light below; to 17 cm (6.5 in). (Note: Birds in Chiapas are a bit darker.)

HABITAT: High elevation forests, including pine and pine–oak forests, in more open sites such as forest edges, clearings; found near the ground.

LOCATIONS: OAX-HI, CHIAP-HI

Plate 67e

White-throated Towhee
Pipilo albicollis
Rascador Oaxaqueño = Oaxacan scratcher
ID: Mid-sized grayish-brown bird with white facial area, whitish chest/belly, and grayish sides; brownish band on throat; dark wings with narrow light bars; dark tail; to 22 cm (8.5 in).

HABITAT: High elevation semi-open sites with scattered trees such as scrubby forest edges, hedgerows in agricultural areas, parks, gardens.

LOCATIONS: OAX-HI

Plate 67f

Collared Towhee
Pipilo ocai
Rascador Collarejo = collared scratcher
ID: Mid-sized olive bird with black face, white eyestripe, and reddish-brown head top; white throat; black chest band; grayish belly; small black bill; to 18 cm (7 in).

HABITAT: Higher elevation forests, including pine and pine–oak forests, forest edges, and semi-open sites with scattered trees; found on or near the ground.

LOCATIONS: OAX-HI

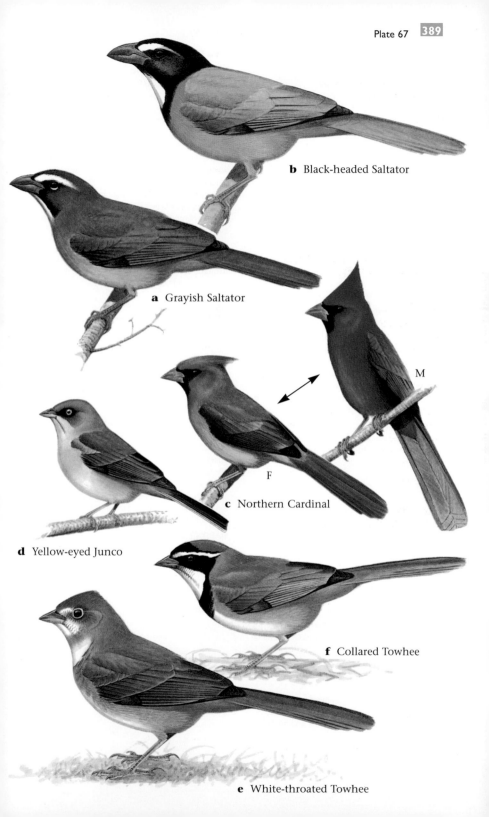

Plate 67   389

b Black-headed Saltator

a Grayish Saltator

M

F

c Northern Cardinal

d Yellow-eyed Junco

f Collared Towhee

e White-throated Towhee

Plate 68a

Green-backed Sparrow
Arremonops chloronotus
Gorrión Dorsiverde = green-backed
sparrow
Gorrión Espalda Verde = green-backed
sparrow
Talero = whip
ID: Small olive-green bird with black and gray striped head; gray chest; whitish throat, belly; small dark bill; flesh-colored legs; 15 cm (6 in).

HABITAT: Low and some middle elevation forests and forest edges; found on or near the ground.

LOCATIONS: CHIAP-LO, TAB, CAM, QRO

Plate 68b

Orange-billed Sparrow
Arremon aurantiirostris
Pico de Oro = golden-bill
Rascador Piquinaranja = orange-billed
scratcher
ID: Small olive-green bird with black, white, and gray striped head, white throat, black chest, whitish belly; grayish sides; small orange bill; 16 cm (6.5 in).

HABITAT: Low and some middle elevation wet forests; found on ground and low in brush in forest interior and more open sites such as forest edges, thickets.

LOCATIONS: OAX-LO, CHIAP-LO, TAB

Plate 68c

Rusty Sparrow
Aimophila rufescens
Zacatanero Rojizo = reddish grass-
eater
Zacatero = grass-eater
ID: Smallish brown bird with black-streaked back; reddish brown on head top; gray neck and facial area; white bar above eye; dark eyestripe and whitish eyering; whitish throat with two black stripes; grayish-brown chest/belly; bill black above, gray below; to 19 cm (7.5 in).

HABITAT: Middle and high elevation forest edges, including in pine–oak forests; also clearings, brushy areas with scattered trees and bushes.

LOCATIONS: OAX-HI, CHIAP-HI

Plate 68d

White-collared Seedeater
Sporophila torqueola
Semillero Collarejo = collared
seedeater
Collarejito = little collar
Dominico
ID: Male is small black bird with white neck and throat; whitish belly and rump; black wings with white bars; black chest band; small black bill; outside breeding season male is olive-brownish with black wings and tail, light brown chest/belly; female is olive-brownish with brown wings and tail, light brown/buffy chest/belly, and brownish bill; to 11 cm (4.5 in).

HABITAT: Low, middle, and some higher elevations; found on or near ground in more open habitats, including brushy forest edges, pastures, gardens, roadsides, marshes.

LOCATIONS: OAX-HI, OAX-LO, CHIAP-HI, CHIAP-LO, TAB, CAM, YUC, QRO

Plate 68e

Bridled Sparrow
Aimophila mystacalis
Zacatonero Bigote-blanco = white-
mustached grass-eater
Zacatonero Patilludo = whiskered
grass-eater
ID: Small bird with gray head with blackish facial area; white stripe below eye and white mark in front of eye; gray chest; white belly; brownish sides; brown back with dark streaks; dark wings with light bars; dark tail and bill; to 16 cm (6 in).

HABITAT: Middle and high elevation semi-open and open scrubby areas with bushes and small trees, thickets, clearings.

LOCATIONS: OAX-HI

Plate 68 **391**

a Green-backed Sparrow

b Orange-billed Sparrow

e Bridled Sparrow

c Rusty Sparrow

F

Non-breeding M

M

d White-collared Seedeater

Plate 69a

Blue-black Grassquit
Volatinia jacarina
Semillero Brincador = hopping seedeater
Marinerito = little mariner
Jaulín Negro = black jaulín
Maromilla

ID: Male is small blue-black bird with smallish black bill and gray legs; outside breeding season male is brown with blue-black wings and tail and has light chest with dark spots; female is brownish with light brown chest/belly with dark streaks; 10 cm (4 in).

HABITAT: Low, middle, and some higher elevation open and semi-open areas including brushy agricultural fields and pastures, roadsides.

LOCATIONS: OAX-HI, OAX-LO, CHIAP-HI, CHIAP-LO, TAB, CAM, YUC, QRO

Plate 69b

Yellow-faced Grassquit
Tiaris olivacea
Semillero Oliváceo = olivaceous seedeater
Mascarita = little mask

ID: Male is small greenish bird with yellow throat and eyestripe; blackish forehead, under eye, and chest; olive-grayish belly; female is olive-colored, duller all over, with hint of yellowish eyestripe and throat; 10 cm (4 in).

HABITAT: Low, middle, and some higher elevations; found in forest edges and open sites such as grasslands, pastures, lawns, fields, roadsides.

LOCATIONS: OAX-HI, OAX-LO, CHIAP-HI, CHIAP-LO, TAB, CAM, YUC, QRO

Plate 69c

Blue-black Grosbeak
Cyanocompsa cyanoides
Picogrueso Negro = black grosbeak
Pico Gordo Bosquero = forest thick-bill
Jaulín Negro = black jaulín

ID: Male is smallish dark blue-black bird with heavy dark bill; female is dark brown with slightly lighter chest/belly; to 18 cm (7 in).

HABITAT: Low and middle elevation forest edges, clearings, agricultural fields adjacent to woodlands.

LOCATIONS: OAX-HI, OAX-LO, CHIAP-HI, CHIAP-LO, TAB, CAM, QRO

Plate 69d

Blue Bunting
Cyanocompsa parellina
Colorín Azulinegro = blue-black bunting
Pico Gordo Acahualero = thick-bill acahualero
Azulejito = little bluebird

ID: Male is small dark blue bird with lighter blue head, small dark bill, and gray legs; female is brown with lighter chest/belly; to 14 cm (5.5 in).

HABITAT: Low, middle, and some higher elevation forests and semi-open sites such as forest edges, scrubby woodlands, thickets, and along waterways.

LOCATIONS: OAX-HI, OAX-LO, CHIAP-HI, CHIAP-LO, TAB, CAM, YUC, QRO

Plate 69e

Blue Grosbeak
Passerina caerulea
Picogrueso Azul = blue grosbeak
Pico Gordo Azul = blue thick-bill

ID: Male is smallish blue bird with black facial area, blackish wings with brown bars, and black tail; female is brown with streaked back, dark wings with lighter bars, bluish shoulder markings, and light throat; to 17 cm (6.5 in).

HABITAT: Low, middle, and high elevation semi-open sites such as forest edges, scrubby areas with scattered trees, thickets, agricultural areas, roadsides.

LOCATIONS: OAX-HI, OAX-LO, CHIAP-HI, CHIAP-LO, TAB, CAM, YUC, QRO

Plate 69 **393**

a Blue-black Grassquit

Non-breeding M

b Yellow-faced Grassquit

M

F

F

M

F

M

c Blue-black Grosbeak

M

F

e Blue Grosbeak

M

F

d Blue Bunting

Plate 70a

Common Opossum
Didelphis marsupialis
Zarigüeya = opossum
Zorro = fox
Tlacuache

ID: A large opossum with yellowish face; blackish or gray back; black ears; yellowish cheek area; to 46 cm (18 in), plus long, almost hairless tail. (Note: Virginia Opossum, *Didelphis virginiana*, is also common in the region and occupies both warm and cool sites; it closely resembles Common Opossum but has whitish cheeks.)

HABITAT: Warm areas, regionwide, in trees and on the ground; common near human settlements; nocturnal.

LOCATIONS: OAX-HI, OAX-LO, CHIAP-HI, CHIAP-LO, TAB, CAM, YUC, QRO

Plate 70b

Central American Woolly Opossum
Caluromys derbianus
Tlacuachillo Dorado = golden tlacuachillo
Tlacuachillo Lanudo = woolly tlacuachillo
Zorrito Platanero = little banana tree fox
Zorrito Cacaotero = little cocoa tree fox

ID: A smaller, very furry opossum, reddish-brown with a grayish face; dark stripe from nose to forehead; gray patch on back; light-colored front feet; last half of tail is hairless; to 30 cm (12 in), plus long tail.

HABITAT: Low, middle and higher elevation wet and dry forests, tree plantations; nocturnal; arboreal.

LOCATIONS: OAX-HI, OAX-LO, CHIAP-LO, TAB

Plate 70c

Water Opossum
Chironectes minimus
Tlacuachillo de Agua = water tlacuachillo
Tlacuache de Agua = water tlacuache
Zorro de Agua = water fox
Perrito de Agua = little water dog

ID: A smaller opossum with webbed toes on rear feet; gray back with broad black/brown bands and a black stripe down its center; head blackish; cheeks and throat whitish; to 30 cm (12 in), plus tail.

HABITAT: Low and middle elevation wet and dry forests and cleared areas; found in water or on ground along fast-flowing watercourses; nocturnal.

LOCATIONS: OAX-HI, OAX-LO, CHIAP-LO, TAB

Plate 70d

Gray Four-eyed Opossum
Philander opossum
Tlacuachillo Cuatro Ojos = four-eyed tlacuachillo
Zorrito Platanero = little banana tree fox
Cuatro Ojos = four-eyes
Tlacuachín

ID: Smaller opossum with gray back and lighter-colored throat, chest, and belly; black face mask; black on top of head; a white mark above each eye; to 33 cm (13 in), plus hairless tail.

HABITAT: Low and middle elevation forests, plantations and other agricultural areas; prefers dense vegetation near water; nocturnal; found in trees and on the ground.

LOCATIONS: OAX-HI, OAX-LO, CHIAP-HI, CHIAP-LO, TAB

Plate 70e

Mexican Mouse Opossum
Marmosa mexicana
Ratón Tlacuache = mouse tlacuache
Zorrito Enano = dwarf fox

ID: Small brown, reddish-brown, or yellowish-gray opossum with blackish eyerings; light belly; light-colored feet; long, mostly naked tail; to 15 cm (6 in), plus tail.

HABITAT: Low, middle and some higher elevation forests and more open sites such as tree plantations and savannahs; found on ground or low in trees; night-active.

LOCATIONS: OAX-HI, OAX-LO, CHIAP-HI, CHIAP-LO, TAB, CAM, YUC, QRO

Plate 70 395

a Common Opossum

b Central American Woolly Opossum

d Gray Four-eyed Opossum

e Mexican Mouse Opossum

c Water Opossum

Plate 71a

Fishing Bat
(also called Greater Bulldog Bat)
Noctilio leporinus
Murciélago Pescador = fishing bat

ID: A large brown or reddish bat; usually one light stripe on back; forward-pointing ears; lips split in the middle, with drooping folds of skin, like a bulldog; no noseleaf; head and body length to 10 cm (4 in), plus tail; wingspan to 60 cm (2 ft).

HABITAT: Lowland forests, pastures, plantations, near water; roosts in hollow trees, caves; forages over water for small fish.

LOCATIONS: OAX-LO, CHIAP-LO, TAB, CAM, YUC, QRO

Plate 71b

Woolly False Vampire Bat
Chrotopterus auritus
Falso Vampiro = false vampire
Chinacate

ID: A large grayish-brown bat with long fur, very large ears, largish noseleaf, and white wing tips; head and body length to 11 cm (4 in); little or no tail.

HABITAT: Low and middle elevation wet and dry forests, plantations; roosts in hollow trees, caves.

LOCATIONS: OAX-HI, OAX-LO, CHIAP-HI, CHIAP-LO, TAB, CAM, YUC, QRO

Plate 71c

Short-tailed Fruit Bat
(also called Seba's Short-tailed Bat)
Carollia perspicillata
Chinaca de Cola Corta = short-tailed chinaca

ID: A small brown or gray bat with short, narrow snout with relatively large, spear-shaped noseleaf; small warts on chin; triangular, pointed ears; head and body length to 6.5 cm (2.5 in).

HABITAT: Low and middle elevation forests, gardens, agricultural areas; roosts in trees, caves, riverbanks, buildings; prefers moist areas; feeds at fruit sources, flowers (especially at the piper plant's candle-like flowers).

LOCATIONS: OAX-HI, OAX-LO, CHIAP-HI, CHIAP-LO, TAB, CAM, YUC, QRO

Plate 71d

Nectar Bat
(also called Common Long-tongued Bat)
Glossophaga soricina
Murciélago Mielero = honey bat

ID: A small brown or gray bat with a long snout; small, spear-shaped noseleaf; short, blunt ears; lower lip with a V-shaped notch lined with small bumps; head and body length to 6.5 cm (2.5 in), plus short tail; wingspan to 28 cm (11 in).

HABITAT: Low and middle elevation wet and dry forests and open areas; roosts in caves, trees, bridges, buildings; forages within forest but also in open areas, dry stream beds, plantations, farms; hovers at flowers to feed on nectar.

LOCATIONS: OAX-HI, OAX-LO, CHIAP-HI, CHIAP-LO, TAB, CAM, YUC, QRO

Plate 71e

Broad-eared Free-tailed Bat
Nyctinomops laticaudatus
Murciélago de Cola Larga = long-tailed bat
Murciélago Coludo = tailed bat

ID: Mid-sized brown or reddish-brown bat with lighter, often buff-colored belly; slightly upturned nose and snout; no noseleaf; very large ears; wings almost transparent; head and body length to 9 cm (3.5 in).

HABITAT: Low elevation forests, forest edges, and more open areas; roosts in trees, caves, rocks, buildings; forages often over water, wetlands for insects.

LOCATIONS: OAX-HI, OAX-LO, CHIAP-HI, CHIAP-LO, TAB, CAM, YUC, QRO

Plate 71 397

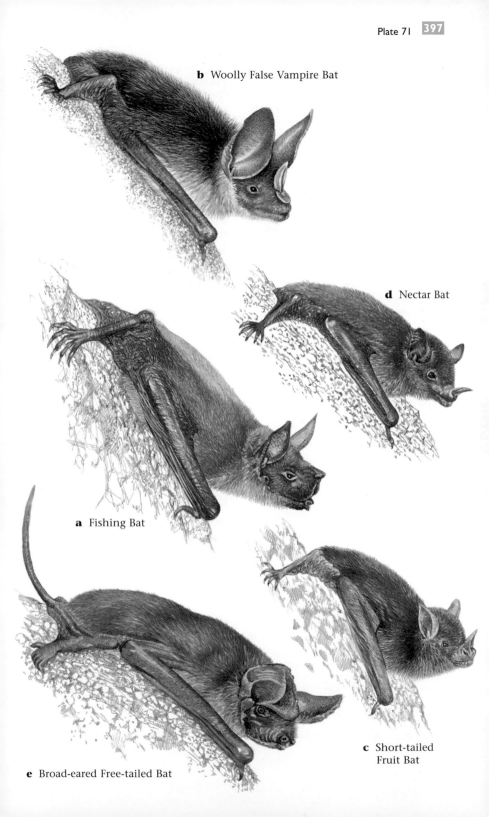

b Woolly False Vampire Bat

d Nectar Bat

a Fishing Bat

c Short-tailed Fruit Bat

e Broad-eared Free-tailed Bat

Plate 72a

Black Mastiff Bat
Molossus ater
Murciélago Coludo = tailed bat

ID: Mid-sized black or dark brown bat with black wings, short fur; no noseleaf; low, rounded ears; hairy feet; head and body length to 9 cm (3.5 in), plus longish, naked tail.

HABITAT: Lowland wet and dry forests and around human settlements; roosts in trees, rock crevices, buildings.

LOCATIONS: OAX-HI, OAX-LO, CHIAP-HI, CHIAP-LO, TAB, CAM, YUC, QRO

Plate 72b

Hairy-legged Bat
Myotis keaysi
Chinaca Café = coffee chinaca

ID: Tiny dark brown bat with paler underparts; small pointed snout without noseleaf; triangular, pointed ears; head and body length to 5 cm (2 in), plus tail.

HABITAT: Low and middle elevation forests, gardens, parks, agricultural sites; roosts in trees, rock crevices, buildings.

LOCATIONS: OAX-HI, OAX-LO, CHIAP-HI, CHIAP-LO, TAB, CAM, YUC, QRO

Plate 72c

Jamaican Fruit-eating Bat
Artibeus jamaicensis
Murciélago Higuero = fig bat
Murciélago Come Frutas = fruit-eating bat

ID: A large, stout-bodied bat, black, brown, or grayish; large head usually with light stripes on face; short, broad snout with spear-shaped noseleaf; V-shaped row of small bumps on chin; head and body length to 10 cm (4 in); wingspan to 40 cm (16 in).

HABITAT: Low and middle elevation wet and dry forests, gardens, plantations; roosts in caves, hollow trees, under palm leaves; feeds on fruit.

LOCATIONS: OAX-HI, OAX-LO, CHIAP-HI, CHIAP-LO, TAB, CAM, YUC, QRO

Plate 72d

Common Vampire Bat
Desmodus rotundus
Vampiro Patas Pelonas = bare-legged vampire bat
Vampiro = vampire bat

ID: A mid-sized, dark brown bat with shiny fur; tips of hair on back often silvery white; short snout with U-shaped, fleshy, skin folds; large, sharp middle incisor and canine teeth; triangular, pointed ears; no tail; head and body length to 9 cm (3.5 in).

HABITAT: Low and middle elevation wet and dry forests, clearings, farm areas; roosts in trees, caves; nocturnal; often flies along riverbeds.

LOCATIONS: OAX-HI, OAX-LO, CHIAP-HI, CHIAP-LO, TAB, CAM, YUC, QRO

Plate 72e

Mexican Funnel-eared Bat
Natalus stramineus
Murciélago Canela = cinnamon bat

ID: Very small pale brown, reddish-brown, or buff-colored bat; belly usually paler; longish narrow snout; no noseleaf; funnel-shaped, triangular ears, small eyes; head and body length to 6.5 cm (2.5 in).

HABITAT: Low and middle elevation wet and dry forests and semi-open sites such as scrub forests and tree plantations; cave roosts.

LOCATIONS: OAX-HI, OAX-LO, CHIAP-HI, CHIAP-LO, TAB, CAM, YUC, QRO

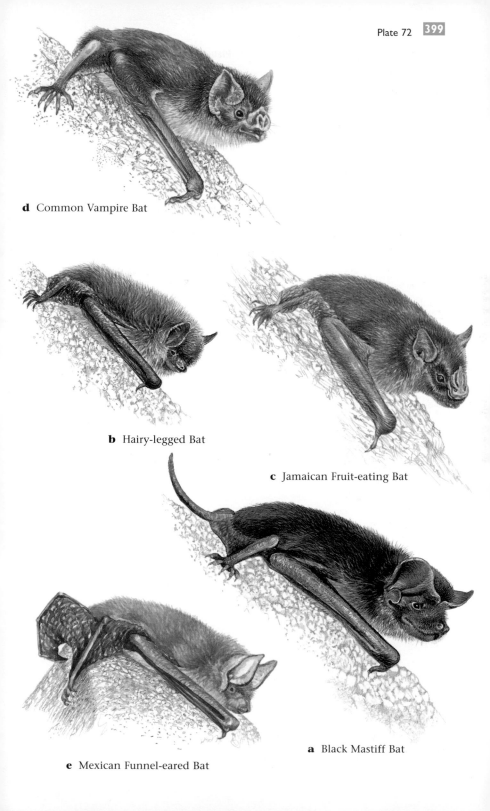

Plate 72 **399**

d Common Vampire Bat

b Hairy-legged Bat

c Jamaican Fruit-eating Bat

a Black Mastiff Bat

e Mexican Funnel-eared Bat

Plate 73a

Yucatán Black Howler Monkey
Alouatta pigra
Mono Aullador = howling monkey
Saraguato

ID: Large, long-haired, all-black monkey; to 55 cm (22 in), plus long prehensile tail; males larger than females.

HABITAT: Lowland wet and dry forests; arboreal, usually in upper reaches of trees.

LOCATIONS: CHIAP-LO, TAB, CAM, QRO

Note: This species listed as endangered, CITES Appendix I, and as threatened, USA ESA.

Plate 73b

Mantled Howler Monkey
Alouatta palliata
Mono Aullador = howling monkey
Saraguato

ID: Large black monkey with a more-or-less noticeable fringe (mantle) of pale or chestnut hair on back, sides, belly; to 55 cm (22 in), plus long prehensile tail; males larger than females.

HABITAT: Lowland wet and dry forests; arboreal, usually in upper reaches of trees.

LOCATIONS: CHIAP-LO, TAB, CAM

Note: This species listed as endangered, CITES Appendix I and USA ESA.

Plate 73c

Central American Spider Monkey
(also called Geoffroy's Spider Monkey)
Ateles geoffroyi
Mono Araña = spider monkey
Chango = monkey
Mico = monkey

ID: A large monkey, brown, chestnut, or silvery; lighter-colored belly; dark or black lower legs, feet, hands, and forearms; pale skin around eyes and nose; to 63 cm (25 in), plus long prehensile tail.

HABITAT: Low and, seasonally, some middle elevation wet and dry forests; arboreal.

LOCATIONS: OAX-LO, CHIAP-HI, CHIAP-LO, TAB, CAM, YUC, QRO

Note: This species is endangered, CITES Appendix I listed.

Plate 73   401

a Yucatán Black Howler Monkey

b Mantled Howler Monkey

c Central American Spider Monkey

Plate 74a

Silky Anteater
(also called Pygmy Anteater)
Cyclopes didactylus
Hormiguero Dorado = golden anteater
Miquito de Oro = small golden monkey
Mico Platanero = banana tree monkey

ID: Small anteater with gray, brownish, or yellowish dense, silky fur; darker on top, with a dark line running from top of head along the back; to 18 cm (7 in), plus thick prehensile tail as long as or longer than body.

HABITAT: Low and middle elevation wet forests; nocturnal; arboreal; especially found among vines and thin tree branches.

LOCATIONS: CHIAP-LO, TAB, CAM, QRO

Plate 74b

Northern Tamandua
Tamandua mexicana
Oso Hormiguero = anteater bear
Oso Hormiguero de Collar = collared anteater bear
Hormigero = anteater
Chupamiel = honey sucker

ID: Mid-sized anteater with long, pointed snout; brown or yellowish and legs with a black "vest" on belly and back that encircles body; 47 to 77 cm (18 to 30 in) long, plus tail as long as body; last section of tail is bare.

HABITAT: Low and middle elevation forests, and more open sites such as forest edges, tree plantations; active day and night; found in trees and on the ground.

LOCATIONS: OAX-HI, OAX-LO, CHIAP-HI, CHIAP-LO, TAB, CAM, YUC, QRO

Plate 74c

Nine-banded Armadillo
Dasypus novemcinctus
Armadillo
Ayotochtli
Mulita = little mule

ID: Gray to yellowish body; hairless back consisting of hard, bony plates; about nine movable bands in midsection; long snout; large ears; scales on head, legs; to 57 cm (22 in), plus long, ringed tail.

HABITAT: Forests, scrub areas, thickets, and grasslands; nocturnal; terrestrial.

LOCATIONS: OAX-HI, OAX-LO, CHIAP-HI, CHIAP-LO, TAB, CAM, YUC, QRO

Plate 74d

Paca
Agouti paca
Tepezcuintle
Tuza Real = royal tuza

ID: Large, pig-like rodent; brown or blackish with horizontal rows of whitish spots on sides; 60 to 80 cm (24 to 31 in), plus tiny tail.

HABITAT: Low and middle elevation wet forests and drier areas near water; nocturnal; found on the ground.

LOCATIONS: OAX-LO, CHIAP-HI, CHIAP-LO, TAB, CAM, YUC, QRO

Plate 74e

Central American Agouti
Dasyprocta punctata
Guaqueque Alazán = chestnut guaqueque
Guaqueque
Agutí

ID: Large, pig-like rodent; reddish-brown, brown, or blackish back and sides; 40 to 62 cm (16 to 24 in), plus tiny tail. (Note: A separate species, the Mexican Black Agouti, *Dasyprocta mexicana*, always black or very dark brown, occurs in lowlands of Tabasco, northern Oaxaca, and northern Chiapas.)

HABITAT: Low and middle elevation forests, plantations, gardens; diurnal; found on the ground.

LOCATIONS: CHIAP-HI, CHIAP-LO, CAM, YUC, QRO

Plate 74   403

b Northern Tamandua

a Silky Anteater

e Central American Agouti

c Nine-banded Armadillo

d Paca

Plate 75a

Mexican Hairy Porcupine (also called Prehensile-tailed Porcupine)
Sphiggurus mexicanus
Puercoespín = porcupine
Zorro Espín = spiny fox

ID: Small brown or black porcupine with long prehensile tail bare at the end; spines ("quills") are largely hidden by long, soft hair; usually 30 to 40 cm (12 to 16 in), plus tail.

HABITAT: Low, middle, and some higher elevation forests and more open areas; nocturnal; mostly arboreal but also found on the ground.

LOCATIONS: OAX-HI, OAX-LO, CHIAP-HI, CHIAP-LO, TAB, CAM, YUC, QRO

Plate 75b

Deppe's Squirrel
Sciurus deppei
Ardilla Montañera = mountain squirrel
Ardillita Selvática = small jungle squirrel
Ardilla de Deppe = Deppe's squirrel
Ardilla Gris = gray squirrel
Moto

ID: Small brownish or grayish squirrel with grizzled appearance; light belly; gray front legs and shoulders; black and white "frosted" tail; to 22 cm (9 in), plus 19 cm (7 in) tail.

HABITAT: Low, middle, and higher elevation forests; found in trees and on the ground.

LOCATIONS: OAX-LO, CHIAP-HI, CHIAP-LO, TAB, CAM, YUC, QRO

Plate 75c

Yucatán Squirrel
Sciuris yucatanensis
Ardilla Arborícola = tree squirrel
Ardilla Gris = gray squirrel
Ardilla = squirrel

ID: Mid-sized blackish or grayish squirrel with grizzled appearance; light belly; black and white "frosted" tail; dark or gray feet; to 26 cm (10 in), plus 23 cm (9 in) tail.

HABITAT: Lowland wet and dry forests, tree plantations; arboreal.

LOCATIONS: CHIAP-LO, TAB, CAM, YUC, QRO

Plate 75d

Red-bellied Squirrel (also called Mexican Gray Squirrel)
Sciurus aureogaster
Ardilla Gris = gray squirrel
Ardilla Negra = black squirrel (black individuals)
Ardilla = squirrel

ID: Mid-sized gray, black, or brownish squirrel, highly variable; often with light, orangish, or reddish-brown belly; long black and white "frosted" tail, often with reddish-brown stripe underneath; some individuals are all black; to 30 cm (12 in), plus 30 cm (12 in) tail.

HABITAT: Low, middle, and high elevation forests, including pine–oak forests; found in trees.

LOCATIONS: OAX-HI, OAX-LO, CHIAP-HI, CHIAP-LO, TAB

Plate 75e

Flying Squirrel
Glaucomys volans
Ardilla Voladora = flying squirrel
Ardilla Planeadora = gliding squirrel

ID: Small gray or grayish-brown squirrel with whitish belly; folded, loose skin conspicuous between arm and leg; to 14 cm (5.5 in), plus long tail.

HABITAT: Pine and pine–oak forests; found in trees; dusk- and night-active.

LOCATIONS: OAX-HI, CHIAP-HI

Plate 75   405

c Yucatán Squirrel

b Deppe's Squirrel

d Red-bellied Squirrel →

e Flying Squirrel

a Mexican Hairy Porcupine

Plate 76a

Jaguarundi
Herpailurus yaguarondi
Leoncillo = little lion
León Miquero = monkey lion
Onza = wildcat
Yaguarundi

ID: Smallish to mid-sized slender cat without spots; gray, brown, or reddish; to 50 to 65+ cm (20 to 25+ in), plus long tail.

HABITAT: Wet and dry forests, regionwide; active day or night; usually seen on the ground, but also climbs.

LOCATIONS: OAX-HI, OAX-LO, CHIAP-HI, CHIAP-LO, TAB, CAM, YUC, QRO

Note: This species is endangered, CITES Appendix I and USA ESA listed.

Plate 76b

Ocelot
Leopardus pardalis
Ocelote = ocelot
Tigrillo = little tiger
Tigre Cangrejero = crab-eating tiger

ID: Medium-sized yellow/tawny cat with black spots and lines; tail shorter than rear leg; to 70 to 85 cm (28 to 34 in), plus tail.

HABITAT: Low and middle elevation forests and more open sites such as forest edges and scrub areas, regionwide; mostly nocturnal; found on the ground or in trees (where it sleeps).

LOCATIONS: OAX-HI, OAX-LO, CHIAP-HI, CHIAP-LO, TAB, CAM, YUC, QRO

Note: This species is endangered, CITES Appendix I and USA ESA listed.

Plate 76c

Margay
Leopardus wiedii
Tigrillo = little tiger
Gato Tigre = tiger cat
Pichigueta

ID: Small to mid-sized yellowish, tawny, or brownish gray cat with black spots and lines; tail longer than rear leg; to 50 to 70 cm (20 to 28 in), plus tail.

HABITAT: Low and middle elevation forests, regionwide; nocturnal; found mostly in trees but also on the ground.

LOCATIONS: OAX-HI, OAX-LO, CHIAP-HI, CHIAP-LO, TAB, CAM, YUC, QRO

Note: This species is endangered, CITES Appendix I and USA ESA listed.

Plate 76d

Jaguar
Panthera onca
Tigre = tiger
Tigre Real = royal tiger

ID: A large or very large cat, yellowish/tawny with black spots; to 1.1 to 1.8 m (3.5 to 6 ft), plus tail; you will know it when you see it.

HABITAT: Low, middle, and higher elevation forests and semi-open areas, regionwide; active day or night.

LOCATIONS: OAX-HI, OAX-LO, CHIAP-HI, CHIAP-LO, TAB, CAM, YUC, QRO

Note: This species is endangered, CITES Appendix I and USA ESA listed.

Plate 76e

Puma
Puma concolor
León del Montaña = mountain lion
León = lion

ID: A large cat, brownish, reddish-brown, or tawny; white throat and around mouth; 0.9 to 1.5 m (3 to 5 ft), plus long, dark-tipped tail.

HABITAT: Low, middle, and higher elevation forests and semi-open sites such as forest edges.

LOCATIONS: OAX-HI, OAX-LO, CHIAP-HI, CHIAP-LO, TAB, CAM, YUC, QRO

Note: This species is threatened throughout its range, USA ESA listed.

Plate 76 407

a Jaguarundi

b Ocelot

c Margay

d Jaguar

e Puma

Plate 77a

Coyote
Canis latrans
Perro de Monte = forest dog
Coyote
ID: Medium-sized and dog-like; coat color varies from grayish to yellow-beige, often with "grizzled" or salt and pepper appearance; to about 75 cm (30 in) long, plus tail (sometimes black-tipped).

HABITAT: Low, middle, and higher elevation semi-open and open areas, including coniferous forests, grasslands, and agricultural sites; active day or night; often seen during early morning.

LOCATIONS: OAX-HI, OAX-LO, CHIAP-HI, CHIAP-LO, TAB, CAM, YUC, QRO

Plate 77b

Gray Fox
Urocyon cinereoargenteus
Zorro Gris = gray fox
Gato de Monte = forest cat
Zorro = fox
ID: Small, silver-gray, dog-like mammal, with reddish ears and shoulders; to 52 to 64 cm (20 to 24 in), plus tail.

HABITAT: Low and middle elevation forests and more open areas, especially forest edges and around wetlands; active day or night; often seen during early morning; found on ground and in trees.

LOCATIONS: OAX-HI, OAX-LO, CHIAP-HI, CHIAP-LO, TAB, CAM, YUC, QRO

Plate 77c

Hooded Skunk
Mephitis macroura
Zorrillo Rayado = lined skunk
Zorrillo Listado = striped skunk
ID: Black skunk with white back, or black with narrow white stripes on each side; very long bushy tail white, black and white, or mostly black; to 38 cm (15 in), plus tail.

HABITAT: Low, middle, and higher elevation semi-open and open sites, particularly around waterways, brushland, agricultural areas; active at dusk and night.

LOCATIONS: OAX-HI, OAX-LO, CHIAP-HI, CHIAP-LO, TAB

Plate 77d

Hog-nosed Skunk
Conepatus mesoleucus
Zorrillo de Espalda Blanca = white-backed skunk
Zorrillo = skunk
ID: Black long-snouted skunk with white back; white usually continues onto bushy tail; 30 to 50 cm (12 to 20 in), plus tail.

HABITAT: Low, middle, and high elevation forests and more open sites such as gardens, agricultural areas; nocturnal; found on the ground.

LOCATIONS: OAX-HI, OAX-LO, CHIAP-HI, CHIAP-LO

Plate 77e

Striped Hog-nosed Skunk
Conepatus semistriatus
Zorrillo de Espalda Blanca = white-backed skunk
Zorrillo Pijón = pijón skunk
Zorrillo = skunk
ID: Black or dark brown skunk with wide white stripe on top of head and neck, dividing to two white stripes along back; bushy white tail; longish bare snout; 30 to 49 cm (12 to 19 in), plus tail.

HABITAT: Low, middle, and high elevation forests and more open sites such as gardens, agricultural areas; found on the ground; nocturnal.

LOCATIONS: OAX-LO, CHIAP-HI, CHIAP-LO, TAB, CAM, YUC, QRO

Plate 77 409

c Hooded Skunk

d Hog-nosed Skunk

e Striped Hog-nosed Skunk

b Gray Fox

a Coyote

Plate 78a

Tayra
Eira barbara
Viejo de Monte = old mountain man
Cabeza de Viejo = old man's head
ID: Medium-sized weasel-like mammal, resembling a large mink; black or brown body; tan, brown, or yellowish head and neck; often a large yellowish spot on throat/chest; to 52 to 70 cm (20 to 28 in), plus long densely furred tail.

HABITAT: Low and middle elevation wet forests, drier areas, gardens, and agricultural sites; found in trees or on the ground; day- and dusk-active.

LOCATIONS: OAX-HI, OAX-LO, CHIAP-HI, CHIAP-LO, TAB, CAM, YUC, QRO

Plate 78b

Neotropical Otter
(also called River Otter)
Lontra longicaudis
Nutria = otter
Perro de Agua = water dog
ID: Short-legged, short-haired brown mammal with whitish throat and belly; first half of tail very thick; webbed feet; 45 to 75 cm (18 to 30 in) long, plus tail.

HABITAT: Found in and around rivers and streams, regionwide; active day and night.

LOCATIONS: OAX-HI, OAX-LO, CHIAP-HI, CHIAP-LO, TAB, CAM, YUC, QRO

Note: This species is endangered, CITES Appendix I and USA ESA listed.

Plate 78c

Grison
(also called Huron)
Galictis vittata
Grisón
Hurón
Rey de las Ardillas = king of the squirrels
ID: A weasel-like mammal with short legs; grayish or "grizzled" above and on sides; black muzzle, throat, chest, and limbs; 45 to 55 cm (18 to 21 in), plus short tail.

HABITAT: Low and middle elevation forests, especially near waterways, regionwide; night- and morning-active; found on the ground.

LOCATIONS: OAX-HI, OAX-LO, CHIAP-HI, CHIAP-LO, TAB, CAM, YUC, QRO

Plate 78d

Long-tailed Weasel
Mustela frenata
Comadreja = weasel
Onza, Oncita, Oncilla = wildcat
Saben, Sebencito
ID: Smallish brown weasel with cream-colored throat, chest, and belly; white stripe through eye area; tip of tail black; 18 to 30 cm (7 to 12 in), plus tail.

HABITAT: Low, middle, and higher elevation forests, drier cleared areas, and agricultural areas; active day or night; usually found on the ground but also climbs.

LOCATIONS: OAX-HI, OAX-LO, CHIAP-HI, CHIAP-LO, TAB, CAM, YUC, QRO

Plate 78   411

a Tayra

b Neotropical Otter

c Grison

d Long-tailed Weasel

Plate 79a

Northern Raccoon
Procyon lotor
Mapache = raccoon
Osito Lavador = little washer bear
Tejón Solitaria = solitary badger

ID: Gray-black back with "grizzled" appearance; whitish face with black mask; gray forearms and thighs; light gray/white feet; pointed muzzle; 45 to 64 cm (18 to 25 in), plus ringed tail. (Note: The Pygmy Raccoon, endemic to Cozumel Island, closely resembles the Northern Raccoon, but is smaller.)

HABITAT: Low and middle elevation forests and open areas, often near water; nocturnal; found on the ground and in trees.

LOCATIONS: OAX-HI, OAX-LO, CHIAP-HI, CHIAP-LO, TAB, CAM, YUC, QRO

Plate 79b

White-nosed Coati
Nasua narica
Coatí
Tejón = badger
Tejón de Manada = pack badger
Pisote
Cuchucho

ID: Light-, dark-, or reddish-brown raccoon-like mammal with grayish or yellowish shoulders; white muzzle, chin, and throat; 45 to 70 cm (18 to 27 in), plus very long, faintly ringed tail. (Note: Cozumel Island Coatis are smaller.)

HABITAT: Low and middle elevation wet and dry forests and forest edge areas, regionwide; day active; found on the ground and in trees.

LOCATIONS: OAX-HI, OAX-LO, CHIAP-HI, CHIAP-LO, TAB, CAM, YUC, QRO

Plate 79c

Kinkajou
Potos flavus
Martucha
Micoleón = lion monkey
Mico de Noche = night monkey

ID: Grayish- or reddish-brown short-haired mammal, sometimes with a darker stripe along back; roundish head with short muzzle; short legs; 40 to 55 cm (16 to 22 in), plus very long, prehensile tail.

HABITAT: Low and middle elevation forests, tree plantations; nocturnal; found in trees.

LOCATIONS: OAX-HI, OAX-LO, CHIAP-HI, CHIAP-LO, TAB, CAM, YUC, QRO

Plate 79d

Cacomistle
Bassariscus sumichrasti
Cacomixtle
Mico Rayado = striped monkey
Goyo
Wilo
Babisuri

ID: A light brown or tawny-brown densely furred mammal with dark stripe down back; dark facial mask surrounding yellowish or whitish rings around eyes; dark lower legs and feet; 39 to 47 cm (15 to 18 in), plus long, bushy, ringed tail.

HABITAT: Low and middle elevation forests; nocturnal; found in middle and upper levels of trees.

LOCATIONS: OAX-HI, OAX-LO, CHIAP-HI, CHIAP-LO, TAB, CAM, QRO

Plate 79 413

c Kinkajou

d Cacomistle

a Northern Raccoon

b White-nosed Coati

Plate 80 (*See also*: Peccaries, Deer, p. 230; Tapir, p. 232)

Plate 80a

Collared Peccary
Tayassu tajacu
Pecari de Collar = collared peccary
Jabalí de Collar = collared wild boar
Puerco de Monte = mountain pig
ID: Gray or blackish pig-like mammal with long, coarse hair; white or yellowish "collar" around shoulders; to 80 to 92 cm (31 to 36 in).

HABITAT: Low and middle elevation wet and dry forests, agricultural areas, regionwide; day active.

LOCATIONS: OAX-HI, OAX-LO, CHIAP-HI, CHIAP-LO, TAB, CAM, YUC, QRO

Note: This species regulated for conservation purposes, CITES Appendix II listed.

Plate 80b

White-lipped Peccary
Tayassu pecari
Pecari de Labios Blancos = white-lipped peccary
Jabalí de Labios = lipped wild boar
Senso
Marina
ID: Black or brown pig-like mammal with long, coarse hair; white lower cheek and throat; to 92 to 110 cm (36 to 43 in).

HABITAT: Low and middle elevation wet and dry forests; day active.

LOCATIONS: OAX-HI, OAX-LO, CHIAP-LO, TAB, CAM

Note: This species regulated for conservation purposes, CITES Appendix II listed.

Plate 80c

White-tailed Deer
Odocoileus virginianus
Venado Cola Blanca = white-tailed deer
Venado de Llano = field deer
Venado de Campo = field deer
Venado Saltón = jumping deer
ID: Mid-sized light-, dark-, or grayish-brown deer with white belly, white under tail, and, often, white chin/throat; males with branched antlers; 1 to 1.8 m (3 to 6 ft) long; about 1 m (3 ft) high at shoulders.

HABITAT: Low, middle, and higher elevation forest edges (inlcuding pine and pine–oak forests) and more open areas, such as pastures and grasslands; active day or night.

LOCATIONS: OAX-HI, OAX-LO, CHIAP-HI, CHIAP-LO, TAB, CAM, YUC, QRO

Plate 80d

Red Brocket Deer
Mazama americana
Temazate
Venado Cabrito = little goat deer
Venado de Montaña = mountain deer
Venadito Rojo = small red deer
ID: Small reddish or reddish-brown deer with dark brown head and neck, and brownish belly; white under tail; males with small, straight antlers; 1 to 1.4 m (3.4 to 4.6 ft); 70 cm (28 in) high at shoulders.

HABITAT: Low, middle and some higher elevation wet forests, forest edge areas, and plantations; active day or night.

LOCATIONS: OAX-HI, OAX-LO, CHIAP-HI, CHIAP-LO, TAB, CAM, QRO

Plate 80e

Baird's Tapir
(also called Central American Tapir)
Tapirus bairdii
Danta
Anteburro
ID: Large mammal, brownish, black, or grayish, with short, often sparse hair; lighter-colored throat and chest; vaguely horse-like head with large, hanging upper lip, or "proboscis;" to 1.8 to 2.0 m (5.8 to 6.5 ft); weight to 200+ kg (440+ lb).

HABITAT: Low, middle, and higher elevation wet forests and swampy areas; active day or night; found on land or in shallow water.

LOCATIONS: OAX-HI, OAX-LO, CHIAP-HI, CHIAP-LO, CAM, QRO

Note: This species is endangered, CITES Appendix I and USA ESA listed.

Plate 80   415

c White-tailed Deer

d Red Brocket Deer

e Baird's Tapir

a Collared Peccary

b White-lipped Peccary

Plate 81a

West Indian Manatee
(also called Caribbean Manatee)
Trichechus manatus
Vaca Marina = sea cow
Vaca Acuática = aquatic cow
Manatí

ID: Very large, hairless, gray aquatic mammal; paddle-like front limbs, each with three nails; to 4.5 m (14.5 ft).

HABITAT: Shallow coastal areas, especially around some offshore islands, coastal lagoons, estuaries; may move inland along waterways as far as northern Chiapas.

LOCATIONS: CHIAP-LO, TAB, CAM, YUC, QRO

Note: This species is endangered, CITES Appendix I and USA ESA listed.

Plate 81b

Atlantic Spotted Dolphin
Stenella frontalis
Delfín = dolphin

ID: Gray or blackish dolphin with pronounced beak; pale spots on body; underparts lighter; dorsal fin about 25 cm (10 in) high; 1.5 to 3 m (5 to 10 ft) long.

HABITAT: Caribbean and Gulf Coasts; coastal and offshore areas.

LOCATIONS: TAB, CAM, YUC, QRO

Plate 81c

Bottle-nosed Dolphin
Tursiops truncatus
Delfín = dolphin

ID: Gray or slaty-blue dolphin with pronounced beak; flippers and flukes often darker; underparts lighter; dorsal fin about 25 cm (10 in) high; 1.8 to 3.5 m (6 to 11 ft) long.

HABITAT: Caribbean and Gulf Coasts; bays and lagoons, offshore areas. (Note: Bottle-nosed Dolphins also occur in the Pacific, but may or may not be the same species that occurs in Atlantic waters.)

LOCATIONS: TAB, CAM, YUC, QRO

Plate 81d

Common Dolphin
(also called Saddleback Dolphin)
Delphinus delphis
Delfín = dolphin
Tonina

ID: Smaller dark brown or blackish dolphin with gray sides and, often, white and cream-colored or yellowish side stripes; light belly; long, prominent beak; large dorsal fin (to 50 cm, or 20 in, wide and 40 cm, 16 in, high); 1.5 to 2.5 m (5 to 8 ft).

HABITAT: Pacific coastal and offshore areas.

LOCATIONS: OAX-LO, CHIAP-LO

Plate 81   417

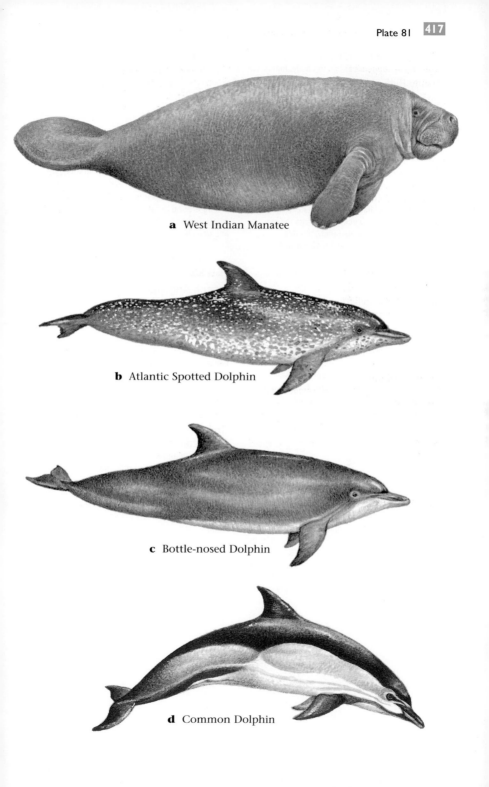

a West Indian Manatee

b Atlantic Spotted Dolphin

c Bottle-nosed Dolphin

d Common Dolphin

Lengths given for fish are "standard lengths," the distance from the front of the mouth to the point where the tail appears to join the body; that is, tails are not included in the measurement.

Text by Richard Francis

Plate 82a
Reef Butterflyfish
Chaetodon sedentarius

Closely related to the angelfish, butterflyfish are also generally monogamous. This species is fairly common and prefers the tops of coral reefs where it can be easily observed, usually in pairs. Very attractive, but not flashy by butterflyfish standards, look for the vertical black bar through the eye. Like many members of its family, the Reef Butterflyfish eats live coral. (to 16 cm, 6 in)

Plate 82b
Spotfin Butterflyfish
Chaetodon ocellatus

This beauty is named for the small black spot on the trailing edge of its dorsal fin. Its snow-white body is fringed by bright yellow dorsal and anal fins. The vertical black stripe that runs through the eye probably serves to deceive predators as to which end is the head. Usually found in pairs cruising over the top of the reef. The best way to get close is to place yourself in their line of travel. They will move away when approached. (to 20 cm, 8 in)

Plate 82c
Longsnout Butterflyfish
Chaetodon aculeatus

This is a rather shy fish, and unlike most butterflyfish tends to be solitary. This fish prefers somewhat deeper water than snorkelers typically explore, and is best observed on SCUBA. As its common name indicates, this species has an elongated snout, which facilitates deep coral probing. The color of this species typically ranges from olive to dusky. (to 10 cm, 4 in)

Plate 82d
Queen Angelfish
Holocanthus ciliaris

One of the most spectacular fish in the Caribbean. This rather shy angelfish must be approached slowly. It is fairly common in the protected reefs preferred by snorkelers, where it can be seen poking around coral heads looking for various invertebrates to eat. These fish are highly territorial and usually occur in pairs. (to 45 cm, 18 in)

Plate 82e
French Angelfish
Pomacanthus paru

Another large angelfish that can be distinguished from the gray angel by the yellow highlights on its scales. It resembles the Gray Angelfish in both habits and temperament. If you are above these fish, they will turn toward the horizontal to better keep an eye on you. (to 35 cm, 14 in)

Plate 82f
Gray Angelfish
Pomacanthus arcuatus

These large angelfish are quite curious and may approach if you remain still. They mate for life and once formed, a pair is rarely separated by more than a few meters. They consume a variety of invertebrates and are quite active during the day. Much less shy than Queen Angelfish. (to 50 cm, 20 in)

Plate 82 419

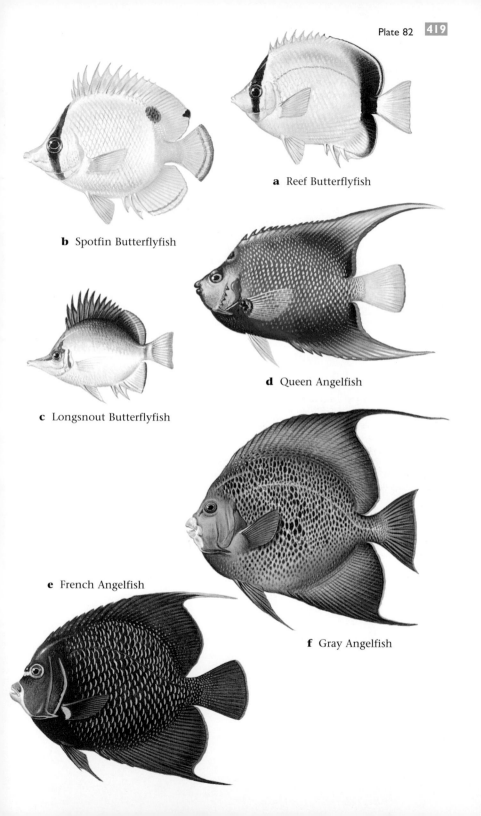

a Reef Butterflyfish

b Spotfin Butterflyfish

c Longsnout Butterflyfish

d Queen Angelfish

e French Angelfish

f Gray Angelfish

Plate 83a
Rock Beauty
Holocanthus tricolor
This is a common but fairly shy species. It patrols well-defined territories on the reef. The quiet snorkeler will be rewarded by the sight of this striking fish. Yellow areas in the front and tail sections are separated by a large area of black. Also notice the blue highlights around the eye. Juveniles are bright yellow with a blue-ringed black bulls-eye on the body toward the tail. (to 20 cm, 8 in)

Plate 83b
Blue Tang
Acanthurus coeruleus
This is one of the surgeonfish, as indicated by the scalpel-like protrusion near the base of the tail, which is deployed in aggressive encounters. Individuals can rapidly change color from a deep purple to powder blue. Abundant and usually in fairly large groups that move restlessly along reef tops grazing on algae. (to 23 cm, 9 in)

Plate 83c
Ocean Surgeonfish
Acanthurus bahianus
This species is also a color-changer (from dark brown to bluish gray). Fairly common and generally found in loose aggregations, along with Blue Tangs and Doctorfish. Grazers, Surgeonfish are approachable but tend to keep a minimum distance from divers. (to 35 cm, 14 in)

Plate 83d
Doctorfish
Acanthurus chirurgus
Very similar in appearance and habits to the Ocean Surgeonfish. Doctorfish can be distinguished by their vertical body bars, which, however, can be quite faint. This is one of the most abundant shallow water species. (to 25 cm, 10 in)

Plate 83e
Bar Jack
(also called Skipjack)
Caranx ruber
A fast-moving predator that courses over the reefs in groups of variable size. Bar jacks can make surprisingly close passes and you may be lucky enough to find yourself in the middle of a swirling school. This species can be distinguished from other jacks by its black stripe, bordered by bright blue, running from its dorsal fin through the bottom half of the tail. (to 60 cm, 24 in)

Plate 83f
Horse-eye Jack
(also called Bigeye Jack, Horse-eye Trevally)
Caranx latus
Another common jack found in open water over reefs, usually in small schools. More skittish than Bar Jacks, the snorkeler will usually only get a brief glimpse of these fast-moving fish. Distinguished by their yellow tails and large eyes. (to 75 cm, 29 in)

Plate 83 421

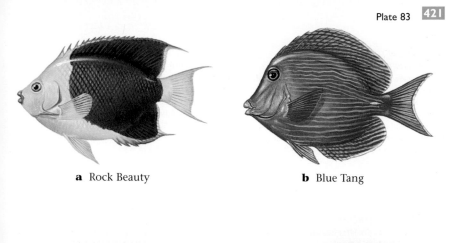

a Rock Beauty

b Blue Tang

c Ocean Surgeonfish

d Doctorfish

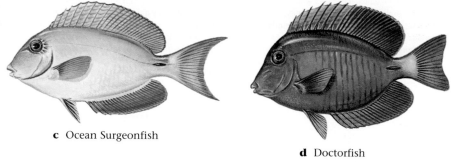

e Bar Jack

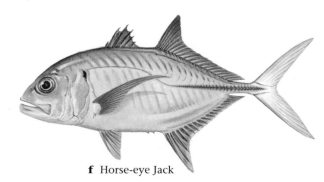

f Horse-eye Jack

Plate 84a

Houndfish
Tylosurus crocodilus
This species belongs to the needlefish family, of
which it is the largest. It is typically found in
shallow water over turtle grass beds or small
patch reefs, where it tends to drift just below the
surface. A favorite food item for Brown Pelicans
in the Yucatán area. This species is fairly shy and
not easy to see from below. (to 1.5 m, 5 ft)

Plate 84b

Great Barracuda
Sphyraena barracuda
This consummate predator has an impressive
array of teeth which it displays while slowly
opening and closing its mouth (to assist
respiration). Not dangerous, but barracudas can
be disconcertingly curious. They may even follow
you around the reef. They exhibit an economy of
movement, generally drifting, but capable of rapid
bursts should prey approach. (to 2 m, 6.5 ft)

Plate 84c

Bonefish
Albula vulpes
One of the species most prized by sports-
fisherpersons. When feeding, usually found in
shallow flats on a rising tide, in the vicinity of
mangroves. Tend to prefer the coral rubble when
they are not feeding. Bonefish are common but
very shy; they must be approached slowly. Not
colorful, they are silvery with no characteristic
markings. Forked tails and an underslung jaw,
somewhat like freshwater suckers. (to 1 m, 3.3 ft)

Plate 84d

Common Snook
Centropomus undecimalis
Another prized sportfish, look for them near
mangroves along the mainland coastline. Wary,
but not as shy as Bonefish; they can be
approached. Snook have a characteristic black
line running through the middle of their bodies.
The shape of their head is rather unique, like a
shallow slide. (to 1.4 m, 4.5 ft)

Plate 84e

White Mullet
Mugil curema
These common fish are found in shallow, open
water over sand or other soft bottom habitats.
They feed on tiny animals found on bottom
detritus or sea grasses. This species has a
characteristic black spot at the base of the
pectoral fin and very large scales. (to 38 cm, 15 in)

Plate 84f

Bermuda Chub
Kyphosus sectatrix
This common species swims in loose schools over
reefs, sometimes quite near the surface. Chubs
have a characteristic oval shape, silvery body,
and dusky-colored fins. (to 76 cm, 30 in)

Plate 84 | 423

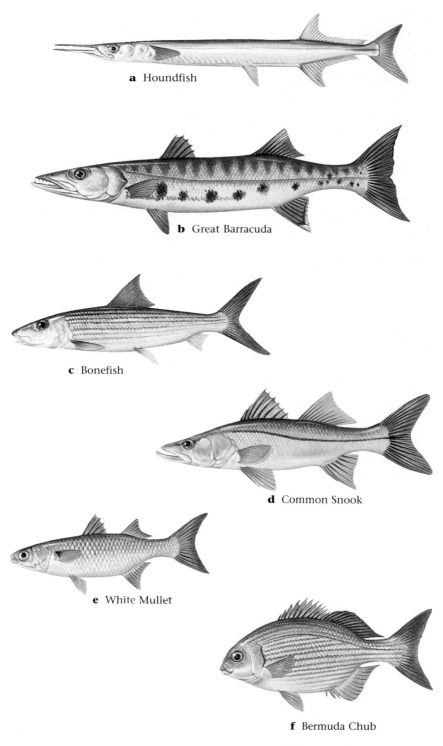

a Houndfish

b Great Barracuda

c Bonefish

d Common Snook

e White Mullet

f Bermuda Chub

Plate 85a

French Grunt
Haemulon flavolineatum

Grunts are named for the characteristic sound they make by grinding teeth-like structures in their throats. They are an important group of predators throughout the Caribbean and closely related to snappers. Grunts are found on reefs, sandy areas and seagrass beds. The French Grunt is one of the common species in this part of the Caribbean. Wavy blue horizontal lines on a yellow body are characteristic. Look for them under ledges. (to 30 cm, 12 in)

Plate 85b

White Grunt
Haemulon plumieri

This grunt species is distinguished by the checkered pattern of blue and yellow on its body, as well as the parallel blue and yellow lines on its long head. Like many grunt species, White Grunts seem to drift in groups of variable size; they retreat to cover when approached. (to 45 cm, 18 in)

Plate 85c

Blue-striped Grunt
Haemulon sciurus

This striking fish has bright blue stripes on a golden-yellow body. Also note the dark dorsal fin and tail. This is one of the most wary of the grunt species; patience and slow movements are required in order to approach. (to 45 cm, 18 in)

Plate 85d

Spanish Grunt
Haemulon macrostomum

This species is not as common as other grunts in the Yucatán reefs. A prominent black stripe runs through the eye to the base of the tail. The body is generally silvery and the fins have yellow borders. (to 43 cm, 17 in)

Plate 85e

Sailor's Choice
Haemulon parra

This handsome grunt has a silvery body with dark highlights. Its fins are quite dusky. The species drifts in small schools over and between reefs. (to 40 cm, 16 in)

Plate 85f

Margate
Haemulon album

This gray grunt prefers the sand flats between reef patches. Behaviorally a typical grunt; fairly wary but sometimes curious, it tends to drift passively, either alone or in small groups. (to 60 cm, 24 in)

Plate 85   425

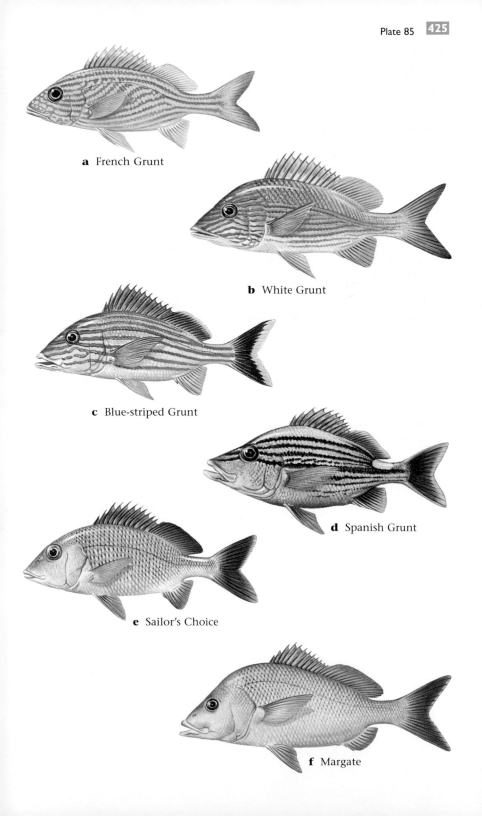

a French Grunt

b White Grunt

c Blue-striped Grunt

d Spanish Grunt

e Sailor's Choice

f Margate

Plate 86a

Atlantic Spadefish
Chaetodipterus faber
You will not confuse this fish with any other. Laterally compressed like a pompano, this species gets its name from its supposed resemblance to a suit of playing cards. The body is silver with dark vertical bars which may, however, rapidly fade. Spadefish occur in slow-moving schools and are fairly approachable. (to 91 cm, 36 in)

Plate 86b

Porkfish
Anisotremus virginicus
This is one of the more abundant species in the Caribbean, and a favorite for snorkelers. Bright yellow lateral stripes behind two striking vertical black bands distinguish this beautiful grunt. As an added bonus, it is quite easy to approach. (to 40 cm, 16 in)

Plate 86c

Mutton Snapper
Lutjanus analis
Most snappers are favorite food fishes for humans, which tends to make them wary. In protected areas though, they can be approached. Look for them in caves and deep crevices. This species has quite variable coloration, from silver to reddish-brown, but almost always with a black spot near midbody. (to 75 cm, 30 in)

Plate 86d

Cubera Snapper
Lutjanus cyanopterus
Difficult to distinguish from other snappers, but with relatively thick lips. This is an extremely shy species. (to 1.5 m, 5 ft)

Plate 86e

Mahogany Snapper
Lutjanus mahogoni
One of the smaller snappers, this species' scales usually have a reddish tinge and more distinct red borders around the fins. It drifts in small groups near cover. (to 38 cm, 15 in)

Plate 86f

Schoolmaster
Lutjanus apodus
This snapper usually occurs in loose schools drifting above the reef; it is quite common but not easy to approach. The body is generally silver and the fins are yellow. (to 60 cm, 24 in)

Plate 86 427

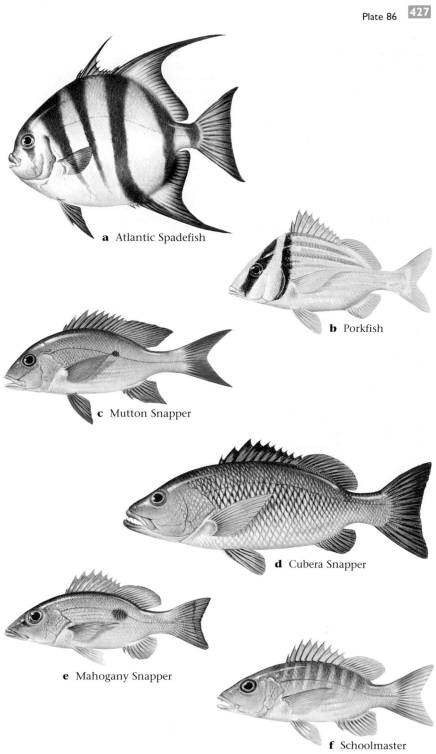

a Atlantic Spadefish

b Porkfish

c Mutton Snapper

d Cubera Snapper

e Mahogany Snapper

f Schoolmaster

Plate 87a
Longfin Damselfish
Stegastes diencaeus
The damselfish are among the most entertaining coral reef inhabitants. Highly territorial and aggressive, especially during the breeding season when the males are tending the eggs. Though diminutive, they will even attack divers, which can be quite comical. This species is dusky brown throughout and is distinguished by its relatively long dorsal and anal fins. The juveniles of this, and most other damselfish, differ markedly from adults in coloration. In this species the juveniles are yellow with distinctive blue stripes dorsally and a prominent blue-ringed black spot near the base of the dorsal fin. (to 13 cm, 5 in)

Plate 87b
Dusky Damselfish
Stegastes fuscus
Very similar to the Longfin Damselfish but with shorter and more rounded fins. Juveniles are bluish with a bright orange swath running from the snout to the middle of the dorsal fin. (to 15 cm, 6 in)

Plate 87c
Threespot Damselfish
Stegastes planifrons
The adults are a fairly bland dusky color but with a small yellow crescent above the eye. Juveniles (shown) are bright yellow with a black spot near the base of the dorsal fin. Another pugnacious species that is extremely active and bold. (to 13 cm, 5 in)

Plate 87d
Cocoa Damselfish
Stegastes variabilis
The adults of this species can be hard to distinguish from other damselfish, but look for a dark spot at the base of the tail. Juveniles (shown) are blue above the eye and yellow below. Not as aggressive as most damselfish. (to 13 cm, 5 in)

Plate 87e
Beaugregory
Stegastes leucostictus
This damselfish prefers coral rubble and sandy areas, and like Cocoa Damselfish, is relatively unaggressive. It is not shy, however. Adults are distinguished from other damsels, such as Longfin and Dusky, by their yellowish-tinged fins. Juveniles (shown) closely resemble Cocoa Damselfish. (to 10 cm, 4 in)

Plate 87f
Bicolor Damselfish
Stegastes partitus
This species defends smaller territories than most damsels. They are aggressive but channel it toward fish of about the same size. This species is easy to identify, with its body divided between a black fore-region and a white rear. Juveniles have a lesser black area and a bright yellow triangular swath originating beneath the chin. (to 10 cm, 4 in)

Plate 87 429

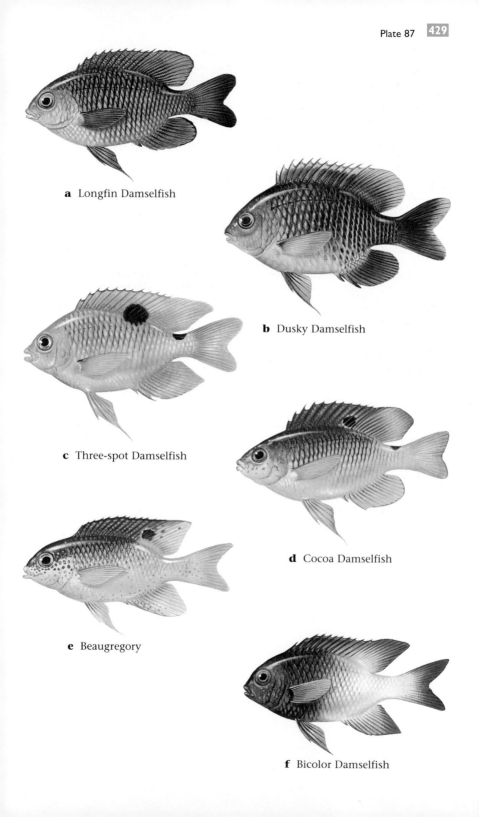

a Longfin Damselfish

b Dusky Damselfish

c Three-spot Damselfish

d Cocoa Damselfish

e Beaugregory

f Bicolor Damselfish

Plate 88a
Yellowtail Snapper
Ocyurus chrysurus
This abundant species is much more streamlined than most snappers. A distinctive yellow line runs from behind the eye to the base of the tail, which is also yellow. It swims above the reefs in loose schools, and is much less wary than most snappers. (to 75 cm, 30 in)

Plate 88b
Yellowtail Damselfish
Microspathodon chrysurus
Large by damselfish standards, this species has a distinctive yellow tail. The juveniles are spectacular in bright light; their dark blue bodies are covered with electric blue dots. Be aware that the juveniles prefer to hang around skin-irritating fire coral. (to 21 cm, 8 in)

Plate 88c
Sergeant Major
Abudefduf saxatilis
Probably the most common damselfish, it can be easily distinguished by its five vertical black bars on a silvery-gray body. Juveniles have more yellow tones, and the adult males turn dark purplish-blue when guarding eggs. They prefer to remain higher above the reefs than most damsels and usually occur in loose aggregations. (to 15 cm, 6 in)

Plate 88d
Blue Chromis
Chromis cyanea
This brilliant blue damselfish with a black nape is extremely abundant in midwater, often in very large groups. The tail is very deeply forked. It is most common slightly below typical snorkeling depths, but look for it over drop-offs. (to 13 cm, 5 in)

Plate 88e
Brown Chromis (also called Yellow-edge Chromis)
Chromis multilineata
Generally tan or brownish-gray with a characteristic black spot at the base of the pectoral fin. The deeply forked tail often has black borders. Similar in habits to the Blue Chromis, with which it often schools. (to 16 cm, 6 in)

Plate 88f
Purple Reeffish (also called Purple Chromis)
Chromis scotti
Not so much purple as varying shades of blue. The juveniles (shown) are a very deep and bright blue. This is a deeper water species that is best observed on SCUBA. Occurs in small groups near the bottom of deep reefs. (to 10 cm, 4 in)

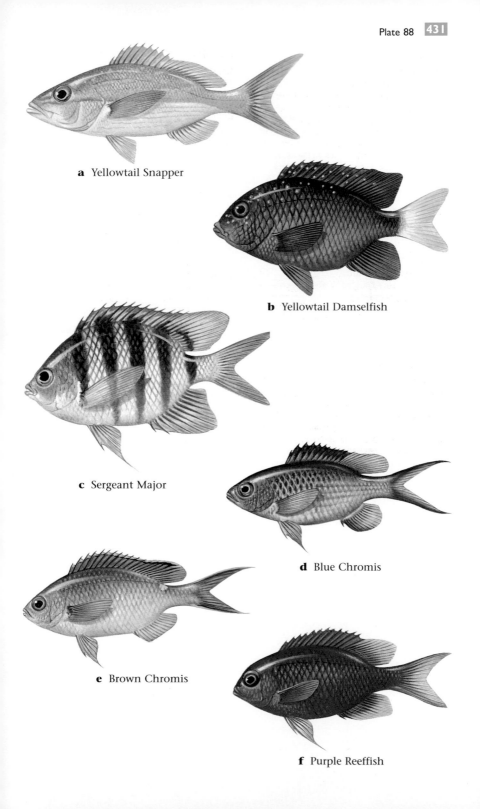

Plate 88 431

a Yellowtail Snapper

b Yellowtail Damselfish

c Sergeant Major

d Blue Chromis

e Brown Chromis

f Purple Reeffish

Plate 89a

Barred Hamlet
Hypoplectrus puella

All hamlets are true hermaphrodites. Their fascinating mating behavior, known as egg-trading, involves the alternate release of eggs and sperm with their partner. This species has brown bars of varying width, interspersed with whitish areas. Bright blue vertical lines on the head. Perhaps the most common hamlet, it prefers the bottom of shallow reefs. All hamlets exhibit an interesting mixture of wariness and curiosity. (to 13 cm, 5 in)

Plate 89b

Indigo Hamlet
Hypoplectrus indigo

Indigo blue bars separated by white bars make this species easy to identify. All hamlets exhibit essentially the same behavior. The Yukatán is a particularly good area to see a variety of these species. (to 13 cm, 5 in)

Plate 89c

Shy Hamlet
Hypoplectrus guttavarius

The name implies that this species is particularly reclusive, although I have not found this to be the case. Sometimes it is very curious. Much of the body is brown to blackish, but head and chest yellowish, as well as all fins. Bright blue lines on snout. (to 13 cm, 5 in)

Plate 89d

Golden Hamlet
Hypoplectrus gummigutta

Yellow body with striking blue and black markings on face. Occurs in deeper water than most hamlets. (to 13 cm, 5 in)

Plate 89e

Yellowtail Hamlet
Hypoplectrus chlorurus

The body of this species often appears black but is actually a dark blue or dark brown. The yellow tail is what distinguishes it from other hamlets. (to 13 cm, 5 in)

Plate 89f

Black Hamlet
Hypoplectrus nigricans

The entire body is black to bluish-brown. Probably the least shy of the hamlets. (to 13 cm, 5 in)

Plate 89 **433**

a Barred Hamlet

b Indigo Hamlet

c Shy Hamlet

d Golden Hamlet

e Yellowtail Hamlet

f Black Hamlet

Plate 90

Plate 90a

Blue Hamlet
Hypoplectrus gemma
Iridescent blue body often with darker eye. This
beautiful species stays close to the bottom and
will remain motionless for long periods while
being observed. (to 13 cm, 5 in)

Plate 90b

Masked Hamlet
Hypoplectrus sp.
This very handsome hamlet has a taupe-colored
body and an upside-down triangular black stripe
below the eye, from whence its name derives.
Some scientists consider all hamlets to be
variants of a single species and ichthyologists
have not yet bothered to provide a species name
for this one, but its color pattern is consistent over
a broad geographic range. (to 13 cm, 5 in)

Plate 90c

Jewfish
Epinephelus itajara
The largest grouper in the Caribbean, this species
is severely threatened by overfishing, particularly
spearfishing. Its appearance is variable but
always splotched or mottled. Shy in unprotected
areas, but much less so where protected. It
spends much of the day inside caves or other
protected areas. Groupers, as well as the hamlets,
are seabass of the family Serranidae. Many are
suspected to be serial hermaphrodites that are
female while young and small, then turn to male
when larger. In some of the largest groupers the
reverse transition is thought to occur. (to 2.4 m,
7.8 ft)

Plate 90d

Nassau Grouper
Epinephelus striatus
Another grouper threatened by overfishing. This
species has five distinctive brown stripes on its
body, but it can change color from a very pale hue
to virtually black. A somewhat steeper forehead
than most groupers. You can often find it resting
on the bottom. (to 1.2 m, 4 ft)

Plate 90e

Yellowfin Grouper
Mycteroperca venenosa
Many groupers exhibit different color phases. This
one varies from black to bright red and white. The
only constants are the yellow edges on the
pectoral fins and the black fringe at the end of the
tail. This is a mid-size grouper, often found resting
on sand. It prefers reef-tops near drop-offs. (to 90
cm, 35 in)

Plate 90f

Tiger Grouper
Mycteroperca tigris
Another mid-sized grouper with highly variable
coloration. The juveniles are usually yellow with
brown bars. (to 1 m, 3.3 ft)

Plate 90 **435**

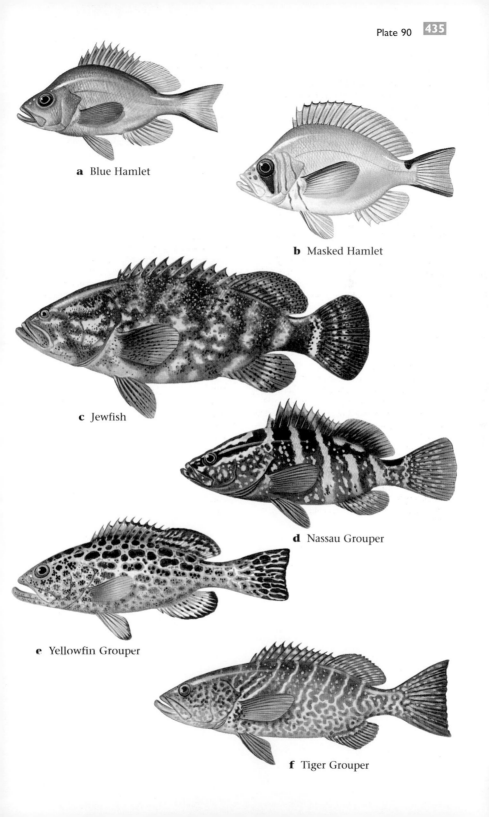

a Blue Hamlet

b Masked Hamlet

c Jewfish

d Nassau Grouper

e Yellowfin Grouper

f Tiger Grouper

Plate 91a
Graysby
Epinephelus cruentatus
This small grouper is one of the most common in the Yucatán area. It can quickly change color. Look for the three distinctive black spots at the base of the dorsal fin. (to 35 cm, 14 in)

Plate 91b
Red Hind
Epinephelus guttatus
A fairly common small grouper with red spots over a white background. It can be found at a variety of depths, including shallow reef patches. (to 67 cm, 26 in)

Plate 91c
Coney
Epinephelus fulvus
This common small grouper exhibits a wide range of color patterns. All individuals have at least a few small spots. It can be quite curious and is more social than most groupers. (to 41 cm, 16 in)

Plate 91d
Tobaccofish
Serranus tabacarius
This small seabass can be found in very shallow water and is not at all shy. Common in the Yucatán region where it prefers areas of coral rubble. Tends to hover near the bottom. The common name derives from the horizontal midbody band, which someone, who must have been smoking something else, thought was the shade of tobacco. (to 18 cm, 7 in)

Plate 91e
Creole-fish
Paranthias furcifer
This member of the seabass family is one of the more common fish in the Caribbean. Occurs in large schools that hover above deeper reefs. Coloration is variable, but often with a purplish hue. Look for the bright red area at the base of the pectoral fin and three spots along the back. (to 38 cm, 15 in)

Plate 91f
Fairy Basslet
(also called Royal Gramma)
Gramma loreto
This beautiful little fish is quite common but wary, and tends to retreat into cracks and crevices when approached. Be patient though, and it will reappear. The purple head region and yellow tail region are about equally divided. (to 8 cm, 3 in)

Plate 91 437

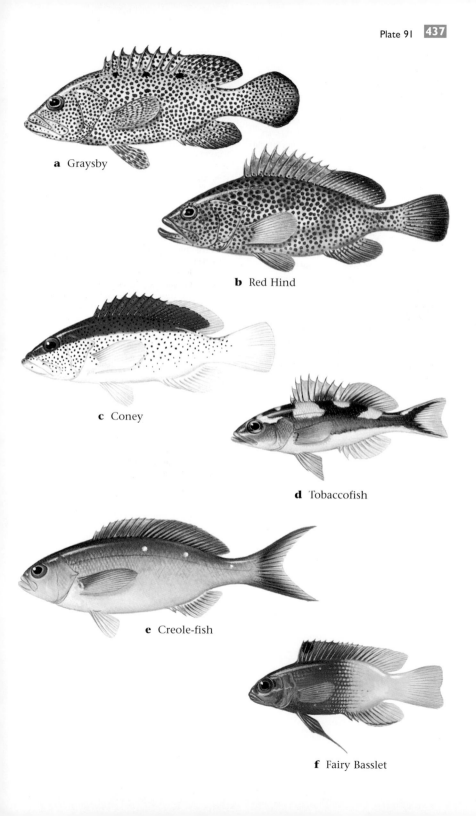

a Graysby

b Red Hind

c Coney

d Tobaccofish

e Creole-fish

f Fairy Basslet

Plate 92

Plate 92a

Blue Parrotfish
Scarus coeruleus

Parrotfish are unique grazers. They feed primarily on algae from rocks and coral, and can munch the hard corals with their powerful beak-like mouth. Watch for the wispy white clouds of coral dust that they occasionally void. Parrotfish are female-to-male sex changers (p. 240) but with a twist. Some males (initial phase males) are born that way, and have a completely different reproductive strategy than the large (terminal phase) males. The latter are solitary and territorial, while the initial phase males tend to congregate with females. Identifying parrotfish is tricky because of the different color phases associated with their development. Terminal phase males are easiest to identify and it is from their coloration that the common names usually derive. In this species, however, all phases come in beautiful shades of blue, from light to quite dark. Terminal phase males have a squared-off head. This is one of the largest parrotfish species. (to 90 cm, 35 in)

Plate 92b

Midnight Parrotfish
Scarus coelestinus

All phases of this species are midnight blue, with brighter blue splashes around the head. Not wary, but will not allow close approach. Parrotfish tend to be quite active and stop only to scrape algae off coral. This is one of the larger parrotfish species. (to 76 cm, 30 in)

Plate 92c

Queen Parrotfish
Scarus vetula

A medium-sized parrotfish, similar in habits to the Midnight Parrotfish. Terminal phase males have blue-green bodies with striking blue and green markings around the mouth. (to 61 cm, 24 in)

Plate 92d

Stoplight Parrotfish
Scarus viride

One of the more common parrotfish in the Yucatán area. This mid-sized species is distinguished by a characteristic yellow spot on the upper part of the gill-cover. Also look for the crescent-shaped tail. (to 50 cm, 20 in)

Plate 92e

Hogfish
Lachnolaimus maximus

This species belongs to the wrasse family, close relatives of parrotfish. A favorite food fish wherever it occurs, they are wary wherever spearfishing is allowed, but can become quite tame where protected. Most wrasse exhibit the complicated reproductive strategy typical of parrotfish, as well as various developmental color phases. This species has only one male type, which has an off-white, grayish, or reddish-brown body, with a black swath over the head. When erected, the dorsal fin has a dramatic appearance, with several long spines in front. (to 91 cm, 36 in)

Plate 92f

Spanish Hogfish
Bodianus rufus

One of the few wrasse species without distinct color phases, this common species is constantly on the move and quite easy to approach. A large purple area on the upper half of the body behind the head is characteristic. (to 40 cm, 16 in)

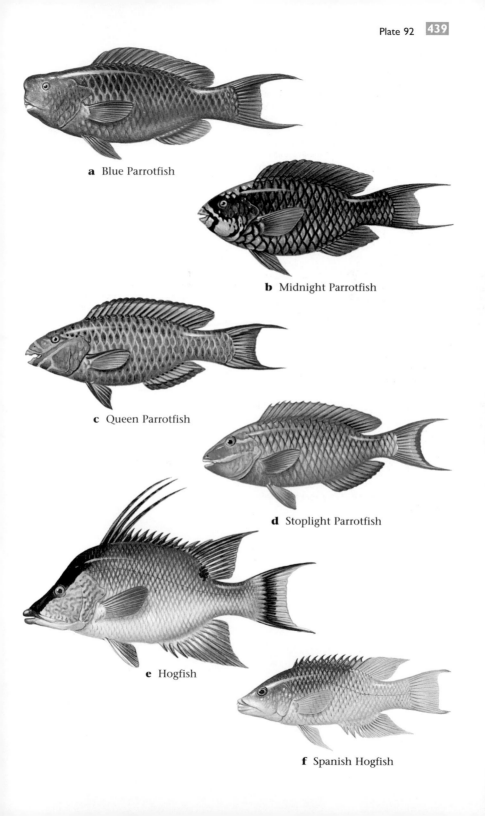

Plate 92 439

a Blue Parrotfish

b Midnight Parrotfish

c Queen Parrotfish

d Stoplight Parrotfish

e Hogfish

f Spanish Hogfish

Plate 93a
Creole Wrasse
Clepticus parrae
This wrasse is one of the most abundant fish in the Caribbean. It prefers the open water over the top of deep reefs where it occurs in large groups, often in the afternoon forming long stream-like schools over the drop-offs. Another wrasse without distinct color phases, the snout area is usually dark purple and the front half of the body dark blue to slate. Older individuals show more yellow and red in the tail region. (to 30 cm, 12 in)

Plate 93b
Yellowhead Wrasse
Halichoeres garnoti
This species exhibits several distinct developmental color phases. Terminal phase males have the yellow head and forebody for which the species is named, and a dark vertical bar at midbody. Initial phase males and females have a dusky back and yellow mid-region. Juveniles are bright yellow with bright blue horizontal stripe. (to 18 cm, 7 in)

Plate 93c
Bluehead Wrasse
Thalassoma bifasciatum
This common wrasse is easy to observe on shallow reefs. Terminal phase males have the bright blue head for which the species is named. Initial phase males and females have a dusky blue color with irregular white stripes. Juveniles have variable color patterns and may act as cleaners (see p. 242). (to 18 cm, 7 in)

Plate 93d
Slippery Dick
Halichoeres bivittatus
This very common species can be found on reefs as well as adjacent sandy areas and turtle grass patches. Terminal phase males are various shades of green with a darker horizontal stripe at midbody. As to the meaning of the common name, don't ask. (to 20 cm, 8 in)

Plate 93e
Squirrelfish
Holocentrus ascensionis
Members of the squirrelfish family are most active at night. By day, you can observe them in protected areas such as rock crevices and inside large barrel sponges preferably in shallow patch reefs. They are not shy and allow close observation. This species is red with white patches and, as do all members of the family, has very large eyes. (to 30 cm, 12 in)

Plate 93f
Blackbar Soldierfish
Myripristis jacobis
This member of the squirrelfish family can often be found in caves or other dark recesses swimming upside down. It can be quite curious and approachable. The body is bright red, with a black stripe behind the gill-cover. (to 20 cm, 8 in)

Plate 93 441

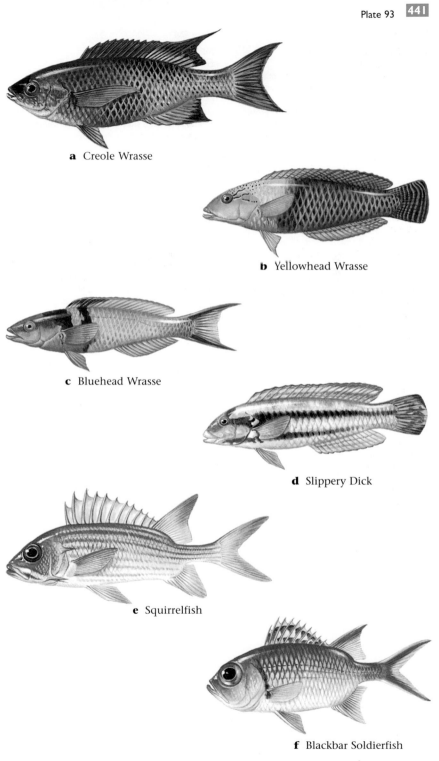

a Creole Wrasse

b Yellowhead Wrasse

c Bluehead Wrasse

d Slippery Dick

e Squirrelfish

f Blackbar Soldierfish

Plate 94a
Glasseye Snapper
Priacanthus cruentatus
Another nocturnal species that can be observed by day in its hideouts. The coloration is variable but usually with silver bars on the back. Unlike other "bigeyes" in the area, this species prefers shallow reefs. (to 33 cm, 13 in)

Plate 94b
Bigeye
Priacanthus arenatus
This species is best observed on SCUBA because it prefers deep reefs where it drifts in small groups. The body is a uniform but variable shade of red. (to 30 cm, 12 in)

Plate 94c
Flamefish
Apogon maculatus
One of the cardinalfish, this species is active at night. By day it can be found in various dark places in a variety of habitats, including reefs and docks. One of the species most often seen on night dives. Body color is salmon to bright red; distinctive features are small white lines above and below the eye and a black spot behind the eye. (to 11 cm, 4 in)

Plate 94d
Neon Goby
Gobiosoma oceanops
Gobies constitute the largest family of fishes, indeed, the largest vertebrate family, yet only the careful observer will be able to enjoy these diminutive fishes. This common goby often acts as a cleaner (p. 242). They often establish cleaning stations at which several congregate waiting for clients. This is a very attractive fish with its black body bisected by an electric blue horizontal stripe. (to 5 cm, 2 in)

Plate 94e
Yellowline Goby
Gobiosoma horsti
This species is much shyer than the Neon Goby. Usually found near sponges on reefs of medium depth. It has a black body and a yellow horizontal stripe running from head to tail, including the top of the eye. (to 4 cm, 1.5 in)

Plate 94f
Cleaning Goby
Gobiosoma genie
As its common name suggests, this species is one of the so-called "cleaner fish." It congregates at established cleaning stations waiting for clients (p. 242). As is typical of gobies that clean other fish, this species is bold and approachable. The upper body is dark with a bright yellow V on the snout which is continuous with two paler body stripes. (to 4 cm, 1.5 in)

Plate 94   443

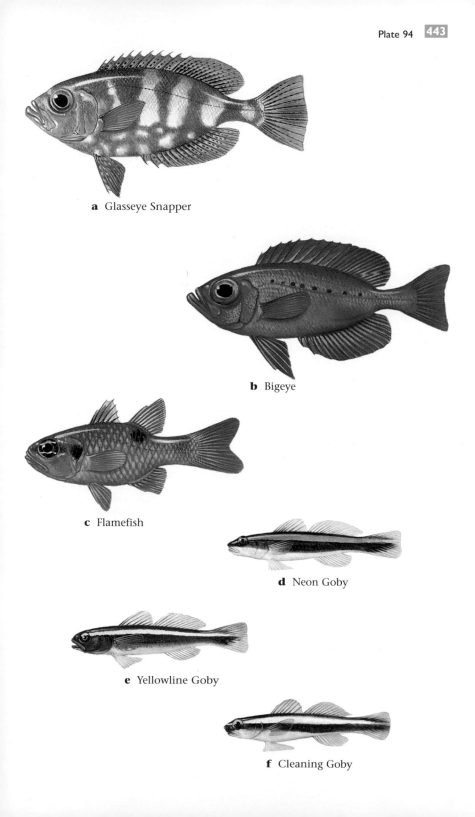

a Glasseye Snapper

b Bigeye

c Flamefish

d Neon Goby

e Yellowline Goby

f Cleaning Goby

Plate 95a

Rusty Goby
(also called Sharknose Goby)
Priolepis hipoliti
The unusual feature of this goby is that it can often be found perched upside down on the roofs of small clefts in reefs or under boulders. This goby is fairly approachable. It has bright orange spots on the dorsal, tail, and anal fins, as well as several dusky body bars. (to 4 cm, 1.5 in)

Plate 95b

Redlip Blenny
Ophioblennius atlanticus
Large, by blenny standards, members of this species are also full of personality. They typically sit on an exposed perch, ever watchful for territorial intrusions. Their prominent eyes, which move independently of one another, allow them to monitor events in all compass directions. Their coloration varies from gray to reddish-brown; the head is usually darker and redder than the rest of the body. The head looks almost flat in profile. (to 12 cm, 5 in)

Plate 95c

Yellowhead Jawfish
Opistognathus aurifrons
These inhabitants of the sand and coral rubble are fun to watch. They typically excavate burrows in the sand and line the entrance with small bits of coral, which tend to migrate from one burrow to another, as they are perpetually stealing their neighbors' goods. They look like small eels, especially when only the upper half of the body is extended above the surface. The body is pale and the head a very pale yellow. (to 10 cm, 4 in)

Plate 95d

Peacock Flounder
Bothus lunatus
This flatfish is most active at night. By day you will see it only if you happen to swim close enough to cause it to swim away; otherwise its camouflage is quite effective. This species also prefers sandy areas or coral rubble. The species is named for the striking blue spots on the body and fins. (to 39 cm, 15 in)

Plate 95e

Splendid Toadfish
(also called Coral Toadfish)
Sanopus splendidus
This spectacular fish seems to be endemic to Cozumel Island, Mexico. The flattened head is especially striking with its densely packed black and white stripes. The ventral fins are entirely yellow; the rest of the fins have an attractive yellow border. Also look for the very prominent barbels around the mouth. This is a shy species, most likely to be found in crevices and other dark recesses, where it is supported by its pectoral fins. (to 20 cm, 8 in)

Plate 95f

Sand Diver
Synodus intermedius
This common reef inhabitant belongs to the lizardfish family. It often lies half buried in the sand, which along with its camouflaged coloration both protects it from predators and allows it to dart out and grab unwary prey. Sand divers exhibit a mottled reddish-brown coloration but can change hue to blend with the background. (to 45 cm, 18 in)

Plate 95 445

a Rusty Goby

b Redlip Blenny

c Yellowhead Jawfish

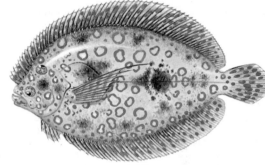

d Peacock Flounder

e Splendid Toadfish

f Sand Diver

Plate 96a

Trumpetfish
Aulostomus maculatus

One of the more distinctively shaped reef creatures, this elongate fish actually looks more like a soprano saxaphone than a trumpet. It is usually seen drifting with head down, which may serve as a sort of camouflage, especially in grass beds or among some gorgonians. It will not tolerate a direct approach but these curious creatures may approach you if you don't flail around too much. (to 1 m, 3.3 ft)

Plate 96b

Sand Tilefish
Malacanthus plumieri

Look for these fish in sandy areas, where their large, inverted, conical burrow entrances are easy to spot. This is a fairly shy species and will retreat to the burrow when approached too closely. But if you are patient, it will eventually emerge and hover over the burrow by undulating its long dorsal and anal fins. (to 60 cm, 24 in)

Plate 96c

Porcupinefish
Diodon hystrix

This member of the puffer family will inflate its body dramatically when threatened, its sharp spines becoming erect during the process. This obviously makes it less easy to swallow for any would-be predator. It does not facilitate movement, however, and it can look quite comical when attempting to swim away in this state, furiously beating its pectoral fins. During the day it occupies various recesses. But you can often spot its head peering outward, and it can then be approached quite closely. (to 90 cm, 35 in)

Plate 96d

Smooth Trunkfish
Lactophrys triqueter

Like puffers, trunkfish manage to negotiate the reefs using their pectoral fins almost exclusively. This species is not wary and allows the diver fairly close inspection. Aside from its peculiar shape and small mouth with those seemingly kissable lips, notice the bulbous eyes, which seem to rotate in various directions like radar dishes. The smooth trunkfish has a dark body with numerous white spots throughout. These spots thin somewhat behind the pectoral fin in older fish and honeycomb markings appear. (to 30 cm, 12 in)

Plate 96e

Queen Triggerfish
Balistes vetula

This is one of the most strikingly beautiful species on the reef. Background body coloration is various shades of purple, blue, turquoise and green, and the head is usually lighter, tending toward yellow; but these fish can rapidly darken or lighten. Irregular black lines radiate from the eye and two striking blue lines run above the mouth. Triggerfish move about primarily by means of the coordinated action of their dorsal and anal fins. Queen Triggerfish prefer reef tops and coral rubble. This species is fairly shy but your patience will be rewarded. (to 60 cm, 24 in)

Plate 96f

Ocean Triggerfish
Canthidermis sufflamen

This is more of an open water species, most frequently observed near drop-offs. However, during the nesting season, it can be found in sandy areas between reef patches. Males create large depressions into which the female lays her eggs, which the male then guards until hatching. This species is almost entirely gray but with a prominent black spot at the base of the pectoral fin. (to 65 cm, 26 in)

Plate 96 447

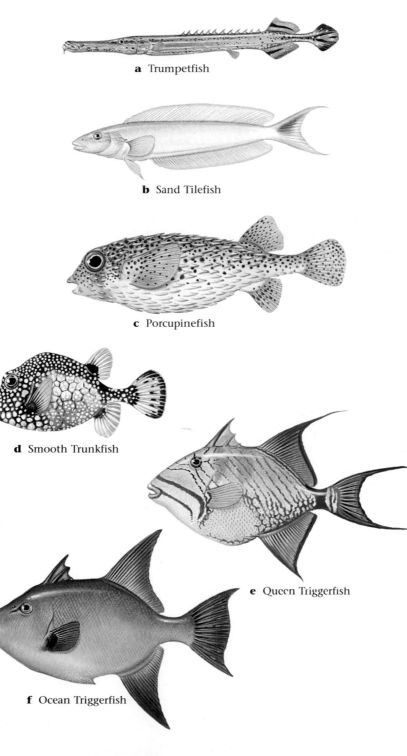

a Trumpetfish

b Sand Tilefish

c Porcupinefish

d Smooth Trunkfish

e Queen Triggerfish

f Ocean Triggerfish

Plate 97a
Black Durgeon
Melichthys niger
One of the more social members of the triggerfish
family, this species occurs in groups of variable
size, sometimes quite large. Quite common and
easy to observe at a distance, they usually do not
permit close inspection. The body is black and
fins blacker, but with prominent pale blue lines
beneath dorsal and anal fins. (to 50 cm, 20 in)

Plate 97b
Scrawled Filefish
Aluterus scriptus
This very odd-looking fish always seems to have a
disheveled appearance because its tail is usually
limp. The body coloration is usually a shade of
yellow-green, and covered with blue and black
spots. Fairly common, it drifts over the reefs,
seemingly without aid of its fins, until it is
approached; it then retreats but again without
much exertion. This fish is a loner. (to 1.1 m,
3.6 ft)

Plate 97c
Yellow Goatfish
Mulloidichthys martinicus
Goatfish look very much like freshwater catfish,
primarily because of their barbels under the lower
jaw. Goatfish prefer sandy areas where they can
be found, usually in small groups, busily probing
the bottom for food. This species has a silvery-
white body with a yellow midbody stripe. The tail
is also yellow. This is one of the more
approachable species and it can be found in quite
shallow water. (to 40 cm, 16 in)

Plate 97d
Spotted Drum
Equetus punctatus
This handsome fish can be found in protected
areas of the reef, under ledges or in various nooks
and crannies, where it rests by day. Drums are
most active at night. The Spotted Drum and the
similar Jackknife Fish have an unusual dorsal fin
that is extremely long and directed upward with a
slight curve. Striking black and white stripes on
the body, and black back and tail with white spots.
Quite unafraid, this species will allow close
inspection if you approach slowly. (to 25 cm, 10 in)

Plate 97e
Green Moray
Gymnothorax funebris
Largest of the morays, this species occurs in
diverse habitats and is fairly common, even in
shallow water. Morays rest by day in crevices
with only their head protruding. They open and
close their jaws in order to breathe, exposing
their teeth. Though they look menacing, they are
actually quite docile and retreat deeper into the
reef when approached. They quickly become
tame, however, and can be observed quite closely
in some protected areas. All morays are active at
night. (to 2.4 m, 8 ft)

Plate 97f
Spotted Moray
Gymnothorax moringa
This species is much smaller than the Green
Moray. It prefers shallow reefs with abundant
rubble, where it can be observed resting by day,
head protruding from its refuge. This species is
more speckled than spotted, seemingly splattered
by black paint. (to 1.2 m, 4 ft)

Plate 97   449

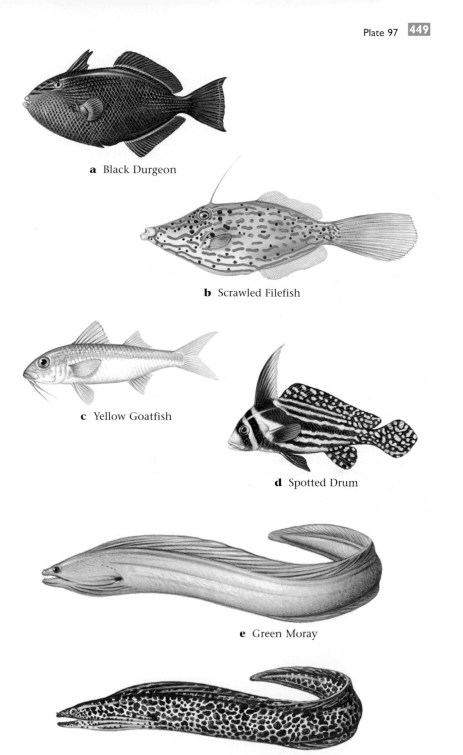

a Black Durgeon

b Scrawled Filefish

c Yellow Goatfish

d Spotted Drum

e Green Moray

f Spotted Moray

Plate 98a

Goldentail Moray
Gymnothorax miliaris
Tiny yellow spots over a dark background distinguish this species from other morays. It prefers shallow to mid-depth reefs and is fairly common. (to 60 cm, 32 in)

Plate 98b

Nurse Shark
Ginglymostoma cirratum
This is the shark species you are most likely to observe. Nurse sharks are quite sluggish by shark standards, spending much of their time resting on the bottom. But they do reach impressive sizes and if one happens to swim by, your heart rate will increase. Nurse sharks seem to be missing the bottom half of the tail. Their heads are also larger than those of most sharks. (to 4.2 m, 14 ft)

Plate 98c

Bull Shark
Carcharhinus leucas
This is the most heavy-bodied of the requiem sharks. Bulls prefer inshore waters and some even migrate hundreds of miles up rivers into freshwater lakes, most notably in Nicaragua. Though fairly common, they are rarely seen. I once watched from shore while a large Bull Shark approached to within 3 m (10 ft) of a group of snorkelers who, nonetheless, remained unaware of its presence. This species should be treated with utmost respect. (to 3.5 m, 11.5 ft)

Plate 98d

Hammerhead Shark
(also called Smooth Hammerhead)
Sphyrna zygaena
This species spends most of its time in open water, but it does cruise the reefs on occasion, especially at night. The bizarre head with eyes stuck at each end ensure that this shark will not be confused with any other. It reaches impressive sizes and, though quite wary, should be treated with care. (to 3.5 m, 11.5 ft)

Plate 98e

Southern Stingray
Dasyatis americana
This common species prefers sandy areas where it lies buried to varying degrees. The name derives from the venomous spine near the base of the tail, contact with which can be exquisitely painful. Because Stingrays prefer shallow water, they are a factor to consider while wading in sandy areas. They are quite unafraid and will not move unless you are almost on top of them. (to 1.5 m, 5 ft)

Plate 98f

Spotted Eagle Ray
Aetobatus narinari
This common and quite handsome species prefers to stay well up in the water column, where it seems to fly through the water with its considerable wing-like fins. Eagle Rays have venomous spines but do not pose any threat to divers. They are quite wary, in fact, and best observed by staying motionless. They often leap out of the water for unknown reasons, and the sound they make upon re-entry is an impressive clap. (to 2.3 m, 7.5 ft)

Plate 98 451

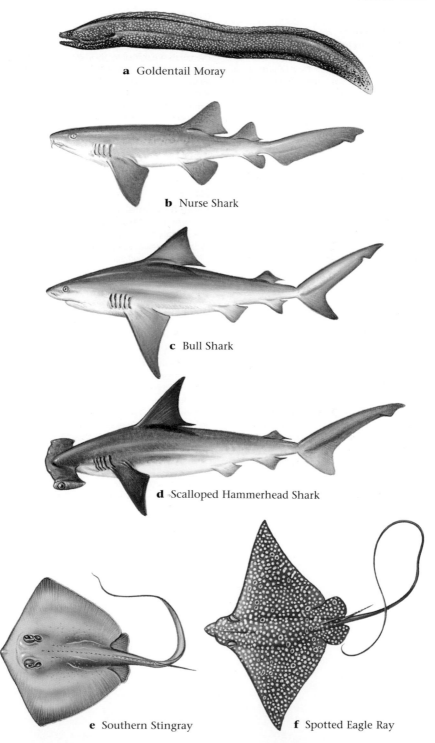

a Goldentail Moray

b Nurse Shark

c Bull Shark

d Scalloped Hammerhead Shark

e Southern Stingray

f Spotted Eagle Ray

Plate 99

Plate 99a

Staghorn Coral
Acropora cervicornis
One of the fastest growing coral species, it
prefers shallow, calm water. Forms dense tangles
of branches in which only the outer tips are alive.
Staghorn coral is fragile and subject to storm
damage, but it recovers more quickly than most
corals due to rapid growth rate.

Plate 99b

Elkhorn Coral
Acropora palmata
Another fast-growing species, it prefers shallow
areas with good water movement and wave
action. This coral can cover vast areas of shallow
bottom, and it is one of the characteristic species
living on the shallow fringing reefs around small
islands along the Yucatán eastern coast. The
branches are flattened like moosehorns.

Plate 99c

Lettuce Coral
(also called Thin Leaf Coral)
Agaricia tenuifolia
This is one of the plate or sheet corals, but in
contrast to most species, the "leaves" are
positioned vertically and resemble leafy lettuce.
Parallel wavy ridges run across each leaf. The
color varies from gray to brown, sometimes with
tints of blue or green. Can form large colonies in
shallow water with plenty of wave action. Its
crevices often harbor small fish and invertebrates
such as brittle stars, urchins, and shrimp.

Plate 99d

Fire Coral
Millepora alcicornis
Fire corals are named for their skin-irritating
capacity. They can cause welts and swelling in
sensitive people, although the effect is usually of
short duration. Pictured here is the branching
species of fire coral but fire corals can become
encrusting on certain substrates. They tend to be
brownish-orange but the color is variable. This
species prefers deeper, calmer water than other
fire corals.

Plate 99e

Large Star Coral
Montastrea cavernosa
This species forms very large coral heads in the
form of mounds or domes. The polyps look like
small blisters of variable hue when retracted. Like
many coral species, the polyps are active at night
and retracted by day. This species is found in a
wide variety of reef habitats.

Plate 99f

Common Star Coral
(also called Boulder Star Coral)
Montastrea annularis
This species is common in a variety of reef
environments. There are several distinct forms
that some consider to be separate species. The
most spectacular form, sometimes called
Mountainous Star Coral, has irregular pillars and
bumps.

Plate 99   453

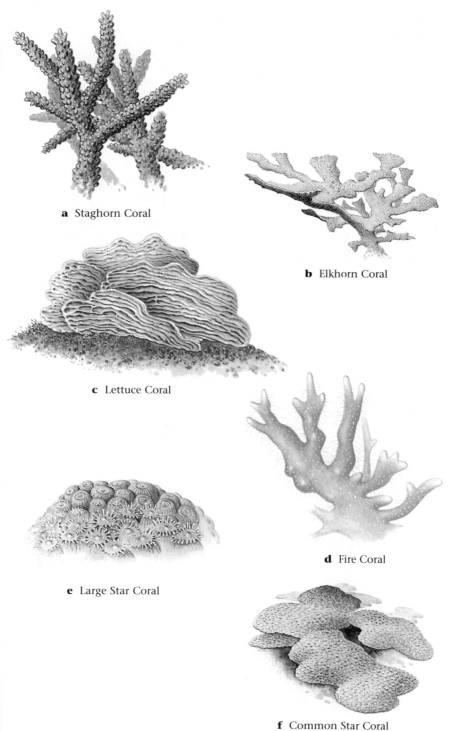

a Staghorn Coral

b Elkhorn Coral

c Lettuce Coral

d Fire Coral

e Large Star Coral

f Common Star Coral

Plate 100a
Porous Coral
(also called Finger Coral)
Porites sp.
Stubby, smooth branches with blunt tips. The coralites are embedded, giving the branches a smooth appearance when the polyps are retracted. However during the day, the polyps are often extended, in which case the branches have a fuzzy appearance. Can be found in most reef environments.

Plate 100b
Common Brain Coral
(also called Smooth or Symmetrical Brain Coral)
Diploria strigosa
A favorite subject for macro-photography because of the numerous wavy ridges that look something like the outside of a human brain. The effect is especially pronounced in the varieties that form rounded heads.

Plate 100c
Flower Coral
Eusmilia fastigiata
This beauty is another favorite of photographers. The coralites are often iridescent greenish-blue, with a lot of space between each. They form small round heads from which the polyps seem to emerge from a central core, somewhat like a hydrangea. The tentacles are extended only at night. This species is found in several reef environments but prefers protected areas.

Plate 100d
Encrusting Gorgonian (a soft coral)
Erythropodium caribaeorum
Colonies form mats which, when the polyps are retracted, are smooth, like oozing leather. When the polyps are extended, they look like dense, fine hair and move in unison with the currents. This species prefers protected shallow reefs.

Plate 100e
Corky Sea Finger (a soft coral)
Briareum sp.
Colonies of this soft coral consist of several (or a single) cylindrical columns arising from a common base. When the polyps are retracted, the columns are reddish-purple and smooth; when the polyps are extended, the rods look like they are covered with yellow-greenish or brownish hairs. This species prefers shallow and calm areas.

Plate 100f
Swollen Knob Candelabrum (a soft coral gorgonian)
Eunicea sp.
This species forms relatively compact colonies with thick branches. It prefers turbulent, shallow water and requires a hard substrate.

Plate 100g
Common Sea Fan (a soft coral gorgonian)
Gorgonia ventalina
This, and other closely related species of sea fans, are one of the more characteristic sights on a Yucatán reef. The colonies form large, flat, fan-like structures that grow on a single plane. Upon close inspection, you will find an intricate vein-like network of branches, emerging from several large veins or branches. Usually a shade of purple, but sometimes yellow. This species prefers shallow reefs with clear water.

Plate 100   455

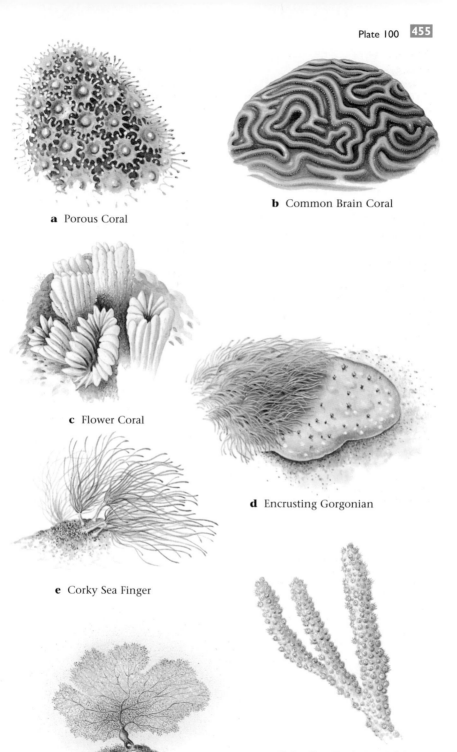

a Porous Coral

b Common Brain Coral

c Flower Coral

d Encrusting Gorgonian

e Corky Sea Finger

f Swollen Knob Candelabrum

g Common Sea Fan

Plate 101a
Black Sea Rod
Plexaura sp.
This species forms bushy colonies that grow in one plane. The stalks are black, which contrasts with the yellowish-brown polyps. It prefers patch reefs in clear water.

Plate 101b
Split-Pore Sea Rod
Plexaurella sp.
Named for the slit-like – as opposed to round or oval – openings that are evident when the polyps are retracted. The structure of the colony can often resemble organ-pipe cactus. This species is common in clear water environments, often in quite shallow areas. Shown with polyps retracted and extended.

Plate 101c
Magnificent Feather Duster
Sebellastarte sp.
This marine worm is the largest of the feather dusters. Most of the worm is hidden from view; only the highly modified head region is visible, most notably the feather appendages that function both for capturing food and as gills for respiration. It will quickly retract when approached too closely. Once it retracts, remain motionless and it may slowly re-emerge. These worms inhabit a wide variety of environments, from patch reefs to pilings.

Plate 101d
Christmas Tree Worm
Spirobranchus giganteus
These worms somewhat resemble the feather dusters but they belong to a separate family, the members of which construct calcareous tubes. The head appendages spiral around a single central core. Their color is variable but usually includes some red or orange with white highlights. They prefer to construct their tubes on living coral but they are not picky as to the type of coral or reef.

Plate 101e
Bearded Fire Worm
Hermodice sp.
This is a fairly active species that often forages in the open. It is covered with tufts of white bristles, interspersed with red gill filaments. Do not touch! The bristles contain a toxin that causes an unpleasant burning sensation, and sometimes a painful wound. When these worms are disturbed, they will display their bristles by way of warning. This species has branched and bushy appendages on the head that look somewhat beard-like. It can be found in a variety of habitats, including reefs, rubble, and grass beds.

Plate 101f
Spiny Lobster
Panulirus sp.
This lobster can reach impressive sizes in areas where it is not hunted. The carapace is brown to tan with two horn-like projections above the eye. The lobsters' very long antennae often project from their hiding places and they always seem to be moving. Being a favorite food item for humans, they are understandably wary and will retreat deeper into their refuges when approached, but will remain facing you. In protected areas they are quite common on the reef.

Plate 101g
Banded Shrimp
Stenopus hispidus
This attractive crustacean seems to be all appendages. Its skinny body and claws are covered with red and white bands. This is one of the cleaning shrimps, and it hangs out at the openings of sponges, waving its antennae to attract its fish clients. It is not particularly wary but will retreat into a protective recess when approached closely. However, it has been known to clean the hands of divers when extended slowly.

Plate 101 457

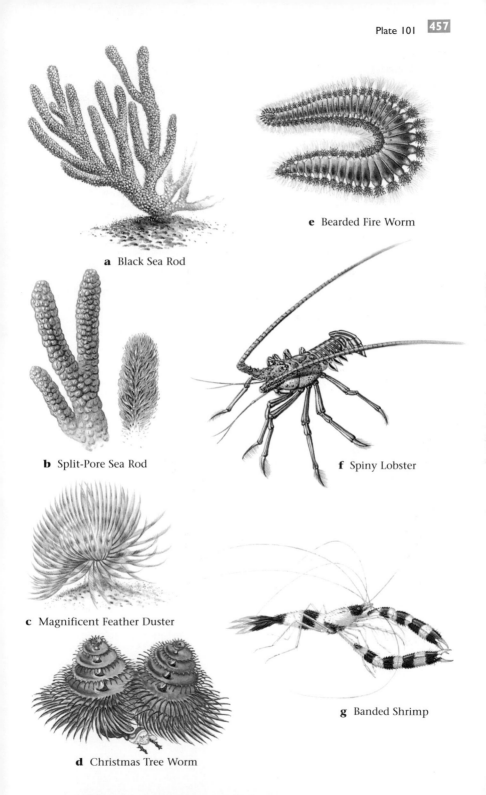

a Black Sea Rod

e Bearded Fire Worm

b Split-Pore Sea Rod

f Spiny Lobster

c Magnificent Feather Duster

d Christmas Tree Worm

g Banded Shrimp

Plate 102

Plate 102a
Queen Conch
Strombus gigas
This huge gastropod was once very common throughout the Caribbean; now its numbers are greatly reduced owing to overfishing, at least in shallow waters. The shells grow in a conical spiral with the outer lip flaring outward. The shell is various shades of orange but often obscured by algae and other encrustations. The snail itself is mottled gray. Its eyes are set at the ends of very long eye stalks. These conchs prefer sandy areas or grass beds between reef patches.

Plate 102b
Flamingo Tongue
(also called Flamingo Tongue Cowrie)
Cyphoma gibbosum
At first glance this appears to be a nudibranch (sea slug) but the creamy white surface covered with orange spots is actually the snail's mantle covering a cowry-shaped shell. The shell can be seen only when the snail retracts its mantle. Most often seen on the gorgonians, including sea fans, upon which they feed. Quite common and found in a variety of shallow water habitats.

Plate 102c
Reef Squid
Sepioteuthis sepioidea
One of the most fascinating reef creatures, and the only squid that frequents the reefs. Reef Squid are intelligent and curious creatures. If you remain almost motionless or swim toward them at an oblique angle, they will allow you to approach closely. Sometimes they will follow divers from a safe distance, observing you with those very large eyes. They will retreat if approached directly, and if they really feel threatened, they will turn on their jet propulsion and disappear in an eyeblink.

Plate 102d
Golden Crinoid
Davidaster sp.
Crinoids, or feather stars, are related to sea stars and sea urchins, though at first glance they look like some kind of gorgonian. They comprise the most ancient group of echinoderms. They have five arms that branch one or more times, so that the terminal arms are always a multiple of five. Each arm has numerous appendages extending laterally to give them their feathery appearance. The arms are used to gather small food particles. Though they appear to be fastened to one spot for life, they can move short distances, and some

species can even swim. This species has 20 feathery arms that are usually yellow-gold to orange in color. The body is usually concealed in a recess of some sort. This species is common but usually occurs below snorkeling depths.

Plate 102e
Blunt-spined Brittle Star
Ophiocoma sp.
Brittle stars are echinoderm relatives of sea stars and sea urchins. They always have five arms emerging from a small central disk. During the day you will find them hiding under rocks and crevices; at night, they are active and can move with surprising speed. The arms break off quite easily, the attribute for which they are named, but regenerate. This species is one of the more common in shallow water. Generally brown to black with blunt spines on the exposed parts of its arms.

Plate 102f
Giant Basket Star
Astrophyton sp.
Basket stars are closely related to brittle stars; their five main arms branch many times giving the appearance of tentacles with a perm. During the day they curl up in a tight ball, usually on a gorgonian. At night, the arms unfurl and are directed toward the current, forming a sort of net by which they capture their planktonic prey. This species is a common reef inhabitant, generally colored orange to brown.

Plate 102g
Cushion Sea Star
Oreaster sp.
Sea stars are the most familiar echinoderms but they are not a particularly prominent component of the Caribbean reefs. This sea star has five short, thick arms. The color is usually some shade of orange to brown, with the dorsal spines forming a net-like pattern. It is common in sand flats and grass beds.

Plate 102 **459**

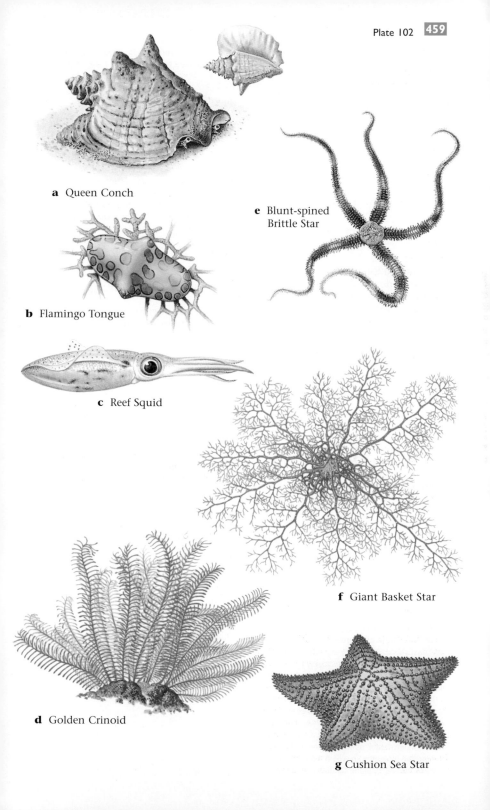

a Queen Conch

b Flamingo Tongue

c Reef Squid

d Golden Crinoid

e Blunt-spined Brittle Star

f Giant Basket Star

g Cushion Sea Star

Plate 103a

Variegated Sea Urchin
Lytechinus sp.

Sea urchins are grazers. Their mouths are on the underside and the spines function to protect these otherwise vulnerable creatures. Urchins are a common cause of injury, because their spines, should they penetrate your skin, are difficult to remove. This beautiful species has short spines and well-defined grooves between plates. Coloration is variable, but usually white or some shade of green. Often camouflaged with sea grass or other debris. Look for it over grass beds or on reefs.

Plate 103b

Long-spined Sea Urchin
Diadema sp.

Formerly abundant throughout the Caribbean, but has experienced a dramatic die-back. This species is one of the main sources of injury for unwary bathers and snorkelers. The long spines easily puncture the skin, often causing infection. This species is typically black. It is found in all habitats.

Plate 103c

Donkey Dung Sea Cucumber
Holothuria mexicana

The common name says it all concerning the outward appearance of this common sea cucumber. Unlike most echinoderms, sea cucumbers lack either spines or arms. They are shaped like a bloated caterpillar and are usually spotted moving slowly over sandy areas between reef patches. Seemingly defenseless, many species contain skin toxins to deter would-be predators.

Plate 103d

Bulb Tunicate
Clavelina sp.

Tunicates, or sea squirts, are among the most abundant marine creatures, and certainly among the least recognized. They are often mistaken for sponges, but they are more closely related to you and me. This is easiest to discern in the larval phase during which they look very much like tadpoles. Pelagic tunicates remain free-swimming for life, but most species settle out and metamorphose into the sponge-like creatures we observe attached to various substrates on the reef. This attractive species is one of the compound tunicates, in which a number of individuals are joined at the base and share a common excurrent siphon. Purplish-blue varieties, known as blue bells, are particularly attractive.

Plate 103e

Sponge
Tethya sp.

Sponges are the most primitive multicellular animals and are an extremely important component of the reef community. They come in a dizzying variety of sizes and shapes. Sponges can be distinguished from non-sponges, such as tunicates, by their large excurrent openings through which they expel the water taken in while filter feeding.

Plate 103f

Branching Vase Sponge
Agelas sp.

This attractive species is common on fairly shallow reefs. Typically there are clusters of up to 30 tubes, each with numerous conical projections. The color varies from lavender to gray. Search this sponge's surface for brittle stars.

Plate 103g

Loggerhead Sponge
Spheciospongia sp.

This species is one of the barrel sponges, large squat creatures with a large central depression. It is often host to shrimps and others that dwell in the canals. Also check the depression for small fish.

Plate 103 461

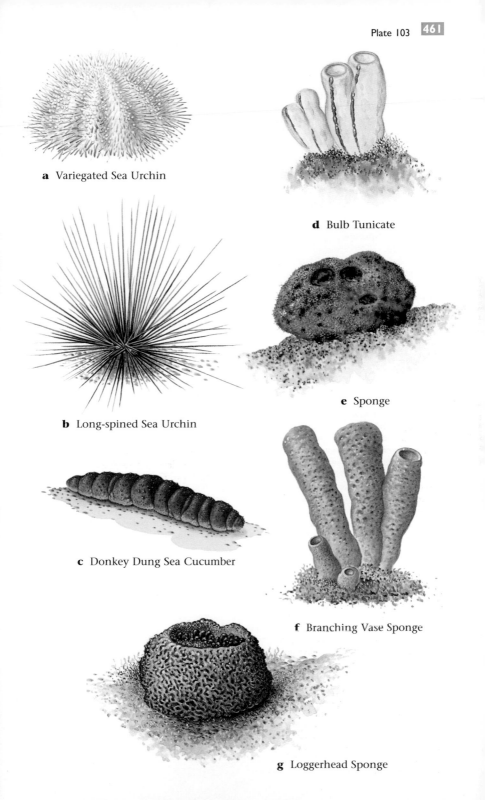

a Variegated Sea Urchin

d Bulb Tunicate

e Sponge

b Long-spined Sea Urchin

c Donkey Dung Sea Cucumber

f Branching Vase Sponge

g Loggerhead Sponge

Plate 104a

Giant Barrel Sponge
Xestospongia muta

These are some of the most spectacular creatures to be found anywhere on the reef. This species grows at depths that require SCUBA. It can grow to over 2 m (6 ft) in diameter and the central depression is large enough to contain a diver, but do not enter as these are fragile creatures that easily break. Large specimens may be over 100 years old!

Plate 104b

Yellow Tube Sponge
Aplysina sp.

This attractive sponge forms clusters of yellow tubes that are joined at the base. This is a soft sponge but don't squeeze it unless you don't mind a purple stain that will last for days. This sponge prefers open water reefs and reef walls. Look for gobies and other small fishes inside the tubes.

Plate 104c

Feather Hydroid
Gymnangium longicauda

Hydroids are related to corals, gorgonians, and anemones, as well as jellyfish. In fact, during the medusa stage of their life cycle, they are hard to distinguish from jellyfish. You are most likely to observe their sessile (non-moving) stage, however, in which they have a fern-like appearance. Each hydroid is actually a colony of numerous individuals. This species is fairly common throughout the Caribbean on the top of reefs where there is some current. Do not touch, as it contains toxins irritating to the skin.

Plate 104d

Moon Jelly
Aurelia sp.

Jellies belong to a large group of animals known as cnidarians, which also includes coral, gorgonians, hydroids, and anemones, all of which possess a specialized structure known as a nematocyst, which is a hook-like barb for injecting toxins. All cnidarians have a medusa stage in which the animal is free-swimming but only the jellies spend most of their lives in this mode. This is one of the more common species in the Caribbean; you are most likely to find them drifting near the surface over the reefs. The moon jelly is almost transparent; the four-leaf clover structure near the top is the reproductive organs.

Plate 104e

Upsidedown Jelly
Cassiopea sp.

This common inhabitant of lagoons and quiet sand flats has a flattened bell and typically orients with arms and tentacles facing upward, hence the name. They have symbiotic single-celled algae from which they derive some of their nourishment. Sometimes they can be observed lying upside down on the bottom, which is thought to facilitate the algae's growth. The closely related Mangrove Upsidedown Jelly is often abundant, not surprisingly, among mangroves.

Plate 104f

Giant Anemone
Condylactis gigantea

Anemones comprise another distinct group of cnidarians; they are very familiar to inhabitants of temperate regions. This species is the largest in the Caribbean and can be distinguished by the distinct swelling at the tip of each tentacle. The main body is usually hidden with only the tentacles visible. Several species of shrimp, including cleaner shrimp, frequently use this anemone for refuge, as do some blennies.

Plate 104g

Turtle Grass
Thalassia testudium

This shallow-water grass grows on sandy areas, forming beds that can cover large areas. This is a flowering plant, though its pale greenish-white flowers are not obvious. Individual blades are flat with rounded tips. Turtle grass beds are an important habitat for many small fishes and invertebrates, and well worth exploring.

Plate 104 463

a Giant Barrel Sponge

d Moon Jelly

e Upsidedown Jelly

b Yellow Tube Sponge

f Giant Anemone

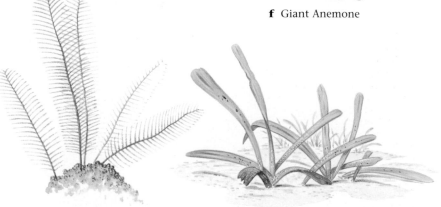

c Feather Hydroid   **g** Turtle Grass

WCS Conservation Work in Latin America

The Conservation Challenge in Latin America

From Mexico to Tierra del Fuego, Latin America is a land of superlatives. Vast tropical rain forests, rivers comprising the largest freshwater systems on Earth, towering mountains and deep oceans are home to animals and plants found nowhere else in the world. Perhaps no other region on Earth presents such a variety of ecosystems and astonishing array of wildlife.

Latin American conservation efforts, however, face difficult social, economic, and environmental challenges. The region's human population has tripled since 1950. South America alone has lost almost a quarter of its forests, while more than 60 percent of Mexico's woodlands have fallen. Increased hunting, fishing, mining, and other natural resource exploitation threaten already stressed ecosystems.

Patagonia's elephant seals travel as far as the South Georgia Islands, staying at sea for up to eight months. (Photograph with permission from William Conway and WCS.)

These problems, compounded by too few trained conservation professionals and a chronic lack of funding for natural resource management, have both pushed many species to the brink of extinction, and reduced the land's ability to support human life.

Working to reverse these trends, the Wildlife Conservation Society has supported conservation work in Latin America since 1909 with its landmark field studies in Trinidad, Venezuela, and the Galapagos Islands. Today WCS conservationists, working mainly through local projects run by nationals, are uniquely positioned to understand local conditions and conservation opportunities. WCS has developed hundreds of innovative conservation projects, from field studies of endangered species such as the Andean mountain tapir, to the protection of

WCS scientists conceived a network of protected areas in Central America called Paseo Pantera — the Meso-America biological corridor. By protecting these areas, the vast ecological diversity of the region will be preserved. Red-eyed treefrogs, tapirs and small hawksbill turtles are among the thousands of species found in the Paseo Pantera. (Photographs with permission from S. Matola, A. Meylan and WCS)

immense areas through the Patagonian Coastal Zone Management Plan and Central America's Paseo Pantera Project. Today, WCS operates more than 100 Latin American projects in 17 countries, from Mexico to Argentina.

All of these projects depend critically on scientific research. Field staff survey wildlife and assess biodiversity to determine how species interact with their habitats and their human neighbors. Projects lasting several years allow researchers to uncover trends and patterns not apparent in short-term research and to build relationships with local communities and governments. Local conservationists are trained to be responsible for the stewardship of their land and Government participation is encouraged.

Crossing Political Boundaries for Regional Conservation

The Biological Corridors of Paseo Pantera

Whether called panther, cougar, mountain lion, puma, or pantera, the New World's premier big cat ranges from Patagonia to northern Canada. The Central American Land Bridge joining North and South America, rose from the sea some three million years ago, allowing the panther and thousands of other species to expand their range and thrive in new territories. To protect this "biological highway," WCS has pioneered a conservation strategy called "Paseo Pantera" – Path of the Panther – to connect an unbroken corridor of parks and refuges throughout Central America.

The pumas and jaguars require huge expanses of unbroken habitat. Biological corridors in Central America are a key solution to preserving the range of these big cats along with many other threatened species. (Photograph with permission from A. Rabinowitz and WCS.)

Just a few decades ago, upland rainforests and dense vegetation covered much of Central America, and mangrove swamps and coral reefs lined its two coasts, forming a chain of natural areas between North and South America. Subsequent human development and population growth has pushed wildlife into dwindling, isolated patchworks of habitat. Working with all seven countries in the Paseo Pantera region, WCS seeks to improve management of existing parks and to restore degraded habitat for migratory wildlife. In Belize, WCS aims to establish new reserves along the Belize Barrier Reef. In reserves in the western Maya Mountains, researchers are developing guidelines to preserve biodiversity. WCS assists Guatemala in the management of the 3.1-million-acre Maya Biosphere Reserve. In Honduras, WCS musters resources to improve the management of protected coastal areas, including the Bay Islands and the Rio Platano Biosphere Reserve – one of the largest protected areas in Central America. In El Salvador, damaged areas are restored and in Nicaragua, plans have been developed to protect Bosawas, Miskito Cat, and the Si-A-Paz. WCS and several other organizations have helped Costa Rica expand Tortuguero National Park to four times its original size. And Panama is investing in conservation projects in the coastal bays of Bocas del Toro.

Manu Reserve in Peru contains incredible diversity of wildlife, including flocks of Scarlet Macaws, shown here at a lick. (Photograph with permission from C. Munn and E. Nycander.)

In 1994, all seven Central American countries signed an agreement affirming the Central American Biological Corridor as a conservation priority – Paseo Pantera will help sustain the region's unique mixture of wildlife well into the future.

Working with Nations to Protect Habitat

Bolivia Creates Two Massive Parks

When wildlife surveys in Bolivia revealed an extraordinary wealth of species in two diverse regions, WCS joined the Bolivian Government, local conservation organizations, and indigenous people to protect these areas. Two massive parks were declared in 1995 spanning 20,000 square miles – an area larger than Switzerland.

The first park, Kaa-Iya del Gran Chaco in southeast Bolivia, protects a vast 8.6 million acres of the Chaco – a unique dry forest and thorn-scrub habitat second only to the Amazon rainforest in size. Agricultural clearing has destroyed most

of the original Chaco, which once covered much of northern Argentina and Paraguay, leaving sparsely populated Bolivia with the last great expanse. Since 1984, WCS has cataloged 46 species of large mammals in Bolivia's Chaco, including giant armadillos, maned wolves, several big cat and primate species, and the rare Chacoan peccary, thought extinct until 1975.

The second park, Alto Madidi in northwest Bolivia, covers 4.7 million acres of glaciers, mountain and lowland rainforests, and Pampas del Heath savannahs. It could be the most biologically rich region in the world with almost 1,200 species of birds, Andean bears, jaguars, giant otters, anacondas, black caimans, and thousands of other animal and plant species.

Today, illegal hunting threatens wildlife. This jaguar skin was found in the Mamiráua Reserve. (Photograph with permission from J. Thorbjarnarson.)

Making these two parks a reality required several years of research and cooperation between the various interested parties. In the huge Chaco park, WCS is supporting the efforts of native peoples to develop management plans, foster ecotourism possibilities and explore other economic incentives to local inhabitants. The Tacana Indians, who have long supported conservation efforts in the Madidi region, will similarly help manage their new park.

Stewardship Through Community Involvement

Mamiraúá Lake Ecological Station

Amazonian Brazil boasts one of the most extraordinary, yet least studied ecosystems in the world – the seasonally flooded forest known as the várzea – an area larger than Florida. Rare uakari monkeys and umbrella birds forage in the canopy, while pink river dolphins, caimans, and Amazonian manatees swim among submerged trees. WCS has recorded over 200 species of fish, nearly 300 species of birds, and an exceptional diversity of trees in this unique environment.

In response to threats such as commercial logging and overfishing, WCS launched one of its most unique Latin American projects: the Mamiraúá Lake Ecological Station, a 2.8-million-acre reserve. Brazilian and international scientists gather information and monitor the resource demands of the 2,000 local inhabitants. Research projects include surveys of wildlife, plants, and fish; and measure the effects of timber extraction, fishing, and subsistence hunting.

Local people have provided crucial support to this initiative, protecting their livelihood from outside exploitation. Ultimately, local participation is the only way to insure future wise stewardship of this flooded Amazonian forest.

In 1995, $1 million was provided for the establishment of the "National Institute for the Várzea," to be built near the reserve in the town of Tefè, the first major Brazilian research institute in the upper Amazon region.

Long-term Commitments

Coastal Conservation in Patagonia

Spectacular concentrations of colonial seabirds and marine mammals flourish along Patagonia's rugged 2,000-mile coastline. The plankton-filled Antarctic current supports southern right whales, elephant seals, Magellanic penguins, and one of the planet's most productive fisheries.

WCS has worked here for over 30 years, promoting the declaration of many coastal reserves, among them Punta Tombo and Punta Leon in Chubat Province.

To combat threats from burgeoning petroleum, shipping, fishing, and tourism developments, WCS has helped produce a Patagonian Coastal Zone Management Plan, a collaborative effort with local organizations to manage fast-growing industries while protecting the region's unique ecology.

WCS monitors Elephant seals and Magellanic penguins as indicator species of the southern Atlantic ecosystem. Chronic oil pollution has reduced penguin

Chronic oil pollution has caused a one-third decline in Magellanic penguin populations at Punta Tombo, Patagonia. WCS is working with Argentinean authorities to move oil tanker routes away from delicate breeding areas. WCS President and General Director William Conway pioneered WCS's work in Patagonia some 30 years ago. (Photograph with permission from William Conway.)

numbers by one-third in the past 15 years and WCS advocates that oil tanker routes be moved farther from the coast. Offshore, WCS examines the impact of whale-watching, which has grown 300 percent in seven years, working with local governments and the whale-watching industry to insure that boats do not disturb the whales and their young. WCS monitors the impacts of commercial fishery "by-catch" – non-target fish species discarded overboard. Fisheries off Patagonia produce up to 50 to 70 percent by-catch, virtually all of which dies on the decks of the fishing vessels, reducing the food supplies of marine birds and mammals.

You Can Make A Difference — Your Membership Helps Save Animals And Their Habitats

Join today! Your membership contribution supports our conservation projects around the world and here at home. For over 100 years, we've been dedicated to saving endangered species — including tigers, snow leopards, African and Asian elephants, gorillas and thousands of other animals.

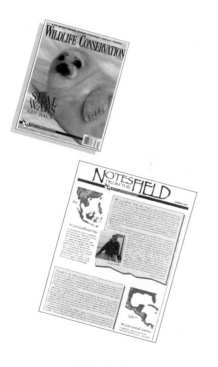

FOR A COMPLETE LIST OF BENEFITS PLEASE TURN THE PAGE ...

YES! I want to help save rare and endangered animals. Here is my membership contribution of:

☐ $25 ☐ $35 ☐ $75 ☐ $150

☐ I cannot join at this time but enclosed is my contribution of $ _____ .

Full name (Adult #1) _____

Address _____

City _____ State _____ Zip _____ Daytime Phone (____) _____

For Family Plus and Conservation members only:

Full name Adult #2 _____

Full name Adult #3 _____ #Children ages 2–18 _____

Your employer may have a matching gift program that can double, even triple, your contribution. Be sure to check with your Personnel or Community Relations Office.

$25 ASSOCIATE
- One year of *Wildlife Conservation*, our award-winning magazine filled with breathtaking photographs and articles that will keep you up to date about our worldwide conservation initiatives.

$35 BASIC MEMBERSHIP
Wildlife Conservation magazine plus the benefits of *full membership* including:
- *Notes from the Field*, a quarterly newsletter with project reports from scientists in the field
- Opportunities to travel with WCS scientists
- A membership card that entitles one adult to unlimited admission for one year to all five of our wildlife parks in New York including the world famous Bronx Zoo, as well as passes for free parking where available.

$75 PLUS MEMBERSHIP
All the benefits of Basic and:
- *Passport to Adventure* … Travel around the world learning about exotic animals without ever leaving home with this interactive, educational package of stickers, fun fact cards, a map and more. Great fun for the entire family!
- A membership card that entitles two additional adults (total of three) and children to unlimited admission for one year to all five of our wildlife parks.

$150 CONSERVATION SUPPORTER
All the benefits of Plus and:
- Limited edition sterling silver antelope pin

JOIN NOW! Mail your membership contribution to:

WILDLIFE CONSERVATION SOCIETY
Membership Department
2300 Southern Boulevard
Bronx, NY 10460–1068, USA

Join using your credit card by calling
1–718–220–5111

or by visiting our website
www.wcs.org

TRAVEL WITH THE EXPERTS!

Wildlife Conservation Society tours take you to wild places with informed escorts who know the country, know the animals and care about wildlife. They have actively participated in establishing national parks or saving endangered species. You'll travel with experts who will share their excitement, wonder and love of wildlife conservation.

From the national parks of Kenya to the rain forest of Peru, WCS travel experts will make your trip an informative, educational and lively adventure.

For more information call or write:
WCS International Travel Program
830 Fifth Avenue
New York, NY 10021 USA
00 1 212–439–6507

Please clip this form and mail it with your gift to Wildlife Conservation Society/2300 Southern Blvd/Bronx, NY 10460–1068, USA.

IMPORTANT: Wildlife Conservation Society is a 401(c)3 organization. Non membership contributions are fully tax-deductible to the extent allowed by law. Membership dues are tax-deductible in excess of benefit value. Magazine is a $12 value, parking passes are an $8 value, *Passport to Adventure* is a $6 value. For a copy of our latest Annual Report you can write to us or to the Office of Charities Registration, 162 Washington Avenue, Albany NY 12231, USA. Your contribution to the Society will be used to support our general programs as described in the Annual Report.

Conservation Through Training

WCS teaches scientific know-how to Latin American students and professionals, to create a core of conservationists, scientists, and decision-makers who will be able to protect the region's wildlife. Local people and park guards are taught how to census and monitor wildlife populations and conservation science is taught to university students and professionals. Since 1989, WCS has held courses in Colombia, Venezuela, Ecuador, and Peru.

The WCS Student Grants Program offers graduate and undergraduate conservationists much-needed funding for research projects to solve conservation problems in Latin America. Since 1987, this program has supported more than 140 projects in Peru, Ecuador, Colombia, Bolivia, and Venezuela.

The Scientific Challenge of Sustainable Use

"Sustainable use" – harvesting natural resources while preserving biodiversity – is often seen as the panacea to unite conflicting environmental and economic interests. However, the exploitation of natural resources, even on a sustainable basis, will inevitably involve some biodiversity loss, a problem under WCS scrutiny. WCS is investigating the effects of selective timber harvesting in Bolivia and Venezuela, for example, where some logging interests claim the practice is sustainable. In Ecuador, the Sustainable Utilisation of Biological Resources (SUBIR) project monitors wildlife populations and works with local communities to develop economic alternatives to destroying habitats.

WCS trains park rangers and conservationists throughout Latin America. These Venezuelan park guards help protect world-renowned regions, including Angel Falls in Canaima National Park. (Photograph with permission from A. Grajal.)

Species Index

General Index

Plate Key Symbols and Codes

 = Lowland wet forest.

 = Lowland dry forest.

 = Highland forest/cloud forest. Includes middle and higher elevation forests, including the pine and pine–oak forests of Oaxaca and Chiapas, and cloud forests.

 = Forest edge/streamside. Some species typically are found along forest edges or near or along streams; these species prefer semi-open areas rather than dense, closed, interior parts of forests. Also included here: open woodlands, tree plantations, and shady gardens.

 = Pastureland/savannah (grassland with scattered trees, shrubs))/non-tree plantations/gardens without shade trees/roadside. Species found in these habitats prefer very open areas.

 = Freshwater. For species typically found in or near lakes, streams, rivers, marshes, swamps.

 = Saltwater/marine. For species usually found in or near the ocean or ocean beaches.

Location symbols (see Maps 2, 3 and 4, pp. 15, 16, 18):

OAX-HI refers to the large central portion of Oaxaca, containing middle and high elevation terrain; it includes the entire area around Oaxaca City.

OAX-LO refers to low elevation regions of Oaxaca, primarily the lowlands along the Pacific Ocean, the northern swath adjacent to the state of Veracruz, and the Isthmus of Tehuantepec. Not all species that occur in the northern lowlands occur in the southern lowlands, and vice versa.

CHIAP-HI refers to the large central portion of Chiapas, containing middle and high elevation terrain; it includes the areas around Tuxtla Gutiérrez and San Cristóbal.

CHIAP-LO refers to low elevation regions of Chiapas, primarily the lowlands along the Pacific Ocean and the swath adjacent to Tabasco and northern Guatemala (the Petén). Not all species that occur in the northern lowlands occur in the southern lowlands, and vice versa.

TAB is Tabasco, the small state north of Chiapas, bordering the Gulf of Mexico (this state encompasses only lowlands).

CAM is Campeche, the western part of the Yucatán Peninsula (this state encompasses only lowlands).

YUC is Yucatán, the northern and central parts of the Yucatán Peninsula (this state encompasses only lowlands).

QRO is Quintana Roo, the eastern section of the Yucatán Peninsula (this state encompasses only lowlands).